CAUSALITY AND EXPLANATION

CAUSALITY AND EXPLANATION

WESLEY C. SALMON

New York • Oxford

Oxford University Press

1998

Oxford University Press

Oxford   New York
Athens   Auckland   Bangkok   Bogota   Bombay   Buenos Aires
Calcutta   Cape Town   Dar es Salaam   Delhi   Florence   Hong Kong
Istanbul   Karachi   Kuala Lumpur   Madras   Madrid   Melbourne
Mexico City   Nairobi   Paris   Singapore   Taipei   Tokyo   Toronto   Warsaw

and associated companies in
Berlin   Ibadan

Library of Congress Cataloging-in-Publication Data
Salmon, Wesley C.
Causality and explanation / Wesley C. Salmon.
    p.   cm.
Includes bibliographical references and index.
ISBN 0-19-510863-9; ISBN 0-19-510864-7 (pbk.)
1. Science—Philosophy.   2. Science—Methodology.   3. Causality
(Physics)   4. Explanation.   I. Title.
Q175.S2337   1997
501—dc21   96-48651

Page v constitutes an extension of the copyright page.

9 8 7 6 5 4 3 2 1

Printed in the United States of America
on acid-free paper

2. Determinism and Indeterminism in Modern Science

Reprinted from Joel Feinberg, ed., *Reason and Responsibility*, 2nd ed., Dickenson Publishing Company, 1971. Copyright © 1970, Wesley C. Salmon. All rights reserved.

3. Comets, Pollen, and Dreams: Some Reflections on Scientific Explanation

Reprinted from Robert McLaughlin, ed., *What? Where? When? Why?* copyright © 1982, D. Reidel Publishing Company, pp. 155–178, with kind permission from Kluwer Academic Publishers.

4. Scientific Explanation: Causation *and* Unification

Reprinted with kind permission from *Crítica*, vol. 22, no. 66 (1990), pp. 3–21.

6. A Third Dogma of Empiricism

Reprinted from Robert E. Butts and Jaakko Hintikka, eds., *Basic Problems in Methodology and Linguistics*, copyright © 1977, D. Reidel Publishing Company, pp. 149–166, with kind permission from Kluwer Academic Publishers.

7. Causal and Theoretical Explanation

Reprinted from S. Körner, ed., *Explanation*, copyright © 1975, Basil Blackwell, pp. 118–145, by kind permission from Blackwell Publishers. (Originally published under the title "Theoretical Explanation.")

8. Why Ask, "Why?"?: An Inquiry Concerning Scientific Explanation

Reprinted from *Proceedings and Addresses of the American Philosophical Association*, vol. 51, no. 6 (1978), pp. 683–705. Copyright © 1978, Wesley C. Salmon. All rights reserved.

9. Deductivism Visited and Revisited

Reprinted from Adolf Grünbaum and Wesley C. Salmon, eds., *The Limitations of Deductivism*, University of California Press, 1988, pp. 95–127. Copyright © Adolf Grünbaum and Wesley C. Salmon.

10. Explanatory Asymmetry: A Letter to Professor Adolf Grünbaum from His Friend and Colleague

Reprinted from John Earman, Allen I. Janis, Gerald J. Massey, and Nicholas Rescher, eds., *Philosophical Problems of the Internal and External Worlds*, by permission of the University of Pittsburgh Press. Copyright © 1993, University of Pittsburgh Press. (Originally published under the title "On the Alleged Temporal Asymmetry of Explanation.")

11. Van Fraassen on Explanation, Philip Kitcher, coauthor

Reprinted from *Journal of Philosophy*, vol. 84, no. 6 (1987), pp. 315–330. Copyright © 1987, Journal of Philosophy, Inc. Reprinted by permission.

12. An "At-At" Theory of Causal Influence

Reprinted from *Philosophy of Science*, vol. 44 (1977), pp. 215–224, University of Chicago Press, publisher. Copyright © 1977, Philosophy of Science Association. All rights reserved.

13. Causal Propensities: Statistical Causality versus Aleatory Causality

Reprinted from *Topoi*, vol. 9 (1990), pp. 95–100, with kind permission from Kluwer Academic Publishers. Copyright © 1990, Kluwer Academic Publishers.

14. Probabilistic Causality

Reprinted from *Pacific Philosophical Quarterly*, vol. 61, nos. 1–2 (1980), pp. 50–74, by kind permission of Blackwell Publishers.

15. Intuitions—Good and Not-So-Good

Reprinted from Brian Skyrms and William L. Harper, eds., *Causation, Chance, and Credence*, Vol. 1, pp. 51–71, with kind permission from Kluwer Academic Publishers. Copyright © 1988, Kluwer Academic Publishers.

16. Causality without Counterfactuals

Reprinted from *Philosophy of Science*, vol. 61 (1994), pp. 297–312, University of Chicago Press, publisher. Copyright © 1994, Philosophy of Science Association. All rights reserved.

18. Causality: Production and Propagation

Reprinted from Peter D. Asquith and Ronald N. Giere, eds., *PSA 1980*, vol. 2, pp. 49–69. Copyright © 1981, Philosophy of Science Association. Reprinted by permission.

20. Scientific Explanation: Three Basic Conceptions

Reprinted from Peter D. Asquith and Philip Kitcher, eds., *PSA 1984*, vol. 2, pp. 293–305. Copyright © 1985, Philosophy of Science Association. Reprinted by permission.

21. Alternative Models of Scientific Explanation, Merrilee H. Salmon, coauthor

Reprinted from *American Anthropologist*, vol. 81, no. 1 (1979), pp. 61–74, by permission of the American Anthropological Association. Not for further reproduction.

22. Causality in Archaeological Explanation

Reprinted from Colin Renfrew et al., eds., *Theory and Explanation in Archaeology*, pp. 45–55. Copyright © 1982, Academic Press. Reprinted by permission.

23. Explanation in Archaeology: An Update

Reprinted from L. Embree, ed., *Metaarchaeology*, pp. 243–253, with kind permission from Kluwer Academic Publishers. Copyright © 1992, Kluwer Academic Publishers.

For Charlotte and Bruce
with love

———

Tribute to Carl G. Hempel
1905–1997

On November 9, while this book was being manufactured, Carl G. Hempel died. As we mourn his passing, we are deeply grateful for his admirable personal qualities and his epoch-making contributions to twentieth-century philosophy of science. Among his many accomplishments, his work on scientific explanations stands out as the benchmark to which all subsequent studies of this topic must be referred. Among the experts he is recognized as the preeminent authority. Readers of this book will not find his name in the index. The reason is simple. His influence, directly or indirectly, pervades almost the entire book.

The only suitable entry would be:

Hempel, Carl G., 3–403 passim.

Preface

Having worked actively on scientific explanation for more than thirty years, I recently discovered that it is a sexy topic. I use the term "sexy" in its nonsexual sense. What I mean is that a huge federally funded project involving billions of dollars was defended on philosophical grounds, namely, that by aiding scientists in explaining natural phenomena, it would lead to deeper understanding of our universe. In 1987 Nobel laureate physicist Steven Weinberg testified before the U.S. Congress regarding funding of the superconducting super collider (SSC) on that basis. Later on (1992), while the continuation of its funding was under consideration by Congress, Weinberg published an important and influential book, *Dreams of a Final Theory,* in which he tried to show why the SSC would be worth the additional investment. Regrettably, this project was scuttled after it was undertaken and after enormous amounts of labor and money had already been expended on it. The final essay in this collection, "Dreams of a Famous Physicist: An Apology for Philosophy of Science," offers a philosophical analysis of this challenging work. Although I thoroughly agree with Weinberg's scientific goals, I take strong exception to his explicitly declared attitudes toward philosophy of science. I find this treatment of scientific explanation deeply flawed.

My point of departure for this whole collection lies in the eighteenth-century Enlightenment, more specifically, in David Hume's epoch-making critique of causality. In the last decade of the twentieth century, we have, I believe, taken significant steps toward an actual solution of the fundamental problems he posed concerning the nature of causality—i.e., toward understanding the kinds of connections that link causes and effects. The initial essay, "A New Look at Causality," offers a preview of the issues developed in greater detail in subsequent essays, especially those in Part III. As I point out in the Introduction, there is an obvious and basic relationship between the concepts of causality and explanation. To a surprising extent, this relationship has been ignored, denied, or severely underrated in much of the twentieth-century philosophical literature on scientific explanation.

Even more surprising to the modern reader, I imagine, is the fact that the very existence or possibility of scientific explanation was denied by many outstanding philosophers and scientists at the beginning of the twentieth century. Today it is widely agreed that one of the chief aims of scientific endeavor—if not *the* principal goal—is to facilitate our understanding of the universe in which we live and of our place in it. To my mind this is one of the greatest philosophical achievements of the century. The fifth essay, "The Importance of Scientific Understanding," elaborates this theme. Let us hope that the lesson is not ignored as we face global problems in the twenty-first century.

The essays contained herein were written over a period of many years, but they are not presented chronologically. Those in Part I are genuinely introductory. They are not simple, but they should be accessible to readers who are seriously interested in the subject. The essays in Parts II, "Scientific Explanation," and III, "Causality," are attempts at substantive contributions to these two subjects. They represent my efforts over a period of two decades (1975–1995) to come to terms with the fundamental problems associated with causality and explanation, including the development of ideas on probabilistic causality that fit harmoniously with my views on causal and statistical explanation. The essays in Part IV, "Concise Overviews," are survey articles containing more technical details and, therefore, more accurate summaries of the topics they treat. They can be seen as highly condensed versions of the main themes of *Scientific Explanation and the Causal Structure of the World* (Salmon, 1984b) and *Four Decades of Scientific Explanation* (Salmon, 1990b). The essays in Part V address specific issues in particular scientific disciplines, namely, archaeology and anthropology, astrophysics and cosmology, and physics. They aim to show that this area of philosophy of science is not irrelevant to the sciences. Because I am not so vain as to suppose that every reader of this book will want to read every essay, brief abstracts of the essays appear at the beginning of the parts in which they appear. I hope these will help the reader pick and choose according to his or her particular background and interests.

Seven of the essays (essays 1, 5, 17, 19, 24–26) are previously unpublished; the remainder appeared in a variety of places. In those that were published elsewhere, I have not hesitated to make slight revisions and corrections to improve the grammar and style, including an effort to make the material reasonably gender-neutral. I have enclosed substantial insertions in square brackets.

When I began serious work on this book, I was literally at sea. While on sabbatical leave from Pittsburgh, I took my laptop computer on the SS *Universe,* encircling the globe in a hundred days. As a member of the immediate family of Merrilee Salmon, who spent the term teaching in the Semester-at-Sea program, I enjoyed a leisure that was both exciting and conducive to work. I am most grateful to her for this opportunity and also to Bill Soffa, academic dean, for many stimulating conversations on philosophy of science. I spent most of the academic year 1995–1996 in Konstanz, Germany, as a recipient of an Alexander von Humboldt Foundation Award that enabled me to do most of the remainder of the project. I should like to express my deep gratitude to the Humboldt Foundation for the support of my research, to Gereon Wolters, who nominated me, and to the

University of Konstanz, which extended all kinds of professional courtesies. By virtue of the Humboldt Award, I had the opportunity to present lectures at several institutions in Germany, where I greatly benefited from informed and stimulating discussion. My warmest thanks go also to my esteemed colleague John Earman, who, during periods between our travels, enlightened me on topics related to determinism and indeterminism in classical and modern physics. Among my many debts to my wife, Merrilee, is the fact that she has given me all of the substantive material on archaeology and anthropology, which makes up an important part of the essays in this book, particularly essays 21–24. I wish also to thank Hana Novak for elegantly rendering all of the figures, Kathy Rivet for indispensable secretarial assistance, and Charlotte Broome for expertly compiling the index. Expressions of gratitude to many other persons are found in the individual essays.

Pittsburgh, Pennsylvania   W. C. S.
May 1997

Contents

CAUSALITY AND EXPLANATION

Introduction

To most people the suggestion that there is a close connection between causality and explanation would come as no surprise. Even if these two concepts do not go precisely hand in hand, their domains have major areas of convergence. In many cases to explain a fact is to identify its cause. Yet, as general concepts, causality and explanation are far from clear. Both have wide ranges of applicability; both are abstract, ambiguous, and vague. They have important theoretical dimensions as well as practical aspects. Similar remarks apply to the closely associated concept of understanding. As a vast philosophical literature testifies, to explicate these three concepts is no trivial task. This introduction is offered as a conceptual map of the broad territory they cover, showing where our concerns fit into the overall picture. It offers a general impression of "the lay of the land."

1. Causality

The concept of causality pervades our thinking about ourselves, about our environment, and about the entire universe we live in.

- It is fundamental to our attempts to gain *intellectual understanding* of the universe and its contents—its physical systems, its living organisms, and its sentient beings. Scientific explanations, from which such understanding derives, are often, if not always, causal.
- Causal concepts are central to our *practical deliberations*. We need to know the causes and effects of depletion of ozone in the upper atmosphere. We need to know whether "secondhand" smoke causes harm to human health.
- Causality is invariably involved in our *technology*, where we attempt to achieve desirable effects while avoiding undesirable ones. Can we produce electrical energy

3

by controlled thermonuclear fusion without encountering problems as serious as those connected with fission reactors?

- *Everyday practical planning* involves causal considerations. We avoid leaving iron tools out in the rain because exposure to moisture causes them to rust. We plant seeds in springtime in order to reap food or flowers later on.
- Causal terminology permeates *ordinary language*. Note how many common verbs express causal efficacy: "break," "fix," "move," "send," "hurt," "help," "make," "antagonize," "comfort," etc.

Causal concepts are ubiquitous: in every branch of theoretical science—physical, biological, behavioral, and social; in the practical disciplines—architecture, ecology, engineering, law, and medicine; in everyday life—making decisions regarding ourselves, our loved ones, other living persons, and members of future generations.

Decision making necessarily involves evaluations, and value judgments cannot be properly rendered without considering consequences. Discussions of values often refer to means and ends. If a corporation contemplates construction of a shopping mall, the end in view is to make a profit. Even if we grant that it is a worthy goal—not a self-evident truth by any means—we must ask whether the building of the shopping mall at that place and at that time will achieve that goal. Will the effect of that enterprise actually be the hoped-for profit? Other consequences should be considered. Will the construction have adverse ecological effects? Will it damage the local economy by putting smaller merchants out of business, or will it benefit the local economy by creating new jobs and increases in local trade? And if the enterprise is actually undertaken, who deserves credit if the effects are beneficial, and who deserves blame if they are detrimental? When we attempt to assign legal or moral responsibility to human actions, causal considerations are paramount. Who actually produced the result? Was it done in an appropriate manner? And did the individual or group deliberately and knowledgeably adopt acceptable means? Questions of this sort give rise immediately to the ancient problem of free will and determinism.

Having arrived at this point, we know that we are deep in philosophical territory. Philosophical investigations of causality have a long history and broad relevance. As the preceding paragraph shows, causality figures prominently in ethics; for similar reasons it is present in legal, social, and political philosophy. It is central to the theory of human action: Are our actions caused by our volitions, or is some special form of agency involved?

Causality has traditionally played a central role in metaphysics and theology. One early example is Aristotle's classification of causes into four types: material, formal, efficient, and final. The first cause argument for the existence of God is another obvious example; so is the design argument—clearly a causal argument. The metaphysical dualism of René Descartes raises serious causal problems concerning interactions between mind and matter. A principle of universal causation is deeply embedded in Immanuel Kant's philosophical system. A principle of sufficient reason lies at the foundation of Leibnizian metaphysics. The time-honored doctrine of determinism is a causal thesis.

Contemporary philosophy of language has offered causal theories of meaning, and contemporary epistemology presents causal theories of perception and evidence. It may be that philosophers whose concerns are confined to formal logic and philosophy of mathematics can avoid entanglement with causal concepts, but in virtually every other area of philosophy they abound. The role of causality in scientific explanation, and in philosophy of science more generally, is a major theme of this book.

The ubiquity of causal concepts both within philosophy and in other areas of human endeavor would not, in itself, demonstrate that causality is a topic of serious philosophical concern. Causality presents pressing philosophical issues because we do not have an adequate and generally accepted understanding of the cluster of concepts it involves. Indeed, the inadequacy of our comprehension of causality was dramatically displayed by David Hume's searching eighteenth-century analysis, and confusion is still rampant. Many commentaries on Hume's discussion have appeared in the philosophical literature, but until now no adequate answers to the problems he raised have been available. Moreover, among the most pressing current problems in philosophy of physics is the role—or lack thereof—of causality in quantum mechanics. We face two problems here. One is the question of causal indeterminacy; the other is the apparent presence of causal anomalies, such as action-at-a-distance, in this domain. These issues leave us with fundamental questions about the form explanations must take in quantum mechanics.

2. Explanation

Having seen that the concept of causality is involved in philosophy of science, in particular in the treatment of scientific explanation, let us now turn our primary attention in that direction. Scientific explanations obviously have enormous practical value. We want to explain airplane crashes in order to find ways of averting such accidents in the future. We want to explain the occurrence of diseases to learn how to prevent or cure them. On the intellectual side, scientific explanations of phenomena are the means to understanding them. However, since both words—"explanation" and "understanding"— are highly ambiguous, it is essential to distinguish a variety of senses. Let us begin with "explanation," turning our attention to "understanding" afterward.

- People often ask for explanations of *meaning*—whether of an ordinary word, a poem, a painting, or another work of art. The meaning of a word may be found in a dictionary. The meaning of a poem may be clarified by calling attention to certain metaphors. The meaning of a painting may be exhibited by reference to the iconography of the period in which the work was created. If the process has been successful, we have achieved understanding of the word, the poem, or the painting.
- Another type of explanation involves learning *how to perform* certain activities. A painter might explain how to achieve an appearance of depth by the use of perspective. An automobile owner's manual might explain how to jack up the car in order to change a tire. A guidebook for tourists might explain how to find a particular building in a foreign city.

Explanations of the foregoing types would not ordinarily be requested by posing why-questions. They are explanations not of *why something occurs,* but rather of *what something means* or *how to do something.*

Explanations of meanings do occur in the sciences; to find the meaning of a technical term, one might consult a scientific handbook. Explanations of how to perform various activities are also found in science; a scientist might explain to a technician how to construct a particular type of detecting apparatus. When we speak of scientific explanation, however, we are not usually referring to these kinds of explanations. For the most part we have in mind explanations of *why certain phenomena occur.* The phenomena may be particular facts or general regularities. For example, there has been considerable interest in recent years in the explanation of the extinction of the dinosaurs some 65 million years ago. The explanation most widely accepted at present involves the collision of a large asteroid or comet with Earth at that time. Although the extinction is a complicated occurrence, it is still a particular fact and not a general regularity. The explanation of the elliptical paths of planets around stars (Kepler's first law) on the basis of Newton's laws of motion is a familiar case of explaining a general regularity. This law applies, of course, to other planetary systems and systems of satellites, not just to our solar system.

The foregoing explanations could be solicited with why-questions: "Why did the dinosaurs become extinct?" and "Why do planets move in elliptical orbits?" However, not all why-questions are requests for scientific explanations.

- Knowing all of the scientific facts pertinent to the death of a child, the parents may still ask why their child was taken from them. This is a request for *consolation,* not for scientific explanation.
- A closely related type of why-question asks for a *moral judgment,* for example, why does a male employee receive a higher salary than a female employee who does the same kind of work at least as well?

Although I do not want to focus too closely on linguistic form, I think that requests for consolation and moral judgment can often be phrased as *why-should* questions. Why should our child have died? Why should the man receive higher pay than the woman? Requests for scientific explanations can usually be phrased as *why-does* or *why-do* questions. "Why should" questions ask for justification; "why does/do" questions ask for factual information.

- A particularly important sort of explanation involves *motives.* When asked why you bought a novel by a particular author, you might reply that it promised to be entertaining reading for a long trip on which you were about to depart. Explanations that appeal to purposes constitute a familiar type; they are often sought and given in everyday life.

Where human behavior is concerned, explanatory appeals to conscious purpose are unobjectionable. Moreover, it may be that much human behavior can be explained in terms of unconscious purposes. Purposes, conscious or unconscious, are also appropriate components of explanations of the behavior of at least some animals other than humans. Male orangutans offer pieces of meat to females in order to receive sexual favors.

There is a serious danger here. Because references to motives are so common and so satisfying where human behavior is concerned, people have sometimes concluded that the *only* form of genuine explanation is in terms of the motives of humans, other animals, other material objects, or supernatural beings. From this point there are two ways to go.

The first is to attribute mental states to things we normally regard as inanimate objects. The result is animism—a blatant form of anthropomorphism. The second is to deny that science has the capacity to explain natural phenomena. This view was rather widely shared in the early part of the twentieth century, and remnants of it are still with us today. Perhaps this attachment to purposive explanations accounts for the frequent claim that genuine explanation cannot be found in science, but only in theology or metaphysics.

When an explanation makes reference to motives, purposes, or ends, we call it *teleological.* Such an explanation involves *final causes* in Aristotle's sense. Aristotelian physics is teleological: nature *abhors* a vacuum and terrestrial matter *seeks* its proper place in the cosmos. Newtonian physics is nonteleological; it operates according to *efficient causes.* The biblical account of the origin of species, which explicitly invokes God's purposes, is teleological. Charles Darwin's evolutionary theory is nonteleological; it explains the species in terms of natural selection. Historically both physics and biology made significant progress by eliminating teleological explanations.

Nevertheless, contemporary biology does employ explanations in terms of *functions.* Consider a famous example. During the industrial revolution pollution from the factories in Liverpool darkened the naturally light bark of the plane trees in that area. The peppered moth, which lives in these trees, had possessed a light color, which served as camouflage to protect it from predators. However, when the color of the bark darkened and the light color was no longer an effective camouflage, the species developed a dark color, which then fulfilled that function. Subsequently, when the pollution was substantially reduced and the bark of the plane tree reverted to its natural light color, the peppered moth regained its former light color. This type of explanation, which seems clearly to appeal to an end—the avoidance of predation—raises the question of the status of functional explanations in the biological sciences. Are they legitimate explanations or merely heuristic "explanation sketches" that require causal underpinning? Or are they already fully causal explanations?

Functional explanations are not confined to the biological sciences. In anthropology and sociology we find that certain practices in various societies are explained in terms of their social functions. A rain dance, performed during a drought, may not have any influence on the weather, but it may enhance social cohesion at a time when the community is under serious stress. It seems to me that Larry Wright's (1976) *consequence-etiology* analysis shows how functional explanations can be understood in terms of straightforward causal relations, thereby qualifying functional explanations as a legitimate subset of causal explanations. (See Hitchcock [1996] for technical details.)

As soon as we enter the area of explanation in the behavioral sciences, we encounter a number of controversial issues related to human behavior. One fundamental question is whether human actions can be explained scientifically, or whether freedom of choice precludes scientific explanation. Another is the question whether human actions can be explained causally or whether—at least as far as intentional behavior is concerned—we have to invoke reasons that are not analyzable in causal terms. Where intentionality is involved, we must ask whether human behavior can be explained without recourse to an account of meanings. Whereas I made a sharp distinction earlier between explanations of meanings and explanations of natural phenomena, it is sometimes said that any adequate understanding of human behavior must involve interpretations of meanings. The reason is that many acts are performed because of their meanings. Religious rites, for example, would often be unintelligible in the absence of their symbolic significations.

Without insisting that all scientific explanations are causal, we can still maintain that knowledge of causal relations enables us to explain a vast range of natural phenomena, and that such explanations yield understanding of the world and what transpires within it.

3. Understanding

As I have already noted, the term "understanding" requires clarification; let us now focus briefly on it. It is taken up in greater detail in essay 5, where figure 5.1 provides a chart for the cluster of concepts it represents. "Understanding" is a broad term that carries many psychological overtones; we can distinguish four major types.

- *Empathic understanding:* In many contexts understanding refers to empathic sharing of feelings. "I understand," spoken to someone who is grieving over the death of a loved one, means that the speaker has experienced the pain of such a loss and, perhaps, experiences that sadness in the present case. To understand another person's behavior is to know that person's motives, values, desires, and beliefs. In a slightly different sense one might claim to understand some particular person when one can predict that person's emotional reactions and behavior. If, however, humans are free agents, predictability may be problematic.

Empathic understanding is based on emotive factors, on feelings and values. The psychological aspect of such understanding is paramount. Its achievement yields psychological comfort—with oneself, with other humans or animals, or with the world. People deeply crave this kind of psychological understanding.

- *Symbolic understanding:* A certain type of understanding relates directly to language; it emphasizes communication and meanings. We speak of people understanding English, Italian, or French. Communication occurs by means of symbols, but not all symbols are linguistic. The lotus design, for example, has deep religious significance in Buddhism. Symbols convey both factual information and emotive content. For this reason symbolic understanding is closely related to all of the other kinds of understanding mentioned here.

One view of human behavior is that, because it is meaningful—that is, purposive or intentional—scientific explanation does not yield understanding of it. Rather, understanding requires interpretation of meanings. This view is closely associated with the idea that human behavior cannot be explained causally because it must be understood in terms of reasons, and reasons are not causes.

- *Goal-oriented understanding:* We can achieve a different kind of understanding by invoking purposes, aims, or goals. This type of understanding splits into two subtypes corresponding to two types of explanation. First, human behavior can often be explained in terms of conscious motives and purposes. For example, I carry water on a desert hike because I expect to be thirsty, and no drinking water will be otherwise available. This is, of course, a *teleological* explanation. The familiarity of such explanations makes them seem especially appropriate when understanding is required. We readily extend such explanations to the behavior of other humans. This kind of understanding tends to blend into empathic understanding, for knowledge of the desires and values of others enables us to know, if not share, their feelings.

In the second subtype *functional* explanations provide understanding. We understand why our blood contains hemoglobin: its function is to transport oxygen from the lungs to other parts of the body, where it is needed for the metabolic processes that sustain the life of the organism.

The possibility of divorcing function from conscious purpose is crucial. If the two are not separated, supernatural purposes are apt to be invoked. As we have seen, functional explanations appear in the biological and social sciences; indeed, they are especially crucial to evolutionary biology. A naturalistic causal interpretation of function enables us to accept functional explanations as legitimate components of natural science.

- *Scientific understanding:* The fourth major type of understanding is linked to scientific explanations in the physical, biological, behavioral, and social sciences. Its cognitive dimension is primary. Scientific explanations must be based on well-established scientific theory and fact; psychological comfort is not at issue. This point deserves emphasis. For example, along with a majority of physicists, I believe that our universe is indeterministic. This conviction, which is grounded in what I take to be a sound interpretation of modern physics, means that some explanations of natural phenomena are irreducibly statistical. Some people feel a deep psychological discomfort with indeterminism. This sort of psychological comfort or discomfort is utterly irrelevant to the correctness of objectively grounded explanations.

The radical ambiguities of "explanation" and "understanding" create almost endless opportunities for obfuscation and confusion. I have gone on at some length describing these various types of explanation and understanding in order to avoid such problems. In particular, we must recognize the possibility of *scientific* understanding, grounded in *scientific* explanation, that is free from considerations of psychological satisfaction and comfort. In so saying I intend neither to disparage these psychological factors nor to denigrate the other kinds of explanation and understanding; I mean only to demarcate one portion of that area as the subject of this book. Explaining meanings and explaining how to do things are essential to many aspects of our lives. Empathic understanding and symbolic understanding are likewise crucial. The total range of human understanding is vast, and I am examining only one part. But it is a part whose importance can hardly be overrated.

Scientific explanation is not a simple matter; the twentieth century has seen many deep differences regarding its nature. We must look at a variety of basic conceptions. We must consider the role of causality in scientific explanation. We must ask why the most influential philosophical theory of scientific explanation in the twentieth century explicitly excluded causal considerations from its account.

The reluctance to introduce causal considerations lies in the problems Hume raised: scientifically minded philosophers have sought to avoid secret powers and mysterious connections. I have tried to advance viable answers to these problems and to integrate them with my treatment of scientific explanation. Nevertheless, I do not claim that all scientific explanation is causal; instead, I distinguish two general types of scientific explanation—one depending on causal and/or mechanical factors, the other emphasizing theoretical unification. Although many philosophers see a conflict between these two conceptions, I find them mutually compatible and complementary. One and the same phenomenon can often be explained in both ways, each providing a different sort of understanding. The present collection of essays, which deal with causality and explanation, is offered in the hope that it will provide some new insight into causality and its role in understanding our world.

Part I

INTRODUCTORY ESSAYS

Causality, Determinism, and Explanation

The essays in Part I provide an elementary overview of the topics covered in this book, and they introduce the major concepts that will be found in the four remaining parts. Although they may not be easy, they should be accessible to serious readers with little or no prior exposure to philosophy. They represent, on a relatively nontechnical level, my most recent thoughts on the main issues treated in the book.

Essay 1, "A New Look at Causality," offers a novel approach, in terms of causal processes and causal interactions, to the fundamental philosophical problems raised by David Hume in the eighteenth century. His classic critique aroused philosophical controversy that remains unabated. In this essay I show how twentieth-century science has opened a new way to attack these issues.

Essay 2, "Determinism and Indeterminism in Modern Science," clarifies these two concepts that figure prominently not only in discussions of human freedom but also in theories of scientific explanation. The question of the status of scientific explanation in an indeterministic world arises repeatedly in subsequent essays. Even if we do not yet have the final word on the truth or falsity of indeterminism, we need to take account of its possibility in framing philosophical theories of scientific explanations.

Essay 3, "Comets, Pollen, and Dreams: Some Reflections on Scientific Explanation," examines three basic approaches to scientific explanation that have been advocated by influential writers in the second half of the twentieth century and are still held. It exposes fundamental differences in these concepts that emerge when they confront explanation in scientific contexts in which statistical laws and functional explanations play major roles.

Essay 4, "Scientific Explanation: Causation *and* Unification," explores the possibility of a rapprochement between two dominant traditions regarding scientific explanation that have generally been seen as mutually incompatible. It shows how progress in the development of both approaches has eradicated many—perhaps all—of the grounds for conflict between them.

Essay 5, "The Importance of Scientific Understanding," shows how scientific explanations enable us to understand the universe in which we live. It exhibits the value of such understanding as we move from the twentieth to the twenty-first century.

1

A New Look at Causality

What is causality? Philosophers have been asking this question for more than two millennia; it is a subject fundamental to metaphysics, epistemology, ethics, philosophy of mind, philosophy of science, and many other fields. However, as we all know, this concept was dealt a severe blow in the eighteenth century by David Hume. Hume sought a necessary connection between cause and effect, but he could not find one either in formal reasoning or in the physical world. We need to take another look.

1. The Problem

Formal reasoning cannot reveal causation because we cannot deduce the nature of an effect from a description of the cause or the nature of the cause from a description of an effect. First, to use one of Hume's favorite examples, suppose that one billiard ball is moving across the table toward a second ball that is motionless. It is impossible to determine a priori what will happen when they meet. There is no contradiction in supposing that the second ball would remain in place while the first returned in the direction from which it came. An unlimited supply of other possibilities can be imagined; for example, the first ball might jump over the second, or the two might vanish in a puff of smoke. Some laws of nature might be violated, but not any law of logic.

We can show quite easily that Hume is right. Consider the first possibility. Suppose that, unknown to us, someone had screwed the second ball firmly to the table. In that case a direct hit from the first ball would result in the effect we imagined: the second ball would remain in place and the first would return from whence it came. We can try the experiment and verify the result. Now, whatever actually happens cannot be logically impossible, so our new description of the situation must be logically consistent. To be sure, in describing the situation we have included some details beyond those Hume specified. However, our description was consistent, and by removing parts to make it agree with Hume's specification, we cannot make it inconsistent. The only way to make it inconsistent would be to *add* something that contradicts our description. So Hume was demonstrably correct in saying that the occurrence of the effect cannot be deduced from the description of the cause.

The same can be said about the possibility of deducing the nature of the cause from a description of the effect. Hume offers another example. Suppose we show a diamond and a piece of ice to someone who is logically competent but who has had no experience of objects of these types. Could such a person, by pure reason alone, deduce that one of these objects is produced by enormous heat and pressure while the other would be utterly destroyed under such conditions? Certainly not. Everyday experience easily sustains Hume's argument. To offer an example—obviously not one of his—suppose you observe that an illuminated ceiling light goes off. This might be because someone turned it off at a wall switch, or because the bulb burned out, or because a circuit breaker was flipped. Further investigation might enable you to find out about the cause, but just from the fact that the light went off and nothing else, it is impossible to deduce which alternative was the actual cause. We know that all of these are logically possible because all of them actually occur on various occasions.

We have examined the inference from cause to effect and the inference from effect to cause, and we have seen that there is no deductive entailment in either direction. If there is any connection in either direction between the cause and the effect, it is not one of logical necessity. Hume has sometimes been criticized for depending too heavily on our psychological ability to conceive or imagine. In the foregoing discussion I have been careful to formulate the arguments in logical rather than psychological terms. We have found that Hume's arguments still hold.

Since deductive logic does not provide the answer to our question about causality, we naturally turn to empirical investigation. Suppose that a child develops a skin rash. This follows shortly after a picnic at which various foods are served in a setting where various kinds of vegetation are present. Perhaps the child ate a large dish of strawberries, but also watermelon and pineapple, and played in a patch of weeds. By inspection we know that the consumption of strawberries was followed by the appearance of the rash, but we do not know whether the one was a cause of the other or whether it was a mere coincidence. We observe two events, the putative cause and the supposed effect. We cannot say a priori that the rash actually had a cause. Perhaps it just occurred, by chance, so to speak. We do not observe a third entity, the power of the one to produce the other, or simply the causal relation between the two. Given only the two events, we do not know whether they are causally related or not.

We can, however, make further observations, even to the point of conducting experiments. We can see whether the rash develops on other occasions when the child eats strawberries but not watermelon or pineapple. We can serve the food indoors, where the child does not come in contact with the plants that were present at the picnic. If we find that the rash occurs regularly after the consumption of strawberries, but does not occur regularly in the other circumstances, we conclude that the eating of strawberries is the cause of the rash. The rash does not appear haphazardly, and it is not regularly associated with any other factor. Hume concludes that it is only by repeatedly observing associated events that we can establish the existence of causal relations. If, in addition to the separate events, a causal connection were observable, it would suffice to observe one case in which the cause, the effect, and the causal relation were present.

Hume concludes from the foregoing considerations that, in situations in which we believe that there is a causal relation, we perceive three features: (1) the temporal priority of the cause to the effect; (2) the spatiotemporal contiguity of the cause to the effect; and (3) the fact that on every occasion on which the cause occurs, the effect follows (constant conjunction). However, according to Hume, we cannot find a physical connection between the cause and the effect; the connection does not exist in the physical world outside of our own minds. Instead, a psychological phenomenon occurs. Since the effect always follows the cause, we are primed to anticipate the effect on the next occasion on which the cause presents itself. The relation between cause and effect is custom and habit. Like Pavlov's dogs, who were conditioned to salivate when a bell rang, we have a conditioned reflex. If there had never been any humans or other intelligent beings, there never would have been causes and effects—that is to say, there never would have been causal relations—in the physical universe. The events would occur, but the causal relation would not exist.

Hume's thesis that we cannot observe any hidden power of a cause to bring about its effect stands in direct opposition to John Locke's earlier claim that we can sometimes perceive just that sort of power, namely, in the case of human volitions. According to Locke, when one decides to raise one's arm, and the arm goes up, one is directly aware of the power of the volition to bring about the action.

Hume rejected Locke's claim on the ground that there is no direct connection between the volition and the action. In order for the volition to lead to the movement of the arm, there must be events in the brain, in the nerves connecting the brain to the muscles of the arm, and in the muscles themselves. If we interpolate this complex series of events between the volition and the motion of the arm, we simply add to the problem. Instead of seeking just the connection between a cause C and an effect E, we need to find the causal connections between the intermediary events $I_1, I_2, \ldots, I_n$. Letting the arrow represent a supposed causal connection or causal power, we may say that instead of locating a single causal connection, $C \rightarrow E$, we must deal with a whole series, $C \rightarrow I_1 \rightarrow I_2 \rightarrow \ldots \rightarrow I_n \rightarrow E$. The problem is not mitigated; it is exacerbated. Similar remarks could be applied to the case of skin rash following the consumption of strawberries. If one had responded to that example by saying that medical science could, in principle, discover the physiological connection between the eating of strawberries and the appearance of the rash, Hume could have responded that such investigations could reveal only links in the causal chain, not the causal power of each link to produce the next one.

After Hume many philosophers sought to avoid the conclusion that the physical aspect of causality involves nothing more than temporal priority, spatiotemporal contiguity, and constant conjunction, and that any further connection or power is simply custom or habit. The story of these attempts is too long to recount here, but a few observations are in order. Usually when philosophers discuss causality, they think of two facts (or types of facts) C and E or of two events (or types of events) C and E between which there is a relation R. Among the problems discussed are whether the terms C and E should be taken to refer to facts or events; this is often treated as a linguistic question. Another question is whether these terms refer to individual facts or events, or whether they should be taken to designate classes of facts or events. Sometimes the principal topic is the logical structure of the relation R, for example, necessary condition, sufficient condition, or a combination of the two. A standard survey of these approaches is offered in John Mackie's *Cement of the Universe* (1974). I want to suggest a different approach, one that emphasizes the physical, as opposed to the logical and linguistic, aspects of causality.

2. Physical Causality

Let us take Hume's challenge seriously: let us try to find a physical connection between cause and effect. This seems possible in the twentieth century, even though it was not in the eighteenth, because the theory of relativity is now available. The first step is to focus our attention on processes instead of events (or facts). We will see that causal processes are precisely the connections Hume sought, that is, that the relation between a cause and an effect is a physical connection (although it may not be the *necessary* connection Hume referred to).

The second step is to distinguish between causal and noncausal processes. According to the theory of relativity, it is not possible to send a signal at a velocity greater than the velocity of light in a vacuum. One often hears it said that nothing can travel faster than light, but this proposition, without qualification, is not true. There are 'things', which I call "pseudo-processes," that can travel at arbitrarily high velocities, but these 'processes' cannot transmit information. Therefore they are not signals.

Consider an example (fig. 1.1). Suppose that we have a building, something like the Roman Colosseum,[1] but round, in the center of which is a beacon that rotates rapidly. Light rays travel in straight lines from the beacon to the wall. Light rays, like other types of electromagnetic radiation, can transmit information from one place to another. For example, one can impose a mark—that is, a modification—on a process of this type; then, without further interventions, this mark will persist in the process for a period of time. These processes are causal; the capacity to transmit marks is an indication of their causal nature. For example, if one puts a piece of red glass in the path of a ray of white light, the light will become red and remain red beyond that point. One could install a red lens in the beacon; if someone did so, the rays would be red, even though the source still emitted white light.

Suppose that we are still in the round Colosseum at night, without the red lens in the beacon. When the white light from the beacon reaches the wall, a luminous spot travels around the wall as the beacon rotates. One can make this spot red at any point by putting a piece of red cellophane at that point; when the luminous spot arrives at that point, it will become red, but immediately after, as soon as it passes that point, it will become white; it will no longer have the red color. It is impossible to apply any mark to the trajectory of the luminous spot that will, without further interventions, persist in the process for a period of time.

For a more everyday example, consider the cinema, in particular, a spaghetti western movie. The actions of the cowboy and his horse are pseudo-processes; the causal pro-

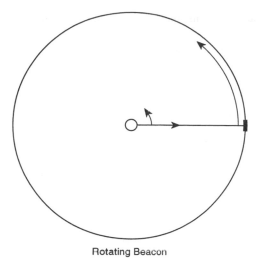

Rotating Beacon

Causal vs. Pseudo-Processes

Figure 1.1

cesses are the light rays that go from the projector, through the film, to the screen. If a spectator, in a state of extreme excitement, draws a pistol and shoots the cowboy, this has no lasting effect on the cowboy, but it creates a hole in the screen. The screen, like all material objects, is a causal process, and the hole will remain in the screen until someone fixes it.[2]

I have been speaking of introducing a mark, but this concept is causal; I must explicate it in terms of a causal interaction. When two processes, causal or pseudo-, meet each other, we have a spatiotemporal intersection. The concept of an intersection is geometrical (in four-dimensional spacetime); it is not a causal concept. We can distinguish two types of intersections: causal *interactions* and noncausal *intersections*. When there is an intersection between two processes in which both are modified, and the modifications persist beyond the place of intersection, this intersection qualifies as a causal interaction. One uses a causal interaction to produce a mark in a process. Recall the two examples of processes in which we introduced marks. When the white light meets the red glass, the glass receives some energy from the light, and the color of the light is changed. When the bullet meets the screen in the cinema, the screen receives a hole and the bullet loses a bit of energy.

We have now explicated the concept of a causal process in terms of a mark, we have explicated the concept of introduction of a mark in terms of a causal interaction, and we have explicated the concept of an interaction in noncausal terms. I would suggest that the causal connection Hume sought is simply a causal process.[3] For example, when I arrive at home in the evening, I press a button on my electronic door opener (cause) to open the garage door (effect). First, there is an interaction between my finger and the control device, then an electromagnetic signal transmits a causal influence from the control device to the mechanism that raises the garage door, and finally there is an interaction between the signal and that mechanism. There are complexities that I have not mentioned in this example, but they involve additional causal interactions and transmission of the types I have just been discussing.

3. Counterfactuals

But there is a problem. I had been expounding something like the foregoing criterion for distinguishing between causal processes and pseudo-processes prior to writing *Scientific Explanation and the Causal Structure of the World* (1984b). On one such occasion Nancy Cartwright presented the following objection to the mark method. Returning to the round Colosseum, let us suppose that a few nanoseconds[4] before the luminous spot reaches the red cellophane, someone were to install a red lens in the beacon. In this case the luminous spot would become red (because of the cellophane at the wall) and remain red (because of the red lens on the beacon) after going past the red cellophane. It would seem that the luminous spot had transmitted a mark.[5] With great philosophical regret I realized that it was necessary to add a counterfactual condition, that is, that the color of the luminous spot would not have changed if the mark had not been introduced into the process. This condition blocked Cartwright's example because, regardless of the presence or absence of the red cellophane, the spot would have turned red and remained red.

The same sort of problem arises in the explication of causal interactions. It was therefore necessary to add a counterfactual condition to the explication of that concept. Roughly speaking, this condition says that two intersecting processes, each of which would have proceeded without modification in the absence of an intersection, interact causally if and only if both are modified at the intersection in ways that persist beyond the locus of intersection.[6]

As a result of the incorporation of these counterfactual conditions, Philip Kitcher (1989) characterized my theory of causal processes and interactions as just another type of counterfactual theory of causality. This critique was disconcerting; but it was not entirely well founded. Because of its emphasis on physical connections, my theory differed fundamentally from explications based on an analysis of causality in terms of forms of conditional statements. However, given well-known problems concerning the interpretation of counterfactual conditionals, their absence from the analysis of causal concepts would be a boon.

4. Conserved Quantities

In a penetrating critique of my theory of causal processes and interactions, Phil Dowe (1992c) proposed an approach to the characterization of causal processes in terms of conserved quantities. According to Dowe, a process is causal if it manifests a conserved quantity, for example, linear momentum, angular momentum, energy, or electric charge. In addition, an intersection of two processes is a causal interaction if there is an exchange of a conserved quantity between them. When there is a collision of two billiard balls

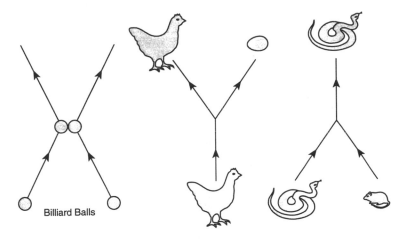

Types of Causal Interaction

Figure 1.2

(Hume's favorite example), the linear momentum of each ball is different after the intersection than it was before. We say that a quantity is conserved if we accept a theory in which there is a law of conservation, for example, the first law of thermodynamics, the law of conservation of energy. If this supposed law were false, energy would not be a conserved quantity, but at present we believe that it is true and that energy is a conserved quantity. We can never have absolute certainty regarding any putative law of nature; therefore, we do not have absolute certainty about any conserved quantity. That fact notwithstanding, we do our best to discover the laws of nature, and consequently, to find out which quantities are conserved.

The first advantage of the theory of conserved quantities over the theory of transmission of marks is the absence of counterfactual propositions. In the theory of mark transmission, we must say that a process is causal if it has the *capacity* to transmit marks, that is, a mark could be transmitted if it were introduced. In addition, as a result of Cartwright's critique, we must say that the process would not have changed in some specified respect if the mark had not been introduced. In contrast, in the theory of transmission of conserved quantities, we can say that a process is causal if it *transmits* a conserved quantity; this analysis does not involve counterfactual propositions. This is a great philosophical advantage because counterfactual propositions notoriously depend on contextual or pragmatic considerations for their truth value. We are looking for objective causal features of the world.

There is another advantage. In the theory that I advocated earlier (Salmon, 1984b, pp. 181–182), I mentioned three types of causal interactions, which I called X, Y, and λ (see fig. 1.2). In that theory I was able to handle only those of type X, but I could not include those of the other two types. And interactions of the other two types are extremely important. When an entity divides into two parts—for example, when an atomic nucleus emits a particle or a hen lays an egg—we have interactions of the Y type. When two entities unite into one—for example, an atom absorbs a photon or a snake eats a mouse—we have interactions of the λ type. The theory of conserved quantities includes, without difficulty, the three types of interactions. It is sufficient that the conserved quantity remain conserved.

I have been talking about *laws* of conservation, but this terminology can give rise to a profound problem. Many philosophers draw a distinction between genuine laws of nature and generalizations that are accidentally true. It is not easy to explicate this distinction. Fortunately this is not necessary for purposes of the conserved quantity theory because the theory does not require *laws* of conservation. It suffices that the proposition stating that energy is conserved be true. The same goes for analogous statements about other conserved quantities. We can completely avoid the problem of laws of nature.

A few years ago I believed that the explication of the concepts of causality in terms of transmission of marks was correct, but I no longer think so. I believe that the mark method is an *extremely useful* tool for the discovery and study of causal processes. A paradigmatic example is the use of radioactive tracers in studying physiological processes. But the mark method does not furnish an adequate explication of the concept. I agree with Dowe that the theory of conserved quantities gives a better explication of the concepts of causal process and causal interaction.

5. Transmission

Nevertheless, it seems to me that Dowe's theory is incomplete—that there is a grave lacuna in the theory of conserved quantities as he has presented it. In fact, I think that the concept of causal *transmission* is a principal part of a satisfactory explanation of the causal structure of the world, but it is not present in Dowe's theory. I believe, however, that we can close that gap—that we can clarify this concept. For this purpose we must return to antiquity, to Zeno's well-known paradox of the arrow.

Zeno said that an arrow in flight cannot move, because at each instant it is precisely where it is, occupying a space exactly equal to itself. In that instant it has no time in which to move. In that place, moreover, it has no space in which to move. Therefore, the arrow cannot move. It is often said that Zeno's problem was his failure to understand the distinction between instantaneous motion and instantaneous immobility, because the infinitesimal calculus did not exist in antiquity. In the calculus, they continue, we can now define instantaneous velocity as the derivative of position with respect to time, that is, dx/dt. Thus, the distinction between being at rest at an instant and being in motion at an instant is simply the distinction between an instantaneous velocity equal to zero and an instantaneous velocity different from zero. But this response is not adequate. To define the derivative it is necessary to consider the limit of the average velocities in periods of time greater than zero—precisely those movements that Zeno held to be impossible (see fig. 1.3). If we consider the position of the arrow at only one instant, without considering its positions at other instants, we cannot determine the instantaneous velocity; in fact, in these circumstances the concept of instantaneous velocity has no meaning. (See Salmon, 1975b, chap. 2.)

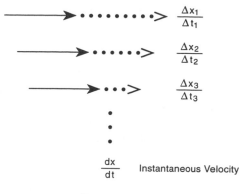

Zeno's Arrow

Figure 1.3

Bertrand Russell (1922b) took account of this fact and, in answer, offered his famous "at-at" theory of motion. According to this theory, motion is no more nor less than being *at* precise points of space *at* precise instants. Movement is a correspondence between the positions and the times; there is nothing more to it. If someone were to ask how the arrow moves from point *A* to point *B,* the answer would be that it is found in the various points between *A* and *B* at different times. If someone were to ask how the arrow gets from one point to the next, the answer would be that in a continuum there is no next point. We describe the motion by means of a mathematical function; a function is a correspondence between two sets of numbers. In the case of movement it involves a correspondence between positions and instants. It seems to me that Russell was absolutely right.

Transmission is a type of movement. When a mark is introduced into a process, the modification moves with the process; for example, a change of color from white to red goes along with a light pulse after the red lens has been placed in its path. According to the "at-at" theory of *transmission,* the mark is transmitted from point *A* to point *B* by being present in the process at every point between *A* and *B* without further interactions with other processes. The same consideration applies to conserved quantities. Let us return for a moment to our Colosseum. In the luminous spot that travels on the wall, we could suppose that there is a constant quantity of energy. However, in this case the energy is not being transmitted along the trajectory of the spot; on the contrary, the energy is present because, and only because, there are many interactions with the light rays from the central beacon. The energy is being transmitted from the central beacon to the wall by the causal processes consisting of the light rays traveling from the beacon to the wall. If there were no external source of energy, there would not be any energy at the wall; indeed, there would be no luminous spot. In contrast, a light pulse, once it has been emitted, carries a fixed quantity of energy without an external source. The same point applies to other conserved quantities.

6. Conditions: Necessary and/or Sufficient

I would like to consider once again the approach to causality in terms of events or facts, one of which is a sufficient or necessary condition of another. This viewpoint has been adopted by many philosophers. Suppose, for example, that a person has a whistle and a dog. When the person blows the whistle, we hear nothing, but in every case the dog comes. We learn inductively that blowing the whistle is a *sufficient* condition for the appearance of the dog. But we do not understand the relation between the cause and the effect unless we know that there is a sound wave that we cannot hear because the frequency is too high for humans but not too high for dogs. The dog comes because it hears the whistle of its master. The wave is the causal process that provides the connection.

Another example involves a tragic accident in California many years ago in which a philosopher died. His car was struck by another car, which skidded on wet leaves on the surface of the street. In those circumstances the interaction among the tires of the other car, the wet leaves, and the surface of the street was a *necessary* condition of the accident. If the wet leaves had not been there, no collision between the two cars would have occurred. The connecting process was the movement of the other car through a red traffic light.

John Mackie's famous theory of causality is more complex and sophisticated than the simple theories of sufficient or necessary conditions (1974, p. 62). According to Mackie, a cause is an INUS condition, where "INUS" is an acronym for a condition that is an Insufficient but Nonredundant (necessary) part of a condition that is Unnecessary but Sufficient. Suppose, for example, that a barn burns down. It might have been caused by a careless smoker dropping a burning cigarette in the barn; it might have been caused by embers from a nearby forest fire falling on the barn; it might have been caused by a stroke of lightning; it might have resulted from spontaneous combustion engendered by fermentation of fresh hay stored in the loft. None of these events is a necessary condition of the fire, but any one of them might be sufficient. However, no fire will occur in any of these cases unless some additional factors are present. For example, if the dropping of the cigarette is to cause the fire, the cigarette must fall on some flammable material such as dry straw, and the incident must go unnoticed by anyone who could have put out the fire before it engulfed the whole building. However, these other conditions would not suffice to start a fire; therefore, the dropping of the burning cigarette was a nonredundant part of this particular sufficient condition.

Similar remarks apply to the other sufficient conditions. Embers falling on the roof of the barn will not produce a fire unless the roof is made of a flammable material and the ember is not doused with water just as it arrives. Lightning does not ignite a fire if the barn is protected by lightning rods. Spontaneous combustion will not occur if salt is used to inhibit fermentation and adequate ventilation permits heat to escape.

Looking more closely at the case of the carelessly dropped cigarette, we see that a fairly complicated network of causal interactions and causal processes is involved. The burning cigarette must travel from the hand of the smoker to a place where flammable material lies. The burning end of the cigarette must interact with the dry straw to ignite a fire. The fire must spread from the straw to the wooden floor or sides of the barn, where another interaction ignites that part of the barn. The heat from the fire must spread, igniting other parts of the barn, until the entire building is burning. Similar closer analyses of the other three sufficient conditions would reveal complex patterns of causal processes and interactions in those cases as well.

Hume offered a regularity view of the nature of causation; its fundamental characteristic is constant conjunction. Of course, as I have already mentioned, conditions of temporal priority and spatiotemporal contiguity are also required. However, it has long been recognized that night follows day and day follows night in just such a regular fashion, yet we do not want to say that day causes night or night causes day. So philosophers who support a regularity view impose further conditions. Mackie's appeal to INUS conditions is a sophisticated regularity account. Regularity accounts, whether simple or complex, follow Hume in eschewing causal powers and causal connections. On the view of causality I am advocating, causal connections exist in the physical world and can be discovered by empirical investigation. In this respect my theory goes beyond any regularity theory. There is more to causality than regularity; the causal connections furnished by causal processes explain the causal regularities we find in the world.

7. Quantum Mechanics

The field of science most problematic for any theory of causality is quantum mechanics. In 1935 the classic article by Einstein, Podolsky, and Rosen appeared; it raises the problem of action-at-a-distance. More recently, John Bell demonstrated a theorem according to which, if the quantum theory is true, then normal causality does not exist in that domain. In addition, in the 1980s Alain Aspect and his group showed by means of experiment that the predictions of quantum mechanics are correct. There are remote correlations that we cannot explain by means of causal processes. In principle, there are perfect correlations between results of measurements made in two places between which there is a large distance (over ten meters). In a series of measurements of pairs of photons, the result of a measurement of one is necessary and sufficient for the result of the measurement of the corresponding photon. Analogous results can, in principle, be established on pairs of particles. According to many physicists and philosophers, we do not have a causal relation precisely because there are no causal processes to furnish the connections between the results at the two points. This is a serious problem because we do not understand the mechanisms that produce these correlations. This situation is discussed in greater detail in "Indeterminacy, Indeterminism, and Quantum Mechanics" (essay 17).

The problem in quantum mechanics is *not* the problem of indeterminism. Unfortunately, on one occasion Einstein said that God does not play dice with the universe. This remark has been quoted frequently, and almost everyone has interpreted that saying as an objection to indeterminism, but—as Einstein said on many subsequent occasions— that was not an accurate expression of his thought. Einstein was worried about "spooky action-at-a-distance," precisely the same problem that I discussed in the preceding paragraph. Also, for me, indeterminism does not present a problem, but action-at-a-distance worries me deeply (see Mermin, 1981, 1985, for fuller details).

It seems to me that indeterminism is actually compatible with causality; in fact, at present several very interesting theories of probabilistic causality are available (see "Probabilistic Causality" [essay 14]; see also Humphreys [1989]; Eells [1991]; and Hitchcock [1993]). In the theory of causality that is based on causal processes and causal interactions, causal relations do not have to be deterministic. For example, when there is an interaction between two causal processes, several different results may be possible, each of which has its own probability. In a collision between an energetic photon and an electron, as investigated by Arthur Compton, there is a range of possible angles for the trajectory of the electron and a range of possible energies; nevertheless, this constitutes a causal interaction (see "Why Ask, 'Why?'?" [essay 8]).

8. Conclusion

In this essay I have sought a physical relation that constitutes a causal connection; I have suggested that causal processes fill the bill. We have seen, contra Hume, that there are causal connections in nature, but that these connections need not be necessary. It seems to me that we have found a correct answer to Hume's question regarding causality. If this is right, causal connections exist in the physical world, and not just in our minds. Moreover, causality is neither logical nor metaphysical; causality is physical—it is an objective part of the structure of our world (see "Causality without Counterfactuals" [essay 16] for further details).

Notes

This is an English translation, somewhat expanded, of my paper, "La Causalità," that was presented at the University of Florence, under the auspices of the Florentine Center for History and Philosophy of Science, December 1993.

1. North Americans might prefer to take the Astrodome as their example.
2. The mark method and the distinction between causal processes and pseudo-processes were given by Hans Reichenbach ([1928] 1957). He referred to pseudo-processes as "unreal sequences."
3. This connection need not be necessary; such connections are discussed in "Probabilistic Causality" (essay 14).
4. A nanosecond is 10^{-9} seconds, that is, a billionth of a second in American terminology (one thousand-millionth of a second in British terminology). (Throughout this book, I use "billion" in the American sense to mean a thousand million.) The speed of light is almost exactly one foot per nanosecond.
5. "An 'At-At' Theory of Causal Influence" (essay 12) was published prior to Cartwright's mention of this counterexample. That essay remains significant because it first introduced the analysis of causal propagation.
6. See Salmon (1984b, pp. 148–155, 171) for the precise counterfactual formulations.

2

Determinism and Indeterminism in Modern Science

[The concepts of determinism and indeterminism are used repeatedly in the following introductory essays without any serious attempt to clarify them. This essay, first published in 1971, embodies my effort to provide a reasonably comprehensive and accurate introductory explication of these notions and their close relatives. Causality and explanation are intimately involved. In his Lakatos Award-winning book *A Primer on Determinism,* my esteemed colleague John Earman complimented it as "a quick and very readable survey of the received philosophical opinion on this topic" (Earman, 1986, p. 3). This is not unmitigated praise; the theme of his book is that "the received philosophical opinion" is thoroughly confused and mistaken. I believe that his assessment is largely correct. He writes:

> Determinism is a perennial topic of philosophical discussion. Very little acquaintance with the philosophical literature is needed to reveal the Tower of Babel character of the discussion: some take the message of determinism to [be] clear and straightforward while others find it hopelessly obscure; some take determinism to be intimately tied to predictability while others profess to see no such bond; some take determinism to embody an *a priori* truth, others take it to express a falsehood, and still others take it to be lacking in truth value; some take determinism to undermine human freedom and dignity, others see no conflict, and yet others think that determinism is necessary for free will; and so on and on. Here we have, the cynic will say, a philosophical topic *par excellence!*
>
> Without any touch of cynicism one may ask what yet another tour of this Babel can hope to accomplish, save possibly to add another story to the Tower. My answer is not at all coy. Essential to an understanding of determinism is an appreciation of how determinism works or fails to work in physics, the most basic of all the empirical sciences; but it is just this appreciation I find lacking in the philosophical literature. . . . Classical physics is supposed by philosophers to be a largely deterministic affair and to

provide the paradigm examples of how determinism works. Relativity theory, in either its special or general form, is thought merely to update classical determinism by providing for Newtonian mechanisms relativistic counterparts that are no less and no more deterministic. And it is only with the advent of the quantum theory that a serious challenge to determinism is supposed to emerge; the challenge is not simply that quantum mechanics is *prima facie* non-deterministic but that "no hidden variable" theorems show that, under plausible constraints, no deterministic completion of the quantum theory is possible.

This picture is badly out of focus. Newtonian physics, I will argue, is not a paradise for determinism; in fact, Newtonian worlds provide environments that are quite hostile to determinism, and some of the alleged paradigm examples of Newtonian determinism are not examples of determinism at all, at least not without the help of props which sometimes have a suspiciously artificial and even question-begging character. The special theory of relativity rescues determinism from the main threat it faces in Newtonian worlds, and in special relativistic worlds pure and clean examples of determinism, free of artificial props, can be constructed. However, the general theory of relativity poses new and even graver challenges, challenges which are currently being addressed on the frontiers of scientific research. The quantum theory, of course, poses challenges of its own; but the first and foremost challenge is not to the truth of the doctrine of determinism but to its meaning in quantum worlds where the ontology may be nothing like that presupposed in the Newtonian and relativistic foundations of the doctrine. (ibid., pp. 1–2)

Having pleaded guilty as charged to these complaints, what possible excuse can I give for republishing my errors? Why not simply refer my readers to Earman's excellent book? The answer is that his book is in large part highly technical, and beyond comprehension to many of the readers I hope to reach. Why not simply omit my essay without comment? Because the issues are so very central to causality and explanation. I have therefore decided to publish the errors, with corrections thereto, in the hope of clarifying some of the issues and persuading those with adequate technical expertise to read what Earman has written. A briefer and somewhat less technical discussion can be found in Earman (1992).]

1. Fate and Destiny

According to a famous legend, the Stoic philosopher Epictetus, who was a slave, broke a vase that his master, who was also a philosopher, treasured. When the master began to beat him, Epictetus protested, "By the philosophy to which we both adhere, it was predestined from the beginning of the world that I should break the vase; I am not to blame and I should not be beaten." His master replied, "By that same philosophy, it was determined for all time that I should beat you," and he continued to do so. This anecdote sums up much of the frustration that people down through the ages have felt when confronted with the problem of free will and determinism. The main purpose of the present essay is to attempt to clarify the notion of determinism and some other concepts closely related to it. Except for a few incidental remarks, I shall leave the problem of free will to other authors.

Determinism is a doctrine that comes in many forms. In ancient mythology, as well as some later religions, it was a crude sort of fatalism. The fates, with conscious intent, decide at the time of one's birth what is going to happen, and nothing anyone can do will make it otherwise. The following passage nicely illustrates the fatalistic view.

> DEATH SPEAKS: There was a merchant in Bagdad who sent his servant to market to buy provisions and in a little while the servant came back, white and trembling, and said, Master, just now when I was in the market-place I was jostled by a woman in the crowd and when I turned I saw it was Death that jostled me. She looked at me and made a threatening gesture; now, lend me your horse, and I will ride away from this city and avoid my fate. I will go to Samarra and there Death will not find me. The merchant lent him his horse, and the servant mounted it, and he dug his spurs in its flanks and as fast as the horse would gallop he went. Then the merchant went down to the market-place and he saw me standing in the crowd and he came to me and said, Why did you make a threatening gesture to my servant when you saw him this morning? That was not a threatening gesture, I said, it was only a start of surprise. I was astonished to see him in Bagdad, for I had an appointment with him tonight in Samarra.[1]

Certain sects of Christianity have maintained that God, who created the world and holds it in his all-powerful control, foreordains exactly what is to happen. This view is known as *predestinarianism*, and it is reinforced by the doctrine of God's omniscience. If God knows with complete certainty and in precise detail what will occur in the future—including whether *you* will go to heaven or to hell—the future is determined to be just exactly what God knows it is going to be. One has no power over one's future and can do nothing to change it. Even one's own acts and apparently free decisions are predetermined by something outside of oneself, over which one has no influence. The feeling of freedom that accompanies many of our decisions and actions is a mere illusion.

Both fatalism and predestinarianism attribute the control of human fate or destiny to some supernatural agency. Most of us, nowadays, reject fatalism as primitive superstition, and few still believe in predestination. Agnostics and atheists find no basis for believing in God at all, and contemporary theists generally believe that God allows humans some measure of freedom. However, it has long been suspected that even a 'hard-headed' scientific worldview would lead to a determinism just as inimical to freedom of choice and action as are fatalism and predestinarianism.

2. Determinism in Classical Physics

In his famous poem *De Rerum Natura*, Lucretius maintains that everything in the universe consists solely of atoms which move about in otherwise empty space, colliding with one another and forming complex arrangements. The earth and the sun, rocks and trees, human beings and other animals—all are just complicated collections of various kinds of atoms. Everything that happens in the universe, including human thought and action, is simply the result of the movements of atoms. Lucretius realized that free will is problematic if we conceive the motions of atoms to be strictly determined by mechanical laws; he writes, "[I]f all movement is always interconnected, the new arising from the old in a determinate order. . . what is the source of the free will possessed by living things throughout the earth?" (Lucretius, 1951, p. 67).

Lucretius tried to resolve the problem by claiming that atoms sometimes swerve spontaneously and without any cause from their otherwise determined courses. Believing that freedom of the will is an established fact, he was led to deny determinism. His argument can be set out as follows:

(1)   If determinism is true, humans do not have free will.
        Humans have free will.
        _____
        Determinism is false.

On the basis of this argument, Lucretius accepted indeterminism as the correct world-view.

Lucretius wrote in the first century B.C., hundreds of years before Isaac Newton formulated the laws that govern the motions and collisions of those tiny lumps of matter the Greek atomists postulated. Before Newton one could have speculated as to whether the laws of mechanics completely determine the motions of material particles; after Newton that question seemed to be closed. From 1687, when the *Principia* (Newton, [1687] 1947) was first published, until about 1900, Newtonian mechanics was tested and retested, confirmed and reconfirmed. Not only did it explain the approximate correctness of Galileo's law of falling bodies and Kepler's laws of planetary motion, but also it accounted for the behavior of the tides and the bulging of the earth at its equator. Moreover, when a delicate laboratory experiment made possible the direct measurement of the gravitational attraction between a large ball of lead and a small one, Newton was found to be right.[2] Newton's laws explained why the orbits of the planets are not perfect ellipses, as Kepler had said, by bringing in the mutual gravitational attractions among the planets themselves instead of considering only the attraction between each planet and the sun. Indeed, when the planet Uranus appeared not to conform to Newton's laws, Neptune was postulated to account for the deviation. Newton's laws enabled astronomers to predict the location of Neptune, and telescopic observation confirmed its existence. These laws led to the discovery of a theretofore unobserved planet.[3]

It is almost impossible to overestimate the impressive success of Newtonian mechanics. As more sophisticated experimental and mathematical techniques were developed to extend the application of Newton's laws to new phenomena, confirming evidence continued to mount. One of the greatest mathematical physicists to contribute to the application of Newtonian mechanics to planetary motion was P.-S. Laplace, who, early in the nineteenth century, wrote:

All events, even those which on account of their insignificance do not seem to follow the great laws of nature, are a result of it just as necessarily as the revolutions of the sun. In ignorance of the ties which unite such events to the entire system of the universe, they have been made to depend upon final causes or upon hazard, according as they occur and are repeated with regularity, or appear without regard to order; but these imaginary causes have gradually receded with the widening bounds of knowledge and disappear entirely before sound philosophy, which sees in them only the expression of our ignorance of the true causes. (Laplace, [1820] 1951, p. 3)

Here is a classic statement of the determinist's position. All events, no matter how large or small, no matter how significant or insignificant, are completely determined by strict laws of mechanics. When people attribute events to final causes (e.g., fate or divine intervention) or hazard (i.e., pure accident or chance), it is only because they are ignorant of the actual facts. The success of Newtonian mechanics offered convincing evidence that all natural phenomena could be explained by the laws of mechanics. As the application of scientific knowledge is pushed further and further, we see that nothing is in principle incapable of explanation on a purely mechanical basis. The argument of Lucretius resulted from the imperfect state of ancient science; if we still accept the first premise, the argument must continue as follows:

(2) If determinism is true, humans do not have free will.
     Determinism is true.
     _____

     Humans do not have free will.

Both of these arguments are logically valid; they differ with respect to their second premises. Newtonian mechanics—so it seemed to Laplace and countless other philosophers and scientists—clearly turned the tide against Lucretius in favor of determinism. Although Lucretius' argument is logically valid, its second premise is not true. What appears to Lucretius to be free will, free choice, or free action is in fact determined, according to Laplace, and any appearance of indeterminacy is only the result of incomplete knowledge of all the causes.

[Although Earman has shown conclusively that Newtonian mechanics is not deterministic, Laplace unquestionably considered it a thoroughly deterministic theory, and so it was generally regarded prior to the twentieth century. I will discuss Earman's arguments at the end of § 5.]

3. Determinism and the Sciences of Life and Mind

If one believed, with Lucretius and Laplace, that there is nothing more than atoms and their motions, determinism seemed unavoidable in the Newtonian era. But not everyone found this materialistic outlook entirely compelling. Descartes ([1641] 1951) had argued persuasively that there are two realms, the physical and the psychological, and that they are quite distinct from each other.[4] One could agree with Descartes that the laws of mechanics, which govern the material world, are strictly deterministic, and still maintain that freedom exists in the mental domain. It is essential to remember the difference between the scientific evidence for determinism in physics and the philosophical speculation that everything is entirely reducible to material atoms and their motions.

Descartes held that only humans, among all the animals, have mental lives; other animals are mere mechanisms. This doctrine reflects the Christian view that only humans have immortal souls. It suffered a sharp setback when Charles Darwin's epoch-making work on evolution in the mid-nineteenth century showed that humans and the other animals are not utterly distinct but closely related. In the face of this result, it might be tempting to suggest that the deepest gulf is not between humans and everything else but, rather, between living and nonliving things. Darwin's work on the origin of species and the descent of humanity did not, after all, explain the origin of life itself. But Darwin's work has an aspect that bears on this distinction as well. Instead of explaining the existence of various species of living things as a result of purposeful "special creation" as recounted in Genesis, he explains it in terms of nonpurposive mechanisms of natural selection. Add to that the chemical synthesis of the 'organic compound' urea from exclusively inorganic substances, and the sharp separation between the biological and the physical realms begins to look less tenable.[5]

In spite of strong indications of continuity between the physical phenomena whose behavior was explained deterministically by Newtonian mechanics and the biological realm of living things, and in spite of humanity's kinship with the rest of the animal kingdom, there still remained the mysterious phenomena of consciousness that seem the almost exclusive property of the human race. One could speculate that chimpanzees, apes, dogs, and horses may have a very primitive mental life, and even, perhaps, a low degree of free will; nevertheless, in humans the conscious aspect is extremely conspicuous (especially to ourselves), and that might be the locus of our freedom. Humans might be so constructed that their physiological aspects are governed by deterministic laws, but their mental lives are still governed by psychological laws that are indeterministic. That is what Descartes had maintained from the outset.

At this point another intellectual giant of the nineteenth century steps into the picture. In an attempt to understand mental illness, Sigmund Freud developed a psychological theory according to which all mental occurrences, even those of the seemingly most trivial sorts, are as strictly caused as are any physical phenomena.[6] Freud postulated unconscious mechanisms that give rise to dreams and neurotic symptoms, and he offered causal explanations of trivia such as slips of the tongue and the pen. Freud's theories were no idle philosophical speculations; they were designed to explain observable phenomena, and they were tested by experience. I do not mean to argue that Freud's theories are still totally acceptable as current theories; neither, for that matter, are Newton's laws. There can be little doubt, however, that he heralded dramatically the possibility that psychological phenomena may be subject to laws just as deterministic as those of Newtonian mechanics. He offers the strong suggestion that our conscious deliberations and 'free' choices can be explained as deterministically as the result of the collision of two billiard balls on a table. By the close of the nineteenth century, determinism seemed well on the way to being a scientifically well grounded view of the entire universe in all of its aspects—physical, biological, psychological, and even social.

4. Determinism and Contemporary Science

Twentieth-century science has in some ways confirmed and extended the grounds for holding a deterministic worldview, and in others it has seemed to undermine determinism. Spectacular progress in the biological sciences has extended enormously the degree to which processes in living organisms can be understood strictly in terms of chemistry and physics. The most striking achievement has been in the field of molecular biology, where the mechanisms of heredity are explained in exclusively chemical terms. The gene is recognized as a large and complex molecule whose properties are fully determined by its chemical structure, and whose capabilities for self-replication are thereby explained.[7]

Protein molecules, the 'building blocks of life,' are known to be constructed out of amino acids. Amino acids have been synthesized, and so have protein molecules and genes. In the near future scientists will very probably have succeeded in synthesizing a viable living organism from inorganic chemicals. [How extraordinarily off the mark was that prediction!][8] These developments constitute an important extension of Darwin's beginnings, and it no longer seems justifiable to deny that the laws that govern the behavior of atoms have complete dominion in the biological realm.

The science of psychology was in its infancy at the turn of the century, but it too has lived up to its nineteenth-century promise. The scientific study of human and animal behavior—from the psychoanalytic, behavioristic, and physiological standpoints—has borne considerable fruit in showing that human experience, feeling, deliberation, choice, and action can be understood in terms of strict psychological laws. It is perhaps too soon to say whether these laws are ultimately reducible to those of physiology, and thence to those of physics and chemistry, but many indications point in that direction. Even Freud believed that the psychoanalytic mechanisms he postulated would eventually be explained in physiological terms. Subsequent neurological studies suggest that it may soon be feasible to explain learning in terms of specific chemical changes that occur in the brain cells, and psycho-pharmacological developments suggest that chemical understanding of feelings and emotions is not too far away. It is certainly plausible, at this point, to suppose that the laws that govern the behavior of atoms also govern our thoughts, feelings, emotions, decisions, and, ultimately, all of our actions.

What sort of picture does this give of a person as a thinking, deliberating, considering, choosing agent? One's life begins when two cells, a sperm and an egg, unite and, following the laws of physics and chemistry, the genes that are present begin to replicate. The individual's heredity, which determines in large measure what one will become both physically and psychologically, is passed on from one's parents through the genes that carry 'the genetic code'. From the beginning, outside influences impinge upon one— even before birth—and these too have a bearing on what one will become and how one will react to further outside influences. Among prenatal influences are, for example, disease virtues such as that of German measles, which may affect the sense organs of the unborn child and deprive it for life of experiences most of us have. When the infant leaves the womb, social factors begin to operate. Again, external causes—vaguely known as 'environmental influences'—become effective. How the person grows depends in part on social factors such as the personalities of the parents and the economic condition of the family, and in part on what one has already become as a result of the hereditary, physiological, and environmental influences that have already operated. Where, if at all, does the individual's genuine choice—freely made—enter the picture? If one grows up to commit murder, is that not just a part of the inexorable causal process in which one is caught up? Is one not just as much a complete victim of heredity and environment as Oedipus was of his fate? Is this not the most reasonable inference from the scientific knowledge that is presently available? Before we try to draw a conclusion, it will be best to take another look at the laws of physics that seem to be fundamental to the whole scheme of things.

As the twentieth century dawned, physics, which seemed so secure, was approaching a crisis. Two great revolutions were about to shake it to its very foundations. One of these revolutions, which consisted in the replacement of Newtonian mechanics by Einstein's special (1905) and general (1916) theories of relativity, did nothing to upset the deterministic character of physics. Newton's laws of mechanics turned out to be not quite correct, so they had to be replaced by some revised laws of mechanics, but ones that were no less deterministic.

The other revolution had a profound bearing on determinism. According to the theories of electromagnetic radiation available at the end of the nineteenth century, a light beam entering a dark box with a small hole will produce inside the box an infinite amount of radiant energy in the ultraviolet region of the spectrum, thus giving rise to a holocaust more terrible than the worst nuclear bomb. This consequence was later aptly called "the ultraviolet catastrophe." Since no such cataclysms occur, something must be drastically wrong with classical physics. In 1900 Max Planck introduced the quantum hypothesis and showed that it yields a far more satisfactory account of blackbody radiation. In 1913 Neils Bohr applied quantization to the orbits of electrons in hydrogen atoms and showed that he could thereby explain the spectral lines emitted by hydrogen gas when it is excited by passage of an electric current. Bohr's theory, unfortunately, did not work at all well for the spectra of helium and the more complex atoms. By about 1926, Werner Heisenberg, Erwin Schrödinger, Max Born, and others had worked out the details of a more satisfactory quantum mechanics, but the theory they produced was fundamentally statistical. The physics of atoms had become indeterministic. For example, it is a consequence of quantum mechanics that atoms of silver, when shot between the poles of a magnet, will be deflected either up or down, but there is no means, even in principle, of determining beforehand which way a particular atom will go. Each one has a 50–50 chance of going either way, and that is all there is to it.[9] Thus, for reasons entirely different from those of Lucretius, modern physicists also attribute indeterministic swerves to atoms in motion.

A natural reaction to examples of this kind is to say that there are real causes that determine which atom will be deflected in which direction, but that we have not yet found them. Some physicists are presently working to find a deterministic theory to replace the current quantum mechanics, one by which it will be possible to explain what now seems irreducibly statistical by means of 'hidden variables'. No one can say for sure whether they will succeed; any new theory, deterministic or indeterministic, has to stand the test of experiment. The current quantum theory does show, however, that the world *may* be fundamentally irremediably indeterministic, for according to the most widely accepted interpretation of the quantum theory, it is.

5. What Is Determinism?

So far the discussion has proceeded as if a number of the fundamental concepts I have been using are clear. Since this is a rather dubious supposition, let us focus attention on some of them in the hope of enhancing our understanding. We will do well to begin with the classic definition of determinism given by Laplace. At this point the aim is not to argue the truth or falsity of determinism but only to say what it means. Laplace writes:

> Given for one instant an intelligence which could comprehend all the forces by which nature is animated and the respective situation of the beings who compose it—an

intelligence sufficiently vast to submit these data to analysis—it would embrace in the same formula the movements of the greatest bodies in the universe and those of the lightest atom; for it, nothing would be uncertain and the future, as the past, would be present to its eyes. (Laplace, [1820] 1951, p. 4)

The intelligence mentioned in this statement has sometimes been called "Laplace's demon," but he never intended to imply that such a demon actually exists—or an omniscient God for that matter. According to a famous anecdote, when Napoleon learned of Laplace's great work, *The System of the World*, he asked Laplace where God fit into the system; Laplace replied, "Sir, I have no need of that hypothesis." What he was trying to do was to capture the import of determinism. To affirm determinism is to maintain that the precise condition of the entire universe at any one instant, together with the laws of nature, logically entail the condition of the universe in its totality at any future instant. Newtonian mechanics is deterministic, for if the precise position and momentum of each and every particle at one moment—say 12:00 noon Greenwich mean time, 15 April 1970—is known, and if the laws of Newtonian mechanics are the true laws of nature, then anyone who could solve sufficiently complicated mathematical equations could deduce with perfect exactitude and rigor the precise state of the universe at any subsequent moment. From these data and these laws, Laplace's demon could calculate any future occurrence. It could ascertain exactly what you will have for breakfast on 15 April 2001, and if you should drop a bit of egg, precisely where it will spot your clothing.

No determinist seriously believes that human beings are at present capable of ascertaining the total future of the universe in all detail, or that we will ever be able to do so. The determinist is saying, instead, that it is possible in principle to make such inferences because the laws of nature and the state of the universe at any one time actually do determine the state of the universe at all future times. The fact that we are unable to make perfect predictions in all cases is, to the determinist, the result of human ignorance and other limitations; it is not because nature is lacking in precise determination. [In fact, as Earman emphasizes, prediction is irrelevant to determinism.]

To what, then, is the indeterminist committed? For the indeterminist, the combination of laws and total state of the universe at one moment do not completely determine the states of the universe at other moments. It is not a failure of our intelligence, a limitation on our knowledge of the laws of nature, or a partial ignorance of the state of the universe at the given moment. Instead, *given* complete knowledge of the state of the universe at some instant, *given* perfectly accurate formulations of the laws of nature, and *given* unlimited ability to solve mathematical equations, the complete state of the universe at some other moment simply does not follow. This is what it means to deny that determinism, as held by Laplace, is true. For example, Lucretius said that the atoms, all originally falling downward through space at a uniform speed, spontaneously swerved from their courses. In our latter-day wisdom we know that space does not, by itself, have a downward direction, and that there is no physical way to distinguish uniform motion through space from rest. Lucretius might just as well have said that the atoms were all sitting there motionless when some of them started dancing around and bumping into one another. Given a precise knowledge of the size, shape, location, and state of motion (rest) of each atom, and given all the laws that govern their motion, there is no way to infer which atom will move, when it will move, what direction it will move in, and what other atoms it will collide with. If you object that there must be *some* reason why one of these atoms moved at the time and in the manner it did, Lucretius will staunchly deny it. It is not just that we do not know the reason—there is no reason!

In this context, I think we can feel the compelling force of the determinist viewpoint. To suppose that atoms start moving about without any cause at all strains our conceptions. It is easy to protest, with the determinist, that there must be some reason; it is tempting to say that the indeterminist is not even offering an intelligible account, let alone a true one. And, indeed, many philosophers have elevated determinism to the status of an a priori truth—one that cannot rationally be denied. It is sometimes called *the principle of sufficient reason*, "a thing cannot occur without a cause that produces it," and sometimes *the law of universal causation*, "everything that happens presupposes something from which it follows according to a rule."[10]

Notice that two very different grounds have been offered in support of determinism. In the first place, it has been regarded as a very general statement that is strongly supported by the success of science in explaining all kinds of phenomena by means of deterministic laws. In the second place, it has been taken as an a priori truth that cannot be rejected without logical absurdity. If it genuinely enjoys the status of an a priori truth, it needs no support from scientific evidence, and science can never conceivably offer any evidence against it.

In view of the results of modern quantum mechanics, it seems inadvisable to regard determinism as an a priori principle. Quantum mechanics, in the form it now has, may not be true, but its truth or falsity is a matter of its correspondence with the facts, not the violation of an a priori principle. Quantum mechanics has shown that science can operate with indeterministic laws without degenerating into unintelligibility or logical absurdity. It seems reasonable to conclude that determinism is not an inviolable a priori principle; rather, its truth or falsity is a very fundamental and general fact about nature that we can hope to establish only more or less certainly on the basis of scientific evidence. If we are tempted to make determinism an a priori principle of reason, it may be because common sense tells us what "stands to reason." Contemporary common sense seems to have assimilated a good deal of the Newtonian worldview, but it has not yet come to terms with the statistical and probabilistic aspects of twentieth-century science.

[Earman illustrates the failure of determinism in classical physics by citing the following situation. Newtonian space and time are infinite, and there is no finite upper limit on the speed at which causal processes—e.g., material particles—can travel. Consider the complete state of the universe at one particular time, say 12:00 midnight Greenwich mean time, 31 December 2000, the final moment of the twentieth century. Call this moment t^*. Since Newtonian physics admits absolute simultaneity, this is a well-defined time slice of the universe—a state of the universe of precisely the sort Laplace envisioned for his demon to use as a set of initial conditions. *If Laplacian determinism were true,* from a complete knowledge of this state and all the laws of nature, the demon could deduce the entire past and future history of the universe.

In order to show that Newtonian mechanics is not deterministic, Earman invites consideration of the following situation. Suppose that at some earlier time t_1, say 12:00

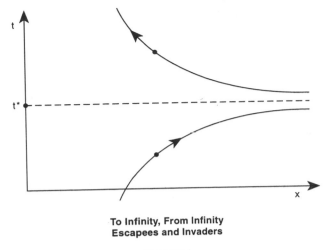

**To Infinity, From Infinity
Escapees and Invaders**

Figure 2.1

midnight GMT, 31 December 1995, there is a particle moving away from us with a rapidly increasing velocity. As we approach t^*, the particle is receding to ever greater distances, so that the trajectory of this particle never intersects the hyperplane that represents the state of the universe at t^* (see fig. 2.1). The particle never travels at an infinite speed, but there is no finite upper bound on its speed. The laws of Newtonian mechanics do not preclude this possibility. Given this situation, the time slice at t^* contains no news of the particle that, roughly speaking, has fled to positive infinity. Knowing the state of the universe at t^* would not enable even Laplace's demon to deduce the total state of the universe at the earlier time t_1.

Since the laws of Newtonian mechanics are time symmetric, the time reversal of the preceding situation is also possible. Consider a time slice t_2, later than t^*, say 12:00 midnight GMT, 31 December, 2005. It is possible that, sometime after t^*, a particle not present anywhere in the time slice t^* approaches us from a great distance traveling extremely rapidly, but slowing down as it gets nearer. Like the preceding particle, it never traveled at an infinite speed, but its speed has no finite upper bound. It would be present in the time slice t_2, but not even Laplace's demon could deduce its presence there from a complete knowledge of the time slice t^* and all of the laws of nature. In this case we have, roughly speaking, an invader from infinity.

Taking both of these possibilities into account, we must conclude that Newtonian mechanics is not deterministic either retrospectively or prospectively.[11] The special theory of relativity does not encounter the foregoing situation because it imposes an upper limit on the speed at which causal influence can be propagated, namely, c (the speed of light in a vacuum). According to this theory, particles cannot escape to infinity, nor can they invade from infinity in the foregoing manner.

The status of determinism in the general theory of relativity is extremely complex; I cannot treat it in detail here. Suffice it to say that when we take a global viewpoint and discuss the entire universe, determinism is doubtful at best; when we take a local viewpoint, determinism surely holds, just as it does for special relativity. Earman writes:

> The usefulness of this triumph of determinism in the small depends upon how small small is. The resultant sense of determinism will be epistemologically useless if the existential clause is filled only by regions so minute as to be irrelevant to typical prediction problems. And in any case, we may have no way of knowing in advance how large or small the region is. Ontologically, determinism in the small does not sustain [the] vision of a world in which the womb of the future contains no ambiguities. The myriad of miniature subworlds within which [this] vision is fulfilled may not join together into a[n] . . . absolute unity in which there is no equivocation or shadow of turning. . . .
>
> While determinism in the small is a certainty in general relativistic worlds, determinism in the medium and the large remains an open question. Additional observational and theoretical results could help to resolve some of the remaining uncertainty; but the ultimate fate of large scale determinism turns on some sticky interpretations of problems about what counts as a reasonable space-time model, and these problems resist narrowly scientific solutions. (1986, pp. 185, 197)

The reader is urged to consult chapter 10 of Earman's book for a rich discussion of the status of determinism in general relativity.]

6. Types of Determinism and Indeterminism

It is traditional to distinguish two kinds of causation, *efficient* causation and *final* causation. Efficient causation has a rather mechanical character, in the sense that effect follows cause without reference to purposes, intentions, or end. If running water erodes the earth from beneath a rock, and the rock rolls down a hill, the whole process is normally regarded as one in which efficient causes are operating mechanically. If the rock crashes into the home of a mine owner who has been exploiting mine workers and people think it is God's punishment, they are treating it as a case of final causation, inasmuch as this account does involve reference to purposes. The view that God created the separate species of living things in order to realize certain of his purposes takes the origin of species to be an example of final causation. Darwin's view, that the species develop by natural selection, regards the same result as the effect of efficient causes. Biological evolution does not have to be considered an instance of efficient causation, however, for theologians can still maintain that evolution is God's way of bringing about the realization of his purposes.

Whether one believes in efficient causation or final causation or a mixture of the two, it is still possible to be a determinist or an indeterminist. Let us adopt traditional terms and say that a person who believes that nature operates only with efficient causes, but never with final causes, is a *mechanist*. Let us say that anyone who believes that there are final causes is a *teleologist*.[12] To be a mechanist or a teleologist is to make a commitment as to *what kinds* of causes there are, but not as to the pervasiveness of causation of either type. A determinist is one who takes a stand on the question of how extensively causes, of whatever type, operate, but not necessarily a commitment on what types of causes there are. We can, consequently, define four distinct positions:

1. *Mechanistic determinism*: Every event is completely determined by causes, and these causes are efficient, not final, causes. Laplace is the classic representative of this position.
2. *Teleological determinism*: Every event is completely determined by causes, and at least some of these causes are final causes. Calvinistic predestinarianism is the most familiar example.
3. *Mechanistic indeterminism*: Events are not completely determined by causes, but to whatever extent they are determined, it is by efficient causes alone. Lucretius, with his indeterministic atomism, would seem to represent this view, as would most modern physicists, who consider quantum mechanics basically indeterministic.
4. *Teleological indeterminism*: Events are not completely determined by causes, but some events are determined to some extent by final causes. An ancient fatalist might represent this view; for instance, the significant events in the life of Oedipus, such as killing his father and marrying his mother, were determined by final causes, but the less important ones, such as the exact positions of the drops of his father's blood, may well have been left to chance.

Scientific progress seems, historically, to be associated with a transition from teleology to mechanism. Aristotle's physics, which dominated the scene for several centuries before Newton, incorporated final causes. Newton's mechanics was entirely non-teleological. Biology before Darwin tended to be teleological, but Darwin, as I have noted, introduced a mechanistic conception of biological evolution through natural selection. Subsequently, even the psychological and social sciences have tended to reject teleological conceptions. The question whether teleological or mechanical conceptions are appropriate is, it seems to me, a matter to be decided by the success or failure of theories and explanations that employ them. The important point is to show how the two types of causation give rise to two types of determinism and two types of indeterminism.

7. Laws of Nature

The laws that are written in law books (also called "statutes") are concocted by humans to *prescribe* how people shall behave. The people who are governed by such laws may conform to them or violate them. The laws of nature, by contrast, *describe* the ways in which various kinds of things in the universe operate, and there is no possibility of violation. If things did not conform to a purported law, it would not be an actual law of nature. Laws of nature, moreover, do not involve a legislator, human or divine, and we should certainly avoid thinking that the existence of laws of nature presupposes a supernatural lawmaker. To fall victim to such an inference would be entrapment by a bad pun.

In science one often hears of Hooke's law, Kepler's laws, Newton's laws, etc. In each case there are one or more statements, propounded by the individual whose name is attached, which *purport* to describe how things like springs, planets, and bullets behave. If these statements do, in fact, state accurately how such things behave, then they express laws of nature. The law of nature itself is a general uniformity or regularity in nature; the statement that is written in the science text seeks to describe this regularity. There is an elementary but crucial distinction between the words used to state a law and the fact of nature that is being described. The word "table," for instance, is a linguistic entity with five letters but neither legs nor a flat surface; the word is not to be confused with a piece of furniture. Similarly, the statement of a law is a linguistic entity, which must not be confused with the regularity that nature actually exhibits. If the sentence in the book is true, it expresses a law; when we assert the statement, we do so because we believe it expresses a law, but we may be quite wrong in thinking so. For example, it was long believed that Newton's so-called laws of motion were true, but we no longer think so; although we still refer to them as "laws," we do not really believe they express genuine laws of nature. We do believe, however, that the speed of light is the greatest speed at which signals of any kind can be transmitted across empty space, and that law is fundamental to Einstein's special theory of relativity.

The doctrine of determinism, as formulated by Laplace, makes essential reference to the laws of nature. It is of utmost importance to remember that such references pertain not to statements found in textbooks but, rather, to the actual regularities that exist in nature. At any given time, of course, we do not know for certain which statements express actual regularities, and any statement we make purporting to express a law of nature may be incorrect, but that does not imply that we cannot speak meaningfully about the actual laws of nature (as opposed merely to our conceptions of the laws of nature). We do not know for certain that a given bottle actually contains scotch whiskey, but we quite properly talk about the contents of such bottles even in the absence of certainty. When I take a drink from such a bottle, it is the contents of the bottle I shall be drinking, not merely my conception—the notion of drinking a conception is sheer nonsense. If it were never permissible to say anything of which we are not absolutely certain, we could never say anything about the physical world.

8. Determinism and Explanation

There are many kinds of explanation, such as explaining the meaning of an unfamiliar word, or explaining how to operate a new camera. Some explanations are answers to the question "Why?" and scientific explanations are frequently, if not always, of that type. For example, suppose a small plane crashed upon takeoff from an airport near Denver on 15 July 1970, and we ask why the crash occurred. A satisfactory answer might point out that the plane failed to clear an obstacle 100 feet high located a certain distance from the end of the runway, and it might cite such relevant conditions as the length of the runway, the type of aircraft involved and the load it was carrying, the altitude of the airport, the air temperature, the wind velocity and direction, and the relative humidity. These specific factors would be related to the crash by general laws; e.g., that increase of altitude, air temperature, and relative humidity increase the distance needed for takeoff. In this type of explanation, two basic kinds of elements are involved, namely, specific conditions obtaining prior to the event to be explained (let us call them *initial conditions*) and *general laws*. The explanation consists in citing the initial conditions and the general laws, and pointing out that the occurrence of the event to be explained follows logically from those premises. An explanation of this type can be schematized as follows:

(3)   Statements of initial conditions
      Statements of general laws
      _____

      Statement that the event to be explained occurs

Such an explanation can be regarded as an argument to the effect that the event to be explained was to be expected, in the light of the initial conditions and the general laws, because its occurrence follows from them.

There is a striking similarity between this characterization of explanation and Laplace's formulation of determinism. Recall that his demon requires (1) knowledge of the condition of the universe at some particular moment, i.e., initial conditions; (2) knowledge of the laws of nature, obviously, general laws; and (3) ability to carry out mathematical deductions, i.e., the ability to establish the validity of the argument. If determinism, as Laplace conceives it, is true, *every future event* is explainable in terms of the laws of nature and some initial conditions. If you want to explain the entire state of the universe at some future time, you would presumably have to take as initial conditions the entire state of the universe at some antecedent time, as well as all of the laws of nature. But to explain some relatively limited and isolated event, such as the plane crash, only some of the conditions obtaining before the crash would be needed (weather conditions in Hong Kong would not be relevant), and some laws of nature would probably be dispensable. In either case, whether you are trying to explain the condition of the whole universe at some time or merely some particular event in it, both laws and initial conditions are required.

In view of the close relationship between determinism and one type of scientific explanation, it is tempting to conclude that events that are causally determined can be explained, and those that can be explained are causally determined. From this point it is easy to take another step and say that when human actions and decisions can be explained, they are determined. One more step leads to the conclusion that to explain human behavior and choices is to show that they cannot be free. To explain human behavior seems to amount to *explaining away* human responsibility! There are, however, a number of dubious steps in this inference.

Whether determinism is true or not, there are many cases in which we do not have enough facts to be able to construct an explanation demonstrating that the event to be explained must have occurred, given the initial conditions and the laws. For example, we say that Susan Jones recovered from her streptococcus infection because she was given penicillin, knowing that not all, but only most, streptococcus infections respond to penicillin. We do not have any set of laws and initial conditions from which it follows that the recovery *must* occur; at best we can show that it is highly probable. It seems there are at least two types of explanation, and they differ in two fundamental ways. The first type, illustrated by the plane crash example, is known as *deductive* explanation; the second type, illustrated by the streptococcus infection example, is known as *inductive* explanation.[13] They differ in the following two ways. First, although both types require the use of general laws, deductive explanations incorporate *universal laws,* which hold without exception, while inductive explanations employ *statistical laws.* For instance, the Bernoulli principle, which is fundamental to aerodynamics, states that *in all cases,* the greater the velocity of flow of a fluid (liquid or gas), the smaller is the pressure it exerts perpendicular to the direction of flow. Universal laws have the overall form "All *F* are *G*." Statistical laws are also generalizations, but instead of saying that something happens in every case, they say that it happens in a certain percentage of cases. The percentage may be specified by a precise number, as in "50.5% of all human babies born are male," or it may be given by a vague word, as in "Most cases of streptococcus infection clear up promptly when penicillin is administered." Second, although each type of explanation consists in an argument, the arguments are deductive (i.e., the conclusion follows with necessity from the premises) and inductive (i.e., the premises confer a high probability upon the conclusion), respectively.

If we understand that schema (3) may represent either an inductive or a deductive argument, both types of explanation conform to it. More explicitly, however, the simplest examples of the two types of explanation can be compared and contrasted via the following two schemas:

(4) All F are G.
     x is F.

     x is G.

(5) Most F are G.
     x is F.
     ================ $[p]$
     x is G.

In each case the first premise is a general law (statistical laws are general in that they refer to a whole class F, but they are not universal in that they do not assert that every member of the class has the property G), the second premise gives the initial conditions, and the conclusion asserts the occurrence of the event to be explained. The single line in (4) signifies a deductive relation between premises and conclusion; the double line in (5) signifies an inductive relation, the number p at the side indicating the degree of probability of the conclusion given the premises. If the probability p attaching to the inductive inference in (5) is near enough to one, we can say that the event to be explained was to be expected in view of the explanatory facts, though it did not necessarily have to happen given these circumstances.

There are still other cases, however, in which we seem to be able to explain occurrences even though the explanatory facts do not make the event very probable—cases, in fact, in which the nonoccurrence of the event is more probable than its occurrence, even in the presence of the explanatory conditions. To cite an example that has been widely discussed, if someone contracts syphilis, and it goes through the primary, secondary, and latent stages without treatment with penicillin, that person may develop paresis. This is one form of tertiary syphilis, but only a small percentage of those who have untreated latent syphilis become paretic. At the same time, the only people who develop paresis are victims of syphilis. If someone develops paresis, we offer as an explanation the fact that they had untreated latent syphilis, even though the probability of a latent untreated syphilitic becoming paretic is considerably less than one half. There are no known characteristics by means of which to predict which cases will develop paresis and which will not.[14]

It is easy to say that explanations of this sort are partial and rudimentary, owing to our lack of knowledge of all of the factors surrounding syphilis and its various manifestations. Such an attitude is probably well founded. Scientific experience indicates that further investigation is likely to provide answers to the question of what makes one syphilitic develop paresis and another not. The explanation provides some understanding of what happened and why, but we have good reason to believe that further research will make possible more complete explanations. The same can be said for the streptococcus infection. Even though the explanation of the cure conferred a high probability upon it, there is good reason to suppose that eventually we will find an objective characteristic of certain streptococcus bacilli that makes them resistant to penicillin. When it has been found, we will be able to tell exactly which streptococcus infections can be successfully treated by penicillin and which cannot. When that information is available, it will be possible to give a deductive explanation of the cure of this particular infection by penicillin.

This discussion of types of explanations and how they can be supplemented has a direct bearing on determinism. If determinism is true, then it is possible in principle to supplement any explanation that is inductive or probabilistic in such a way as to transform it into a deductive explanation. Whenever we use a statistical generalization in an explanation, according to a determinist, it is because our knowledge is incomplete, not because the basic laws of nature are genuinely statistical. On the deterministic view, any reference to chance or probability is, as Laplace remarked, merely an expression of our ignorance of the true laws of nature.

The indeterminist, by contrast, is committed to saying that there are at least some events for which it is impossible to provide deductive explanations; the best we can hope for is some kind of statistical explanation. While some indeterminists might agree that the statistical character of the laws cited in the medical examples is a reflection of the incompleteness of biological science, they could still maintain that in physics there are events that are not amenable to deductive explanation. Lucretius, if he were here and could talk our jargon, might explain the spontaneous movement of an atom by saying that there are various kinds of atoms—large and small, rough and smooth—and that the small, smooth ones have a certain probability of jumping even though they are not bumped by other atoms. Such characteristics are the only ones that are relevant to whether the atoms engage in spontaneous movement, so the best explanation we can give is in terms of such probabilities. If we were to tell him that there *must* be *some* reason why this small, smooth atom rather than another started to move at that moment, we would merely be expressing a deterministic prejudice.

Leaving this historical fiction, we find a similar situation in modern physics. The atoms of certain elements are unstable, and they suffer radioactive decay. The uranium atom, for example, may decay by emitting an alpha-particle from its nucleus. The nucleus constitutes a strong enclosure, and the alpha-particle races frantically back and forth, bumping into the wall of the nucleus about 10^{21} ($=1,000,000,000,000,000,000,000$) times per second, and on the average an alpha-particle makes it out in 4.46 billion years. In other words, it has about one chance in 10^{38} of getting out any time it bombards the barrier of its nuclear prison.[15] When we ask why a particular uranium atom decayed in this manner at this particular time, the answer is that an alpha-particle "tunneled out" of its nucleus. When we ask why the alpha-particle escaped on that particular trial, having failed on countless other occasions, the answer is simply that there is a probability of about 10^{-38} of such an outcome on any given bombardment of the wall. That is all there is to it. Perhaps you want to say that there must be some reason for the success on this trial and the failures on the others, but we do not yet know what it is. According to the most common current interpretation of quantum mechanics, however, that is not the case. We are, according to that view, dealing with an irremediably indeterministic process.

The situation in quantum mechanics arises out of what seems to be a pervasive feature of the atomic and subatomic world. It has been described by an unfortunate phrase, "the uncertainty principle." When one speaks of uncertainty, it is natural to suppose that there is something to be known but we do not know it for sure. Thus, it has sometimes been said that there is an inescapable uncertainty if one attempts to ascertain the values of both the position and momentum of a particle, and similarly for energy and time. If we ascertain the position of an electron with great precision, we will be unable to ascertain its momentum very exactly, and conversely. There is a limit to the joint precision with which two so-called complementary parameters can be known. This way of speaking, as well as many popular attempts to explain the uncertainty principle, strongly suggest that the electron has, at any given moment, an exact position and an exact momentum, but we are not able to find out what both of those values are. This is a serious misinterpretation of the uncertainty principle, as I explain in "Indeterminacy, Indeterminism, and Quantum Mechanics" (essay 17). We should say instead that particles such as the electron and our alpha-particle are actually in physical states that are not characterized by exact values of position and momentum, energy and time. We *can* ascertain the state of the particle, but the state, together with all of the pertinent laws of nature, does not provide the basis for deterministic prediction or deductive explanation of events such as the alpha-particle's tunneling out of the uranium nucleus. Even Laplace's demon could not reliably predict the time at which a particular uranium atom would experience radioactive decay.

9. Explanation and Relevance

If the world is actually indeterministic in the way modern physics suggests, you might infer that some things cannot be explained. Such a conclusion would, I think, be unjustified.[16] It is true that some events could not be explained deductively, but the supposition that there is no other kind of explanation is simply another aspect of the deterministic view. If we embrace indeterminism, we must adopt a suitable conception of explanation to go along with it. For the indeterminist, some events will have to be explained statistically—I do not say "inductively," because I shall be suggesting a different sort of statistical explanation. Moreover, it looks as if we will have to come to terms with events that are extremely improbable: 10^{-38} is a very small number. Shall we conclude that only events with high probabilities can be explained—that those with low probabilities are inexplicable? This result will be forced upon us if we think that explanations, deductive or statistical, must be *arguments* showing that the event to be explained *was to be expected*, for that requires high probability if deductive certainty is lacking. I am inclined to believe, however, that this way of characterizing statistical explanation is inappropriate. The key to an alternative approach will be the concept of *statistical relevance*. (See "A Third Dogma of Empiricism" [essay 6] for further details.)

Suppose a life insurance company is considering issuing a policy to a particular person, Frank Smith, and suppose that at the premium set, the company will make a profit if he lives for at least ten years. The company must decide whether to sell him life insurance at that rate, and so they would like to know whether he will survive for at least a decade. From mortality tables they can find the probability that an unspecified American will live that long, but they know in addition that he is male and 37 years old. Again, the mortality tables will furnish the probability of a 37-year-old American male's living ten years longer. His age and sex are relevant because the probability of survival for a male is different from that for a female, and the probability certainly varies with age. In order to make the decision, the company will secure further evidence about him, e.g., his state of health, his occupation, his personal habits, his marital status, and his hobbies. We know, for example, that the probability of survival is different for heavy cigarette smokers than for nonsmokers, different for diabetics than for people in normal health, different for steeplejacks than for clergymen, and different for married men than for bachelors. Any specification of characteristics of Frank Smith that alters the probability of his living to the age of 47 is *statistically relevant* to the case at hand. Characteristics that do not change the probability are irrelevant. Examples of irrelevant characteristics would be the color of his eyes (but not of his skin), whether his social security number is odd or even, and whether his first child is a boy or a girl.

The insurance company would like to know whether Frank Smith will live another ten years, and whether or not Laplace's demon could predict that fact with certainty, the insurance company cannot. Hence, they must be content with probabilities, and indeed, that is the entire basis of their business. In making decisions as to whom to insure, they try to take into consideration the statistically relevant factors, and they try to avoid getting involved with irrelevant ones.

The same considerations, I believe, enter into statistical explanation. When we ask why Susan Jones's streptococcus infection cleared up quickly, we mention the fact that she was given penicillin, for that is a highly relevant fact. The probability of a streptococcus infection's going away promptly is quite different depending on whether the patient received penicillin or not. When we ask for an explanation of the fact that some individual contracted paresis, the fact that that individual had latent untreated syphilis is cited, for the probability of someone's developing paresis is very different, depending on whether the person ever arrives at the condition of untreated latent syphilis. If we find such explanations incomplete, it is because we reasonably believe that there are additional relevant factors, as yet unknown, that have a bearing on the probability of recovery from streptococcus infection or the occurrence of paresis.

Now, it might occur to you that an incredible variety of factors could be relevant to, say, the contraction of paresis. Whether John Doe's parents are of Latin or Anglo-Saxon extraction might have some bearing on his attitudes toward sex, and hence on the likelihood of his contracting syphilis, and finally on the chance of his becoming paretic. His socioeconomic status might also be relevant in a number of ways, including the probability of his seeking medical treatment should the symptoms of a venereal disease appear. Nevertheless, although such factors may be indirectly relevant in the absence of more detailed information about his medical condition, they become irrelevant in the light of further information. Once it is known that the victim has contracted syphilis, the probability of his picking up a venereal disease is irrelevant. Once it is known that he has arrived at the stage of latent untreated syphilis, the likelihood of his seeking medical treatment in the early stages of the disease is irrelevant. The more immediate conditions, so to speak, screen off the relevance of the more remote ones.[17]

The determinist and the indeterminist alike, in attempting to explain an event, are trying to assemble a *total set of relevant conditions*. By a total set of relevant conditions I mean a set of conditions that cannot be supplemented in any way that would change the probability of the given outcome. This aim is achieved more readily than you might offhand suppose. If you have a universal law of the form "All F are G"—for example, all copper conducts electricity—then the probability of a piece of copper's being an electric conductor is one, and nothing can be added to change that. If you add that the piece of copper was formed into a penny, the probability of its conductivity is still one. If you add that it was originally mined in northern Michigan, the probability of conductivity is still one. Unless the general statement was false in the first place (in which case it did not express a genuine law), what is true of all copper is true of any specific type of copper. We have, indeed, found a total set of conditions relevant to conductivity. Similar considerations apply to negative universal generalizations such as "No whales are fish," the probability in such cases being zero instead of one.

The determinist is very happy with the total sets of relevant conditions that are embodied in universal laws, for these are just the kinds of laws that are needed for deductive explanations. When the laws are statistical, the determinist feels, the explanations are incomplete because there are further relevant conditions to be found. The determinist maintains, in other words, that the only way to achieve a total set of relevant conditions is to find universal laws. The indeterminist takes a different view, maintaining that there are other ways of arriving at total sets of relevant conditions. When asked why an atom experienced spontaneous radioactive decay, the indeterminist might answer that it is an atom of uranium 238, and that it has a half-life of about 4.5 billion years (which is a convenient way of expressing its probability of disintegration). To say merely that it is a uranium atom would not be sufficient, for the different isotopes of uranium have different half-lives, but once the isotope has been specified, nothing further is relevant. It does not matter whether the atom is in a block of pure metallic uranium 238, whether it is alloyed with other uranium isotopes or other metallic elements, whether it is in chemical compound with other elements (e.g., an oxide), whether it is in a magnetic field, or whether it has been blessed by the pope. In such cases, according to the indeterminist, there is a certain probability of spontaneous decay, and nothing we can add has any bearing on that probability. If the determinist says that there must be some further relevant factor that has not yet been found, the indeterminist could appropriately reply, "Perhaps it would be nice if there were, but what guarantee have we that nature is so accommodating to our wishes?"

If indeterminism is true, it does not follow that there are events that are incapable of being explained. To offer an explanation, as I have suggested, is to assemble a total set of relevant conditions for the event to be explained, and to cite the probability of that event in the presence of these conditions. This view of explanation, unlike the standard account of deductive and inductive explanation, does not see an explanation as an argument showing that the event was to be expected on the basis of the explanatory facts. The explanation is, rather, a presentation of the conditions relevant to the occurrence of the event, and a statement of the degree of probability of the event, given these conditions. That degree of probability may be high, middling, or low, but whatever its size, it is an index of the degree to which we would have been justified in expecting it.

A point of clarification must be added lest complete misunderstanding arise. The general laws, be they universal or statistical, that provide the relevant conditions may themselves be explained on a different level, so to speak. If we invoke the general law that all copper conducts electricity, this provides a total set of conditions relevant to the fact that a particular piece of copper, such as a penny placed behind a blown fuse, conducts electricity. However, that does not exclude the possibility of explaining electrical conductivity itself in terms of the behavior of electrons. The fact that such further explanation is possible does not mean that the original explanation of the conductivity of the penny was incomplete; it means only that facts adduced to explain other facts may in turn be explained on a more general or theoretical level.

10. Causes versus Statistical Correlations

In recent years, evidence of a significant statistical correlation between cigarette smoking and various diseases has been widely publicized. The tobacco industry, in its frequent protests that "no causal connection" has been found, has emphatically reiterated the distinction between causal connection and "mere statistical correlation." While I believe that the statements on behalf of the cigarette manufacturers are wrong, and that extremely strong evidence of a causal connection between cigarette smoking and disease has been presented, that is not the major point here. We are interested in determinism and in explanation, and each of these concepts seems to have a deep causal component. When we think of determinism, we think of causal determination, and when we ask "Why?" the natural answer is "Because . . . " To ask why the airplane crashed is to ask what caused the crash.

A persistent statistical correlation   that is, a genuine statistical-relevance relation— is strongly indicative of a causal relation of some sort. Consider some examples. Both fever and characteristic types of spots are symptoms of measles. The fever does not cause the spots and the spots do not cause the fever, yet there is a marked statistical relevance of the one to the other. The reason, of course, is that they are distinct effects of a common cause, and the common cause explains the statistical relation. In similar fashion, there is a high degree of statistical relevance between the drop in barometer reading and the occurrence of a storm, but neither causes the other. Both the storm and the falling barometer are the result of meteorological conditions that barometers are designed to indicate. The main danger in confusing statistical correlation with genuine causation is the danger of confusing symptoms with causes. In medicine, engineering, social work, politics, and other practical pursuits, we know the futility of treating the symptoms when we want to correct the conditions giving rise to them.

In discussing the search for total sets of relevant conditions, I mentioned the fact that some relevant conditions can render others irrelevant by what is called "screening off." The screening-off phenomenon is basically a matter of causal proximity. The measles infection is more closely related to both the fever and the spots than are the spots and fever to each other. The barometer reading is more remote from the storm than is the set of atmospheric conditions responsible for the storm. Primary syphilis is causally more remote from paresis than is secondary or latent syphilis.

What do these causal relations amount to? It seems that the world is full of processes that go on in a relatively continuous way. Billiard balls roll around on tables, bouncing off the cushions and colliding with one another, according to the laws of classical mechanics. Light rays are propagated in accordance with the laws of optics. Springs can be extended and contracted as described by Hooke's law. When the temperature of a gas is increased without changing the size of the container, the pressure increases. These are processes that are governed by universal laws of the kind found in classical physics and used in deductive explanations. If everything that happens in the world follows from antecedent conditions by processes that conform to such laws, we say that the universe is *causally* deterministic. In this case we could say with Laplace, "We ought then to regard the present state of the universe as the effect of its anterior state and as the cause of the one which is to follow" ([1820] 1951, p. 4). If, however, the causal processes are governed by laws that have an irreducibly statistical character, such as we find in contemporary quantum mechanics, then the world is causally indeterministic.

11. Free Will and Indeterminism

Suppose indeterminism, of the sort suggested by modern quantum mechanics, is true. No one knows for sure whether it is, but it might be, and it is interesting to see what bearing that would have on the problem of human free will.

There is good evidence that radiation of the sort emitted in radioactive decay of unstable nuclei can have profound effects on genetic structure and can induce mutations. Suppose that the father of a child was in the vicinity of radioactive materials just prior to its conception, and that a chance disintegration of an unstable atom emitted a gamma-ray that altered a gene that was passed on to the child. Suppose, to make the case dramatic, that the genetic damage of the gamma-ray results in the child's becoming a congenital criminal, although normal character and personality would have developed if that atom had not disintegrated just when it did. Would we be inclined to say that this person's criminal acts are done freely, because of the chance occurrence in heredity, while noncriminal acts would have been unfree if chance were unable to influence genetic makeup? Hardly.

But, you might say, the indeterministic event was not part of this person. It happened before conception, it came from outside of the father and the child, and its results were passed on (suppose) in a fully deterministic manner. Very well. Suppose someone eats food that, unknown to that person, is contaminated with radioactive material. One of these unstable atoms decays, indeterministically, at a vital place in the body. As a result, cancer later develops. Is there any element of freedom introduced because the chance event took place inside the body? Hardly.

But, you might continue, the onset of cancer does not involve any element of thought, deliberation, decision, or choice, and these are vitally involved in freedom. That seems to be a sound point. Suppose, therefore, that you are trying to make up your mind about experimenting with marijuana. If determinism were true, your heredity, your environment, and the physiological processes in your nervous system would totally determine the outcome of your deliberation. If you decided to go ahead and try it, the decision would be a causally determined result of the chemistry of your brain at that moment. Under these circumstances, you might seriously doubt that the choice is free. Suppose, however, that determinism is not true. At the crucial point in your brain is an unstable atom. Its relation to the decision process is something like a trigger mechanism. If that atom disintegrates at the proper moment, it will start a process that will lead causally to the decision to smoke pot. If it does not disintegrate, you will decide against it. Does the decision now seem free? Hardly.

These science fiction speculations are designed for one purpose: to raise the question whether the problem of free will is really connected with determinism in the way it seems to be. Having seen that determinism seems to raise very serious difficulties in connection with freedom of choice and action, we are tempted to jump unreflectively to the conclusion that all will be rosy if we just abandon determinism. When we go on to postulate indeterminism, however, the net result seems to be absolutely no progress at all in the direction of free will. The problem is just as difficult and puzzling—if not more so—under the assumption of indeterminism than it was in the context of determinism. It appears that we can construct the following argument:

(6)   If indeterminism is true, humans do not have free will.
Indeterminism is true.
_____
Humans do not have free will.

We do not know for sure whether the second premise of this argument is true, but modern quantum mechanics makes it at least plausible. That, however, is not the crucial point. Argument (6) can be combined with argument (2) as follows:

(7)   If determinism is true, humans do not have free will.
If indeterminism is true, humans do not have free will.
Either determinism is true or indeterminism is true.
_____
Humans do not have free will.

This argument is a dilemma, and it is logically valid. Moreover, its third premise is necessarily true, for indeterminism holds if determinism does not, and conversely. There are two avenues to follow from here. One can accept all three premises and draw the conclusion that freedom of will, freedom of decision, freedom of choice, and freedom of action are all illusory. The other avenue, and by far the more promising one, I believe, is to reexamine the first premise of arguments (1) and (2), which is the same as the first premise of (7). This premise, which was accepted so facilely at the beginning, has taken us down the long path to argument (7), which might aptly be called "the dilemma of free will." Perhaps the premise is not as self-evident as it appeared at the outset. It may turn out that the question of whether the breaking of the vase by Epictetus was causally determined is far less important than the question of how many vases he, and other slaves, broke after his beating. Legend does not, as far as I know, provide a clear answer to this latter question.[18]

Notes

1. From the play *Sheppey,* by W. Somerset Maugham (London: William Heinemann, 1933; copyright, 1933, by W. Somerset Maugham).

2. The so-called "torsion-balance experiment," first performed by Henry Cavendish in 1798. All previous confirmations of Newton's gravitational theory involved either one or two bodies of astronomic proportions: the influence of the earth on falling bodies, the mutual attraction between the sun and the planets, the influence of the moon on the tides. The Cavendish experiment detected the gravitational attraction between two ordinary medium-size terrestrial objects.

3. The explanation of the "perturbations of Uranus" by the planet Neptune was accomplished in 1843 by John C. Adams, and independently about two years later by U. J. J. Leverrier. Neptune was observed and identified as a planet by J. G. Galle in 1846. Leverrier also determined that there was a small deviation in the path of Mercury, and he postulated a planet Vulcan to explain it, but Vulcan was never found. The deviation was eventually explained by Einstein in his general theory of relativity.

4. In saying that his arguments were persuasive, I do not mean to ignore the severe difficulty of the problem of interaction between mind and matter to which his mind-body dualism led. This problem becomes even more acute if one admits that there is a great deal of interaction between mind and matter, and simultaneously wants to claim determinism for the physical realm and indeterminism for the psychological realm.

5. The synthesis of urea was accomplished by Friedrich Wöhler in 1828.

6. See especially *The Interpretation of Dreams* (1900) and *Psychopathology of Everyday Life* (1901) in Brill (1938). Although Freud lived and worked well into the twentieth century, many of his most significant ideas were developed before the turn of the century.

7. See Asimov (1962) for an accurate and readable popular account of the most important developments in molecular biology. Watson (1968) is a fascinating biographical account of the discovery of the structure of the DNA molecule by one of its co-discoverers.

8. [Since this article was written, there has been a dramatic change in scientific views regarding the environment in which life was supposed to have arisen. For an excellent nontechnical discussion, see Cairns-Smith (1985). The view discussed is thoroughly mechanistic.

A press announcement in *Science News,* 10 June 1995, p. 367, reported successful sequencing of the entire DNA of two bacteria. That means that the genetic codes for these organisms have been completely deciphered. The DNA of viruses had previously been sequenced, but viruses are not capable of self-reproduction outside of a living cell. The bacteria are stand-alone organisms that do not require the genetic material of another organism for reproduction.]

9. This is the famous Stern-Gerlach experiment, and it has fundamental importance in quantum theory. See "Indeterminacy, Indeterminism, and Quantum Mechanics" (essay 17) for details of this experiment.

10. These formulations are due to G. W. Leibniz and Immanuel Kant, respectively.

11. [For discussions of the status of determinism in classical thermodynamics, electrodynamics, and fluid dynamics see Earman (1986, chap. 3, § 10–16).]

12. [An excellent discussion of teleological aspects of science can be found in Wright (1976).]

13. The clearest and most exhaustive technical discussion of these two types of explanation (which are called "deductive-nomological" and "inductive-statistical," respectively) is given in Hempel (1965b). Hempel also provided a masculine version of this example of inductive explanation.

14. According to Clark and Harris (1961, p. 44), "72 out of 100 untreated persons [with latent syphilis] go through life without the symptoms of late [tertiary] syphilis, but 28 out of

100 untreated persons were known to have developed serious outcomes [paresis and others] and there is no way to predict what will happen to an untreated infected person."

15. See Gamow (1961, pp. 111–115). Gamow was responsible for the theoretical explanation of this phenomenon in 1928.

16. I have offered a detailed and technical account of explanation (Salmon, 1971). The present discussion of explanation and relevance is a highly oversimplified version.

17. This concept of *screening off* is of crucial importance in the discussion of explanation and statistical relevance; it is discussed at length in the article cited in the note 16.

18. To my mind, the best approach to the problem of the relation of free will to determinism is given by Stevenson (1944, chap. 14).

3

Comets, Pollen, and Dreams

Some Reflections on Scientific Explanation

Now we know
The sharply veering ways of comets, once
A source of dread, nor longer do we quail
Beneath appearances of bearded stars.
　　　　　—Edmund Halley, "Ode to Isaac Newton"
　　　　　　　(quoted in Cajori, 1947, p. xiv)

1. Introduction

The Newtonian synthesis, which provided the basis for all of classical physics, produced far-reaching changes in our ways of looking at the world. Laplace, who made significant contributions to the development of classical physics, was one of its most eloquent champions. Like Halley, he found the Newtonian explanation of comets an inspiring example of the power and value of modern science.

Despite the fact that classical physics still has wide applicability to various sorts of phenomena, we no longer believe it to be literally true. Nevertheless, it seems to me, certain philosophical views concerning the nature of science that arise directly out of a Laplacian conception of the world continue to exert an enormous influence on current thought about scientific explanation. A caveat should be issued at once. I shall *not* be arguing the historical thesis that Laplace's writings had a direct influence on contemporary philosophers; instead, I shall maintain that the general viewpoint expressed by Laplace, which pervaded much of nineteenth-century thought, has carried over into the twentieth century and permeates much of contemporary philosophy of science.

Laplace was, of course, a firm advocate of mechanistic determinism; accordingly, he believed that biological phenomena and human behavior are as rigidly determined by the laws of Newtonian mechanics as are the motions of comets and atoms. Only our lack of knowledge prevents us from seeing that fact. Many nineteenth-century scientists, in the biological and social sciences as well as the physical sciences—steeped in the tradition of classical physics—believed that all of the phenomena in the world can ultimately, in principle, be reduced to classical physics. It was, I think, this Laplacian conception of mechanical determination, bolstered by at least a century of additional spectacular success of classical physics, that provided the model for scientific explanation most widely accepted by philosophers and scientists in the *twentieth* century. Stated succinctly, the claim is that, *with the aid of suitable initial conditions, an event is explained by subsuming it under one or more laws of nature.* This is hardly more than a translation into more up-to-date terminology of Laplace's colorful statement:

> Given for one instant an intelligence which could comprehend all of the forces by which nature is animated and the respective situation of the beings who compose it— an intelligence sufficiently vast to submit these data to analysis—it would embrace in the same formula the movements of the greatest bodies of the universe and those of the lightest atom; for it, nothing would be uncertain and the future, as the past, would be present to its eyes. ([1820] 1951, p. 4)

Such an intelligence, Laplace must have believed, would exemplify the highest degree of scientific understanding, and would be able to provide a complete scientific explanation of any occurrence whatsoever. (See "Why Ask, 'Why?'?" [essay 8] for further discussion of the demon's ability to explain.)

In the closing months of the nineteenth century, Max Planck provided the basic building block—the quantum of action—of a new science that would undermine and supersede classical physics. Neither Planck nor anyone else at that time could foresee the fundamental conceptual revolution physics was destined to experience in the first quarter of this century. By 1926, Werner Heisenberg and Erwin Schrödinger had formulated the basic theory of quantum mechanics, and Max Born had furnished the statistical interpretation, which has subsequently become the standard physical interpretation of quantum theory. These developments, to say the very least, cast serious doubt on the whole conception of Laplacian determinism. Physical science has by now fairly well absorbed the shock of supposing that the physical world may be fundamentally and irreducibly statistical, though some physicists will staunchly resist this interpretation of quantum mechanics. It is not clear, however, that philosophy of science—as expounded by scientists as well as philosophers—has digested this development, along with its repercussions for concepts such as scientific explanation. It was not until 1962—an astonishing delay—that *any* systematic attempt was made to explicate statistical explanation, and it appears to me that the resulting analysis was far from satisfactory.[1] I suspect that too close an adherence to the Laplacian ideal may have been responsible for some of the difficulties.

In addition to considering statistical explanations, we shall find it necessary in the course of the discussion to take a careful look at functional explanations. Scientific progress has, rightly I believe, tended to purge science of teleological principles. Aristotelian physics, in which nature *abhorred* a vacuum and bodies *sought* their natural places, has been totally superseded by the mechanical physics of Newton.[2] Darwinian evolution, with its principle of natural selection, has replaced the doctrine that species were specially created by God to fulfill divine *purposes.* This laudable attempt to remove purposive and anthropomorphic explanatory principles from science has, I think, made many scientists and philosophers wary of functional explanations, and has encouraged the notion that in fully mature sciences, functional explanations are eliminated in favor of other types. This has led some philosophers to characterize functional explanations as "explanation sketches" or "incomplete explanations" (Hempel, 1962b, pp. 16–19). Although it can be shown quite clearly, I believe, that certain types of functional explanation need not involve any anthropomorphic or teleological elements, philosophers and scientists have not universally been convinced of their scientific legitimacy. Nevertheless, pious hopes to the contrary notwithstanding, important classes of explanations in some sciences are functional explanations, and they are by no means patently reducible to explanations of any other type.

The purpose of the present essay is to reexamine the nature of scientific explanation from the standpoint of contemporary science. I shall pay careful attention to our heritage of Laplacian determinism—with its obvious bearing on scientific explanation—but I shall also try to see how these conceptions have to be modified in the light of more recent developments. As my foregoing remarks have indicated, I shall devote considerable attention to statistical and functional patterns of explanation. In so doing, I shall be raising issues that are matters for consideration by scientists in a wide range of fields, from anthropology to zoology—touching psychology, quantum physics, and sociology, among others, along the way.

2. Laplacian Explanation (Comets)

2.0. Misconceptions

An important benefit of Newton's explanation of comets was to render them less terrifying. This result is achieved, it had sometimes been suggested, by transforming the unfamiliar into something familiar.[3] Describing a comet as a planet-like object with a highly eccentric orbit does help to classify it with better-known objects, and this, it is claimed, is what makes it more understandable.

Appealing as it may seem, this conception of explanation can hardly be considered adequate. It is easy to cite many examples in which the opposite occurs: the familiar is explained by invoking highly esoteric considerations. The outstanding instance is the Olbers paradox—why is the sky dark at night? No fact could be more familiar than the darkness of night, but any adequate explanation of that phenomenon will involve intricate cosmological considerations. Another familiar fact is that offspring resemble their parents in certain respects; its explanation takes us into the chemistry of the DNA molecule and the 'genetic code'. A third example is Freud's explanation of dreams— familiar occurrences to most people—in terms of unconscious wishes, which at the time (if not now) were unfamiliar to the point of being far-fetched. I do not mean to assert that the Freudian explanation of dreams is correct, but the fact that it attempts to explain the familiar by means of the unfamiliar is no obstacle to its acceptability.

A closely related notion requires that explanations must make ultimate reference to conscious aims and purposes if they are to provide *genuine understanding*. Such explanations are teleological. We are familiar with the motives that explain many of our own actions; the demand is sometimes made that any explanation of any other phenomenon must refer to the purposes of the Creator of the world, or perhaps to some purpose that is inherent in nature itself. This view probably lies at the heart of the claim, often made in earlier times, that science in and of itself can provide only description, not explanation. Such a view of explanation has been severely criticized for its blatant anthropomorphism, and I doubt that it enjoys much support among contemporary scientists. At the same time, those sciences such as biology, sociology, and anthropology, which seem to make extensive use of functional explanations, have sometimes encountered serious problems in showing that they were not ipso facto involved in teleology. As I remarked earlier, I think that careful analysis can draw a viable distinction between those functional explanations that are teleological and those that are not. But it remains to be seem what role, if any, functional explanations can play in the overall scheme of scientific explanation.

Having considered some common misconceptions of the nature of scientific explanation, let us attempt to arrive at more adequate formulations. The plural "formulations" is quite deliberate and very important. I shall offer three characterizations of Laplacian explanation which, in that context, may seem to differ only terminologically. When, however, we move on to consider modifications of the Laplacian view demanded by developments in twentieth-century science, the differences take on crucial logical importance.

2.1. The Epistemic Conception of Scientific Explanation

Suppose that we attempt to explain a particular occurrence, such as a lunar eclipse, by citing certain laws that, together with suitable antecedent conditions, entail that the eclipse occurred at a particular time.[4] In this case we can plausibly say that the explanation is a valid deductive argument, with premises consisting of law-statements along with other statements that describe the initial conditions, and with the explanandum-statement as its conclusion. This explanation could be described as an argument to the effect that the event to be explained was to be expected by virtue of the explanatory facts. I shall refer to this view as the *epistemic conception* of scientific explanation. Given an event that, when it occurred, might or might not have been expected, we explain it by showing that it could have been predicted if we had been in possession of the explanatory facts prior to the occurrence. This prediction would have involved a deduction of the explanandum-statement from the explanans-statements. On this view we can say that there is a relation of *logical necessity* between the laws and initial conditions on the one hand and the explanandum on the other.[5]

2.2. The Modal Conception of Scientific Explanation

Under the same circumstances we can say, alternatively, that because of the lawful relations between the antecedent conditions and the explanandum-event there is a relation of *nomological necessity* between them. I shall call this view the *modal conception* of scientific explanation (see, for example, von Wright, 1971, p. 13; and Mellor, 1976). Given the particular set of initial conditions, and the laws of nature, the explanandum-event had to occur. *Nomological necessity,* it might roughly be said, derives from the laws of nature in much the same way that *logical necessity* rests upon the laws of logic. Viewing the matter this way, one can deny that an explanation is an argument, but still maintain that the explanation is the sort of thing that shows that the explanandum-event had to occur, given the initial conditions. In the absence of knowledge of the explanatory facts, the explanandum-event (the eclipse) was something that might not have occurred for all we would know; given the explanatory facts, it had to occur. The explanation exhibits the nomological necessity of the explanandum-event, given the explanatory facts. Although a deductive argument can be constructed (as in the foregoing account) within which a relation of logical entailment obtains, an explanation need not be regarded as such an argument, or any kind of argument at all.

2.3. The Ontic Conception of Scientific Explanation

There is still another way to look at such explanations. The term "law" is used sometimes to refer to a scientific statement describing a regularity in nature, and sometimes to refer to the regularity itself (see note 4). Construing the term "law" in either sense, we can say that to relate an explanandum-event to some antecedent conditions by means of laws is to fit the event to be explained into an intelligible pattern. When I call the pattern "intelligible," I do *not* mean to suggest that it possesses any kind of 'rational necessity', and I do *not* mean to suggest that such patterns can be known a priori. The point is simply that we have formulated the law-statements in terms that we understand, or equivalently, that we have seen and identified the lawful regularity described by the law-statement. In view of the universal character of the laws involved in such explanations, we can also say that, given certain portions of the pattern of events and the lawful relations exhibited by the constituents of the pattern, other portions of the pattern must have certain characteristics. Looking at explanation in this way, we might say that to explain an event is to exhibit it as occupying its (nomologically necessary) place in the intelligible pattern. Because of its emphasis on existent physical relationships, this view may be called the *ontic conception* of scientific explanation.[6]

2.4. Laws: Universal versus Statistical; Causal versus Noncausal

These three ways of thinking about scientific explanation may seem more or less equivalent—perhaps with somewhat differing emphases—as long as we are talking about the kind of explanation that involves appeal only to *universal* laws. A striking divergence will appear, however, when we consider explanations that invoke *statistical* laws. In the Laplacian framework, all of the fundamental laws of nature are strictly universal; in twentieth-century science we must at least entertain the possibility that some basic laws of nature are irreducibly statistical.

Before making the transition to consideration of the nature of scientific explanation in contexts where statistical laws must be taken into account, I must acknowledge one factor in the Laplacian conception that did not appear in any of the three accounts. Its neglect would be a glaring omission in any discussion of this sort of explanation. I refer to the relation of *causation,* which certainly played a large role in Laplace's considerations.

It may be tempting at first blush to suppose that the laws of nature are always causal laws, and that explanation in terms of laws is ipso facto causal explanation. This view seems implicit in Laplace's discussion, and it has been voiced more or less explicitly by a variety of authors.[7] A moment's reflection reveals, however, that many law-statements do not express causal relations; many lawful regularities in nature are not direct cause-effect relations. Night follows day and day follows night, but day does not cause night and night does not cause day. The ideal gas law

$$PV = nRT$$

relates pressure, volume, and temperature for a given sample of gas, and it tells us how these quantities vary as mathematical functions of one another, but it says nothing whatever about causal relations among them. Kepler's laws of planetary motion describe the orbits of the planets, but they offer no causal account of these motions. Each of these regularities—the alternation of night and day; the quantitative relationship among temperature, pressure, and volume of an ideal gas; and the regular motions of the planets—can be explained causally, but they do not express causal relations, and they do not afford causal explanations of the events that are subsumed under them. I shall return to the causal explanation of regularities later on.

3. Statistical Explanation (Pollen)

In 1827, when the botanist Robert Brown first noticed the random dance of microscopic particles of pollen suspended in a fluid, he interpreted it as evidence of their intrinsic vitality, though further observations of other kinds of particles convinced him that this phenomenon had no connection with life. He could not have guessed that he had witnessed rather direct visual evidence of the statistical behavior of molecules of the fluid in which the particles were suspended. That interpretation had to await the publication of one of Einstein's three epoch-making papers of 1905. At that juncture it was still possible to claim that the apparently random agitations were rigidly determined—just as Laplace had maintained—by the motions of tiny particles that strictly obey Newton's laws of motion. But as the quantum theory developed in the first quarter of the century, the idea of a deterministic underlying structure became more and more difficult to defend. By now, a large percentage of those who interpret quantum theory maintain that quantum phenomena are fundamentally and irreducibly statistical in character. To consider a well-worn example, the radioactive decay of a uranium nucleus by spontaneous ejection of an alpha-particle is governed entirely by probability. Given two such nuclei, one of which decays while the other does not, the statistical interpretation simply says that there is a certain probability for each of them to decay, and *there is no further factor* that determines that one will decay and the other will not. This is not a matter of human ignorance; it is a fundamental indeterminacy in the world. I *do not* mean to assert dogmatically that this is the correct interpretation; I *do* believe it has to be entertained seriously. Under these circumstances, it seems to me, we need a concept of scientific explanation that can accommodate indeterminacy—a concept of explanation that can handle the irreducibly statistical cases. For if anything is evident as a result of the physics of the twentieth century, it is that quantum theory has enormous explanatory power.

Let us consider some examples of statistical explanation that are more commonplace. Suppose John Jones has a streptococcus infection from which he recovers quickly after being treated with penicillin (Hempel, 1965a, p. 381). We would naturally explain his quick recovery on the basis of this treatment. However, most, but not all, streptococcus infections respond to penicillin, so we cannot say that he *had* to recover; we can only say that the penicillin treatment rendered his quick recovery highly probable. This explanation falls somewhat short of the Laplacian ideal of showing that the explanandum-event was necessary in the light of the explanatory facts, but it does approximate that ideal in showing that the explanandum-event *was to be expected* with high probability, given the explanatory facts. In admitting such an explanation, we allow for a little looseness or 'play' in the system of lawful connections.

Unfortunately, not all cases of explanation obligingly give us high probabilities. If John Smith develops paresis, it is explained by the fact that he contracted syphilis (more precisely, syphilis in the latent stage that has not been treated with penicillin).[8] The incidence of paresis among cases of latent untreated syphilitics is not high; it is less than 50%. This appears to be a case in which an explanation of the explanandum-event—the occurrence of paresis—can be given, but it does not render that event highly probable, or even more probable than not. Given that Smith has latent untreated syphilis, one should predict that he will *not* develop paresis. What the explanation does afford, however, is a set of conditions that are relevant to the occurrence of paresis, and (at least in our present state of medical knowledge) we can offer no others. We know that no person who does not suffer untreated latent syphilis will contract paresis, but among those who do have untreated latent syphilis, there is no known way of predicting which ones will manifest this form of tertiary syphilis and which will not.

I could continue offering examples of statistical explanations in which the explanandum-event is not highly probable in the light of the explanatory facts—cases in which what is involved in the explanation is quite clearly a suitable assemblage of factors relevant to the occurrence or nonoccurrence of the event to be explained. Such assemblages of relevant factors may yield probabilities that are high, middling, or low. The degree of probability is not what counts; the important consideration is to identify the factors that are statistically relevant. If, for example, we want to explain why a boy became a juvenile delinquent, we may find that he came from a broken home, lived in a neighborhood with a high delinquency rate, fell within a certain socioeconomic class, etc., which makes delinquency highly probable. Another adolescent, who comes from a different home environment, different neighborhood situation, a different socioeconomic background, etc., may have a low probability of becoming delinquent but nevertheless does. The same factors are relevant in the low probability case as in the high probability case, and in my opinion the two explanations are equally adequate. Each appeals to precisely the same probability distribution over the same set of factors relevant to juvenile delinquency.[9]

There is an obvious but fundamental point behind these considerations. If, in a well-specified set of circumstances, a given outcome is highly probable, but not necessary, then in some of these cases the improbable will occur. Even if a coin is heavily biased for heads, it will occasionally land tails-up. The explanation is exactly the same in both types of cases; this outcome resulted from a toss of a coin with a certain high probability for heads and a correspondingly low probability for tails. If tails does occur, we might remark on its unlikelihood, but this is by way of "gloss." It is not part of the explanation (Jeffrey, [1969] 1971).

In the examples of coin tossing, delinquent behavior, onset of paresis, or recovery from strep infection, we believe, quite reasonably, that the cases are not *irreducibly* statistical. We feel very deeply that with additional knowledge of scientific laws, or more specific information about the particular cases, we could say why *this* toss resulted in a tail rather than a head, or why *this* child became delinquent while another in similar circumstances did not. We are apt to feel, consequently, that our explanation is not complete or fully adequate unless we can say why a particular instance constituted an occurrence rather than a nonoccurrence of a given outcome. Indeed, this has sometimes been elevated to the status of a *criterion of adequacy* for scientific explanations in general, namely, that one and the same explanation cannot adequately explain either the occurrence or the nonoccurrence of a given type of event in the same circumstances. But this is a principle we must relinquish, I believe, if we are to make sense of scientific explanation in a genuinely indeterministic setting. The fact that it is difficult, if not well-nigh psychologically impossible, to give it up is a measure of the degree to which the Laplacian conception of the world permeates our thinking, even if it is well over a half-century out of date.

Let us return to the quantum mechanical example. When an alpha-particle forms in a uranium nucleus, it races to and fro inside, repeatedly crashing against the potential barrier that constitutes the wall of the nucleus. In the overwhelming majority of instances it bounces back, but on rare occasions it "tunnels through." All of this can be explained by a quantum mechanical wave function, but that wave function yields only a very low probability (of the order of 10^{-38}) that the alpha particle will escape. Precisely the same wave function explains both the reflections and the penetrations of the barrier; the only difference is that it assigns a high probability to the one and a low probability to the other.

Some philosophers have maintained that statistical laws give us grounds for prediction, or for assigning fair betting odds, but not explanations. Their reason for denying the possibility of irreducibly statistical explanations is that these do not confer any kind of necessity upon the explanandum-event (von Wright, 1971, p. 13). We have, I believe, reached the crunch between the Laplacian conception of explanation, which reflects the deterministic world picture of classical physics, and the statistical conception of explanation, which is more harmonious with contemporary physics.

When the three general conceptions of scientific explanation were elaborated in the Laplacian context, it will be recalled, they all seemed pretty much equivalent to one another. When we look at them in the indeterministic context, that situation changes remarkably. The first two conceptions, epistemic and modal, involved necessity. The epistemic appealed to the *logical necessity* with which a conclusion follows from the premises of a valid deductive argument. The modal conception invoked the *nomological necessity* with which the explanandum-event is related to the explanatory facts by virtue of universal laws of nature. If either of these formulations is taken as canonical for all acceptable explanations, then necessity is built into the concept of explanation from the outset. If we accept that conclusion, then indeterminism in the physical world would render scientific explanation impossible or unintelligible. The third conception—the ontic conception—does not have this consequence.

I do not think that we are forced to accept any such drastic conclusion. I am convinced that statistical explanations are admissible, and quite possibly indispensable, in contemporary science. In the next sections I shall try to sketch the sense in which statistical explanations, even when the associated probability values are low, provide genuine understanding of the phenomena in question. I shall then say more about the three general conceptions that emerged from the discussion of Laplacian explanation.

4. Causality in Explanation

According to the ontic characterization, it will be recalled, an explanation was described as an exhibition of the fact to be explained in its place within the natural patterns of the world. These patterns are based on the lawful regularities that structure the world. Within the Laplacian framework these regularities were seen as strict causal laws, but that deterministic feature was not essential to the characterization. It may be, as modern physics suggests, that the laws are statistical at bottom, and the patterns may be probabilistic ones. If this does represent the actual structure of the world, then many (if not all) events will have to be viewed as probabilistic outcomes of stochastic processes. The pattern of the world is then to be viewed as a series of probability relations. It would be a grievous mistake to think that this sort of thing is not a pattern or to suppose that we cannot know or understand it. I should like to attempt to sketch some of the important characteristics of such understanding.

It is customary to make a sharp distinction between causal relations and statistical or probabilistic relations. This dichotomy, it seems to me, should be called into question. Suppose a brick is hurled with great force at a windowpane; as the pane shatters, we have no doubt that the cause is the impact of the brick. Suppose, instead, that the window is struck by a golf ball traveling with only moderate speed. Under these circumstances, let us say, the windowpane will break in 90% of such cases, but not in the other 10%. The motion of the golf ball up to the point of contact with the window is a causal influence, propagated through space, and it produces the effect of shattering in $9/10$ of the situations in which it is present. No one would hesitate, I should think, in concluding that it was the impact of the golf ball that caused the breakage in any case in which breakage occurred. The contact of the golf ball with the window obviously has a large influence on the probability of the window's breaking at that particular time; there is nothing like a 90% chance of the window's breaking just in the normal course of things, say, as a result of internal stresses, the rumble of a passing truck, the explosion of a gas heater in a house three doors down the block, etc. There are, in other words, probabilistic or stochastic influences which—to borrow a phrase from Leibniz—incline but do not necessitate. I see no reason to refrain from calling such influences "causal," even though they are not deterministic. The fact that we may believe that a deterministic explanation *could be given* if more detailed information were available is no objection. The main point remains; we need not commit ourselves to determinism in order to hold that there are causal influences in the world (see "Probabilistic Causality" [essay 14]).

Causality has had a bad press in philosophy ever since Hume's devastating critique, first published early in the eighteenth century. As is well known, Hume analyzed causal relations in terms of spatiotemporal contiguity, temporal priority, and constant conjunction. He was unable to find any "necessary connection" relating causes to effects, or any "hidden power" by which the cause "brings about" the effect.[10]

Hume's classic account of causation is rightly regarded as a landmark in philosophy; it was, I believe, unjustly ignored or unappreciated by writers such as Laplace. Nevertheless, it seems to me, Hume did overlook one fundamental aspect of causal processes, namely, that they are capable of *transmitting information*. This feature is crucial, I believe, in assessing the role of causality in scientific explanation. In order to understand this point, it will help to introduce a distinction between *causal processes* and *pseudoprocesses*. That this distinction escaped Hume's attention is not surprising, for it has emerged from consideration of Einstein's special theory of relativity (first enunciated in another of his 1905 papers). A basic consequence of that theory is that no *signal*—that is, no process capable of transmitting information—can travel faster than light. For example, radio signals and sound waves are obviously capable of transmitting information; radio waves travel at the speed of light, as do all other types of electromagnetic waves, and sound travels at a much lower velocity. Certain *pseudo*-processes can, however, travel at arbitrarily high velocities, not limited by the speed of light. If, for example, a rotating spotlight is mounted in the center of a circular room, the spot of light it casts upon the wall can travel at as great a speed as you like, depending on how fast the light rotates and how far the walls are from it. There are, to be sure, a number of causal processes involved in this example—the mechanism that rotates the spotlight, the process by which the filament is made to emit light, and the transmission of light *from* the spotlight *to* the wall. All of these processes are subject to the speed limit imposed by nature (as Einstein conceived it) upon all causal processes. The movement of the spot along the wall—though it manifests a high degree of regularity—is *not* subject to such limitation, but it is *incapable* of transmitting information. If, for example, a red filter is placed near the source in the beam of light that travels from the spotlight to the wall, the spot on the wall will be red; the beam of light carries that mark or information from the point at which the filter is interposed along the beam to the wall. If, however, a red filter, interposed near the wall, makes the spot on the wall red, the red mark will not be carried along by the spot that sweeps around the wall. The spot traveling along the wall does not carry information with it; it constitutes a pseudo-process, not a causal process. This example is analogous to the scanning pattern on a TV screen. Electrons are shot from a source at the back of the tube toward the screen; the lateral to-and-fro pattern of electrons impinging on the screen is a pseudo-process. Information is transmitted from the back of the tube to the screen; it is not transmitted along the lines scanned across the screen. It was this ability to transmit information, which distinguishes causal processes from pseudo-processes, that Hume overlooked.

The ideal gas law was cited as a noncausal law; it does not describe any causal processes. Suppose I have a container of some gas (e.g., helium) with a movable piston. If I compress the gas by moving the piston, without altering the temperature of the gas, we can infer that the pressure will be increased. This increase in pressure can be explained causally on the ground that the molecules, traveling at the same average velocities, will collide with the walls of the container more frequently when the volume is decreased by moving the walls closer together. The quantitative relation between pressure and volume (at constant temperature) is not a causal relation; the motions of individual molecules obeying mechanical laws and colliding with the walls of the container are causal processes. This situation is, I believe, rather typical: a noncausal regularity is explained on the basis of underlying causal processes. In a similar fashion, it seems to me, Newton's laws of motion and gravitation, which are causal laws, explain such noncausal regularities as Kepler's laws of planetary motion, Galileo's law of free fall, and the regular ebb and flow of the tides. The regular behavior of the tides had, of course, been known to seafarers for centuries before Newton; indeed, the relationship between the tides and the position and phase of the moon was familiar to mariners prior to Newton, but these mariners did not suppose that they *understood* the rise and fall of the tides on the basis of this lawful relationship.

We can imagine a child on the beach noticing the waves gradually working their way toward the sand castle it has constructed. Alarmed, it asks why this is happening. A very primitive explanation might consist in informing it of the regular way in which the tides advance and recede. Though citing such a noncausal regularity might temporarily satisfy childish curiosity, the "explanation" can hardly be considered scientifically adequate— mainly, I am suggesting, because of its lack of reference to any causal influence. The causal explanation of the noncausal regularity does, in contrast, seem to qualify as a reasonable explanation (though not necessarily one that leaves nothing further to be explained).

Additional examples of a similar sort can be taken from biological or social sciences. The efficacy of inoculation against smallpox was known for centuries before the advent of the germ theory of disease, and before anything was known of the mechanism of immunization. The phenomenon of immunity was *understood* only after the underlying causal processes had been discovered. A well-known correlation between slum environment and reading disabilities in young children may exist, and may be said, in a crude way, to explain why a particular child from the slums cannot read. A reasonably adequate *understanding* of this phenomenon emerges only when we have exhibited the causal relations between economic deprivation and failure to learn to read.

5. Functional Explanation (Dreams)

Early in this essay I referred to the important role played by functional explanations in a rather broad range of sciences. Freud's dream theory is a particularly striking example of explanations of this type, as are many other explanations in psychoanalytic theory. Such explanations also occur in many other biological and behavioral sciences. Consider a simple biological example. The jackrabbits that inhabit the hot, arid regions in the southwestern part of the United States have extraordinarily large ears. If we ask why they have such large ears, the answer is *not* "the better to hear you with, my dear." Instead, the large ears constitute an effective cooling mechanism. If the body temperature begins to rise, the numerous blood vessels in the ears dilate, and warm blood from the interior of the body circulates through them. The animal seeks out a shady spot, heat is radiated from the ears, and the body temperature is reduced. The jackrabbit has these large ears *because* they constitute an effective mechanism for temperature regulation.

Animals that live in environments like that of the jackrabbit must have some method for dealing with high temperatures in order to survive. There are, of course, many devices that can fulfill this function. Some animals, such as the kangaroo rat, develop nocturnal habits, enabling them to avoid the heat of the day. Other animals, such as humans, perspire. Dogs pant. From the fact that a given type of animal survives in the desert, we can infer that it must have some way of coping with great heat. Thus, it can be shown deductively—or at least with high inductive probability—that such animals will have *some mechanism or other* that enables them to adapt to the extreme temperatures found in the desert. It does not follow, of course, that the jackrabbit must have developed large radiating ears, or even that it is highly probable that it would do so. Thus, if we want to explain why the jackrabbit has this particular cooling device—as opposed to explaining why it has some mechanism or other that fulfills this function—it seems implausible to claim that we can do so by rendering the presence of large ears either deductively certain or highly probable in view of the available explanatory facts.[11]

The study of social institutions by anthropologists, sociologists, and other behavioral scientists furnishes further examples of functional explanation. According to A. R. Radcliffe-Brown, social customs can be explained by considering their function or role in society, just as the presence of the heart in mammals is explained on the basis of its function in circulating blood. "Every custom and belief in a primitive society," he writes, "plays some determining part in the social life of the community, just as every organ of a living body plays some part in the general life of the organism" ([1933] 1967, p. 229). One does not, of course, need to subscribe to Radcliffe-Brown's extreme view that *all* social explanation is functional in order to agree that functional explanations of social phenomena are *sometimes* appropriate.

A classic example of a functional explanation of a social custom is Radcliffe-Brown's study of the joking relationship between a young man and his maternal uncle among the Bathonga in Africa ([1952] 1965, chap. 4). When the uncle (his mother's brother) is absent, the nephew comes to his hut, carries on a lewd conversation with the uncle's wife, demands food, steals a prized possession of the uncle, and generally deports himself in a disrespectful manner. Such behavior toward any other relative of an older generation, such as a paternal uncle, would be out of the question, and it would be severely censured if it ever did occur. The maternal uncle, however, is expected to take the nephew's pranks in good humor—without anger, disapproval, or any attempt at retaliation.

Radcliffe-Brown pointed out that among these people kinship relations form a crucial element of the social structure. The disrespectful treatment of certain older relatives by members of the younger generation plays an important role in maintaining the stability of the kinship system. Through detailed analysis Radcliffe-Brown attempted to show how the joking relationship serves to ease the tensions that naturally arise in kinship systems of the sort found among the Bathonga. Such kinship systems are not entirely different from our own, and the tensions to which he referred are similar to the kinds of in-law problems with which we are familiar. As Radcliffe-Brown explicitly notes, however, in other cultures the function of easing such tensions is fulfilled by other means, such as avoidance of contact with the in-laws. In this case, as in the case of the jackrabbit's ears, a certain function must be fulfilled if a system is to survive. In the case of the jackrabbit, the system is a living organism; in the case of the Bathonga, the system is a social institution. In both of these cases there are functional equivalents—alternative mechanisms that could fulfill the function in question. The same is true of Freud's theory of dreams: many different dreams are capable of fulfilling the same unconscious wish. For this reason it seems implausible to try to maintain that the existence of one particular mechanism is either certain or highly probable in a given situation.

It has sometimes been claimed that functional explanations are always illegitimate, or at best incomplete. According to this view, as the biological and behavioral sciences mature and develop, functional explanations will be replaced by explanations of other sorts. Functional explanations, according to this view, may have heuristic value in the early stages of scientific investigation, but they should ultimately be superseded by nonfunctional explanations. For example, it may be true, *as a matter of fact,* that functional explanations in biology will eventually give way to explanations of a purely physico-chemical sort, but I do not believe that we should commit ourselves to this viewpoint on an a priori basis. From a philosophical standpoint, it seems to me, functional explanations may be just as admissible as explanations of any other sort. As long as they play a crucial role in various branches of contemporary science, I do not think they should be ruled out on logical grounds.

Why are functional explanations regarded with widespread suspicion? There seem to be three principal reasons. First, functional explanations have been viewed as teleological and anthropomorphic. This consideration should not deter us, for as I mentioned early in this essay, functional explanations have been purged of teleological elements in areas such as evolutionary biology. This is illustrated by the examples already mentioned. The jackrabbit does not consciously choose big ears to keep its body cool. Likewise, the Bathonga did not consciously choose the joking relationship as a way of easing in-law tensions, and humans do not consciously choose the dreams that are to fulfill their unconscious wishes. Moreover, as is obvious, none of these accounts requires an appeal to the aims of any supernatural agency.

Second, it has sometimes been objected that functional explanations violate a time constraint by explaining the presence of a mechanism in terms of attainment of a *subsequent* goal, rather than on the basis of *preceding* conditions. This objection is also ill founded. It is because large ears have proved effective *in the past* in controlling body temperature that jackrabbits now have large ears. Even if a particular jackrabbit never required the use of a body-cooling mechanism (if, for example, it were transported to a zoo in a cool locale), the large ears could still be given the same functional explanation (as might be done on a descriptive placard at the zoo). The joking relation among the Bathonga existed when Radcliffe-Brown studied it because (if Radcliffe-Brown's account is correct) it has *previously* succeeded in easing in-law tensions in that society. The occurrence of dreams (if Freud is right) is explained by the *past* success of other dreams in preserving sleep against the disturbance of unconscious unfulfilled wishes. This answer to the problem of temporal orientation of functional explanation is treated effectively by Larry Wright (1976, chap. 1).

Third, what amounted to the 'received' theory of scientific explanation for several decades was unable to accommodate functional explanations as such. According to this view, an explanation is an argument to the effect that the fact to be explained was to be expected, either with deductive certainty or with high inductive probability, on the basis of the explanatory facts. Because typically there are functional equivalents—alternative mechanisms that could fulfill the same function—functional explanations do not, in general, render the explanandum expectable, either with deductive certainty or with high probability. The fact that the received view of scientific explanation cannot account for functional explanations may, however, reflect more adversely on this philosophical theory of scientific explanation than it does on functional explanations themselves. If the received view is correct, there are no legitimate functional explanations in science. Some would say the received view is correct, so there are no legitimate functional explanations. But one person's *modus ponens* is another person's *modus tollens*. Others would say there are legitimate functional explanations in science, so the received view is not correct. If we can develop a philosophical theory of scientific explanation that does admit those sorts of functional explanations that do appear to be widely accepted in various branches of science, that fact should, it seems to me, count significantly in favor of the alternative philosophical theory. In the concluding section of this essay I shall try to show that the ontic conception of scientific explanation holds promise of providing just such a theory of scientific explanation.

6. The Three Conceptions Revisited

In the context of Laplacian determinism I characterized three general conceptions of scientific explanation: epistemic, modal, and ontic. In that context the distinctions among the three may have seemed somewhat artificial, but as I remarked, the differences are striking when viewed from the possibly indeterministic standpoint of contemporary science. Some of these features have already been mentioned, but let us bring them together in order to form a coherent overall picture. These considerations are summarized in the following table.

Deterministic	Indeterministic
1. *EPISTEMIC*	
Logical necessity	High *inductive* probability
Argument/deducibility	Inductive support
Nomic expectability with certainty (vs. unexpected)	Nomic expectability with high probability
2. *MODAL*	
Nomological necessity	
Lawful connection with explanatory facts	Statistical explanation impossible
Had to happen (vs. might or might not have happened)	
3. *ONTIC*	
Fitting into intelligible pattern	Fitting into intelligible pattern
Pattern structured by strict causal relations (vs. haphazard/unrelated to other natural occurrences)	Pattern structured by probabilistic causal relations
	Probabilities need not be high

1. The *epistemic conception*: If determinism is true, then it is always possible in principle to provide a deductive explanation of any event, that is, to show that it is logically necessary relative to the explanatory facts. If indeterminism is true, then some events are not fully determined by antecedent conditions and laws of nature, so it is not possible, even in principle, to provide deductive explanations of the Laplacian variety. In view of this fact, some proponents of the epistemic approach have loosened the requirements sufficiently to admit that events that are not fully determined may still be explained if their occurrences can be rendered highly probable in terms of statistical laws of nature. In particular, Hempel, the leading advocate of the epistemic conception, developed a pattern of explanation, known as the inductive-statistical (I-S) model, that plays precisely this role in the theory of scientific explanation (Hempel, 1962a).[12] In making the transition from the deterministic context to the indeterministic context, the fundamental logical relation of deductive entailment is replaced by the relation of high inductive probability. Thus, the requirements for a satisfactory scientific explanation are relaxed in such a way as to allow for the possibility of irreducibly statistical explanations, provided that the event-to-be-explained can be rendered highly probable in view of the explanatory facts. Under these circumstances, we can still say that the event-to-be-explained *is to be expected,* with high probability rather than deductive certainty, in view of the explanatory facts.

2. The *modal conception*: If indeterminism is true, then there are some events with at least some aspects that are not physically necessitated by antecedent conditions on the basis of laws of nature. With respect to such features of events of this sort, it is simply impossible to show that they *had to occur,* and hence they defy scientific explanation. I can see no way in which the modal conception can be transformed to enable it to handle explanation in indeterministic contexts. To replace physical necessity with some sort of probability relation would be to relinquish the modal conception and to move either to the epistemic conception or to the ontic conception.[13] The adherent of the modal conception faces a severe dilemma. Either one makes an a priori commitment to determinism, or one has to deny that quantum mechanical explanations, as they are usually construed, qualify as legitimate scientific explanations. Neither alternative seems tenable.

3. The *ontic conception*: According to this conception, events are explained by showing how they fit into the physical patterns found in the world. In the Laplacian context of classical physics, it appeared that these patterns were strict deterministic patterns; in the light of contemporary physics, it now appears that some, at least, of these patterns are inherently statistical. But this fact poses no obstacle to the construction of scientific explanations. Statistical patterns are bona fide patterns.

Carbon 14 atoms, for example, decay in a statistically regular way, and this regularity provides the basis for the technique of radiocarbon dating, which has proved to be a valuable tool for archaeologists. Other radioactive atoms decay in accordance with different statistical patterns. The half-life of carbon 14 is 5715 years; the half-life of tritium (hydrogen 3) is 12.26 years; the half-life of uranium 238 is 4.46 billion years. Among other things, these regularities imply that there is a very high probability that a given tritium atom will decay in a period of 5715 years, there is a 50–50 chance that a given carbon 14 atom will decay in the same period, and there is a very small probability that a given uranium 238 atom will decay in that same period. One important point to be emphasized in this context is that some events fit into statistical patterns with very low probabilities. For example, there is current speculation to the effect that the proton is not absolutely stable but decays with a half-life of the order of 10^{30} to 10^{32} years. To gain perspective on the time scale involved, it should be recalled that the total age of the universe since the primordial "big bang" is now thought to be about 10^{10} years. Thus, the probability of a given proton's decaying within the next year is truly minute. Experiments are now being designed, however, with the aim of detecting proton decays. Even though the probability of any given proton's decaying is very small, there is a reasonable chance of detecting such an event if a large enough collection of protons is examined for a few years. Since events fit into statistical patterns with high probabilities in some cases, with middling probabilities in other cases, and with small probabilities in still other cases, the size of the probability of the explanandum-event has no bearing on the possibility of providing a statistical explanation of it.

The situation regarding statistical explanation can now be summarized. The modal conception does not allow for statistical explanations of particular events.[14] This view seems untenable in the light of the patent explanatory success of contemporary statistical theories in the sciences. The epistemic conception admits statistical explanations of particular events, provided that the associated probabilities are high enough. How high is high enough? That, I believe, is a profoundly embarrassing question (see "A Third Dogma of Empiricism" [essay 6]). The ontic conception allows statistical explanations of any events that occur within a definite statistical pattern, regardless of the size of the associated probability.

In attempting to make a decision between the epistemic and the ontic conceptions of scientific explanation, the question of whether it is possible to explain events whose occurrences are intrinsically improbable emerges as a crucial one. As a proponent of the ontic conception, I am inclined to give an affirmative answer. There are two main reasons.

First, to maintain that highly probable events can be explained while improbable ones cannot seems to involve a strange and arbitrary lack of parity. Consider, for example, a famous genetic experiment conducted by Gregor Mendel. In a particular population of pea plants, he showed that there is a probability of ¾ that any given plant will bear red blossoms and a probability of ¼ that it will bear white blossoms. Assume that ¾ is large enough to qualify as a high probability; if it is not, the example can easily be modified to furnish a higher value. Then, according to the epistemic conception, we can explain the occurrence of a red blossom in that group of plants, but we cannot explain the occurrence of a white blossom. It seems obvious to me, however, that under those circumstances we understand the occurrence of a white blossom just as adequately as we understand the occurrence of a red blossom. As Jeffrey ([1969] 1971) has persuasively argued, the fact that one occurs with a higher probability than the other is beside the point.

Second, as I tried to argue in the preceding section, if functional explanations are to be considered admissible, we will have to allow the possibility of explaining facts that do not have high probabilities. For those who are dubious about the force of the argument from symmetry given in the preceding paragraph, this argument may be decisive. It is beyond the scope of this essay to attempt to provide a detailed account of functional explanation, but the fact that functional explanations do seem to be considered acceptable in various branches of contemporary science strongly suggests that our conception of scientific explanation ought to be broad enough to accommodate explanations of this sort (see "Alternative Models of Scientific Explanation" [essay 21]). Of the three conceptions I have discussed, only the ontic appears to be capable of this.

In my earliest article on scientific explanation (Salmon, 1965), I attempted to develop a theory based on relations of statistical relevance; this led to the elaboration of the statistical-relevance (S-R) model (Salmon, 1971). Statistical explanations constructed along the lines of this model could accommodate events whose probabilities are low, medium, or high. In more recent writings I have attempted to supplement the statistical relevance model with considerations of causal relevance. ("Causal and Theoretical Explanation" [essay 7]). Recalling the claim made in a previous section—that causal relations can also be statistical—we see that the statistical and the causal considerations which have been discussed in this essay can be brought together to form a unified theory of scientific explanation.

Statistical and causal relations constitute the patterns that structure our world—the patterns into which we fit events and facts we wish to explain. Causal processes play an especially important role in this account, for they are the mechanisms that propagate structure and transmit causal influence in this dynamic and changing world. In a straightforward sense, we may say that these processes provide the ties among the various spatiotemporal parts of our universe. We have here, I believe, an answer to Hume's question about the nature of the connections between causes and effects. They are the channels of communication by which the physical world transmits information about its own structure. When we recognize these causal processes and the role that they play in unifying the patterns into which facts and events fit, then we have gone a long way toward scientific understanding of our world and what goes on within it. The ontic conception thus constitutes a causal conception of scientific explanation that seems to be in harmony with twentieth-century science. In recognizing the statistical aspects of causal relations, it provides an appropriate advance beyond the Laplacian ideas which have, until recently, had an almost inestimable influence upon our thought about scientific explanation.

Notes

The material in this essay is based on work supported by the National Science Foundation (USA) under Grant no. SOC-7809146. I should like to express my gratitude for this support, and to thank colleagues and students, too numerous to cite individually, in Australia and America, for many valuable comments and criticisms. I am also grateful to the University of Melbourne for making possible an extended visit to Australia, where these ideas were discussed under circumstances highly conducive to constructive intellectual work.

1. This attempt was made by Carl G. Hempel (1962a). My first criticism of this approach was given in my (1965).

2. Even if Isaac Newton himself preserved teleological elements in his worldview, they were eliminated by subsequent practitioners of classical mechanics such as Laplace.

3. This conception of scientific explanation is expressed in Holton and Brush (1973, p. 185).

4. The term "law" has two quite distinct meanings in the context of discussions of scientific explanation: on the one hand, it sometimes refers to a regularity that exists in nature; on the other hand, it sometimes refers to a statement that such a regularity obtains. When the distinction between these two meanings is important, and when the context does not make entirely clear which sense is intended, I shall use the phrase "law of nature" to refer to the natural regularity itself, and I shall use such phrases as "law-statement" or "scientific law" to refer to the linguistic entity. In order to qualify as a law-statement, a statement must be true. Statements that have all other characteristics of law-statements, but that may fail to be true, are known as lawlike statements.

5. This is, as a matter of fact, the most widely accepted view of scientific explanation, at least in contexts where universal laws are available for explanatory purposes. For a clear and thorough discussion of this epistemic approach, both in the deterministic and in the indeterministic contexts, see Carl G. Hempel (1965b). I attack this 'received view' in "A Third Dogma of Empiricism" (essay 6).

6. Although Hempel is to be identified primarily as a proponent of the epistemic conception, he does offer the following characterization of scientific explanation in the concluding paragraph of his major essay: "The central theme of this essay has been, briefly, that all scientific explanation involves, explicitly or by implication, a subsumption of its subject matter under general regularities; that it seeks to provide a systematic understanding of empirical phenomena by showing that they fit into a nomic nexus" (1965a, p. 488). I have expressed a similar idea (1977a, p. 162).

7. Most notably, it was suggested in the classic article by Hempel and Oppenheim ([1948] 1965). Hempel has subsequently rejected this notion (see 1965b, p. 352).

8. This example is due to Michael Scriven (1959, p. 478). In the United States the term "paresis" specifically designates one form of tertiary syphilis.

9. This example, along with its accompanying analysis, was contributed by James G. Greeno ([1970] 1971, pp. 89–91).

10. It may be that Hume's critique is mainly responsible for the fact that such contemporary authors as Hempel have explicitly denied that scientific explanation *must* have any causal component (see Hempel, 1965a, p. 352).

11. We can, of course, explain why this particular jackrabbit has large ears on the grounds that it inherited this trait from its parents, both of which had big ears. But if we ask why this trait is present in the species, the answer may be that it originated on the basis of some sort of chance mutation. The fact that it was perpetuated and propagated is due to natural selection on the basis of its survival value. This point is discussed by Baruch Brody (1975).

12. An improved version is given in Hempel (1965a, pp. 381–403).

13. D. H. Mellor (1976) tries to make this transition, but this appeal to degrees of possibility and necessity strikes me as insufficient for the purpose. It seems to me that the replacement of physical necessity with high *inductive* probability leads to the epistemic conception, while the replacement by high *physical* probability leads to the ontic conception.

14. Hempel (1965a, pp. 380–381) discusses the deductive-statistical (D-S) model in which *statistical regularities* are explained by deductive subsumption under broader statistical law. Statistical explanations of this sort pose no difficulties for the modal conception, but such explanations cannot explain *occurrences of individual events*.

4

Scientific Explanation

Causation and *Unification*

For the past few years I have been thinking about the philosophy of scientific explanation from the standpoint of its recent history. Many of these reflections have been published in *Four Decades of Scientific Explanation* (1990b), a condensed version of which is given in "Scientific Explanation: How We Got from There to Here" (essay 19). They have, I believe, provided some new insight on some old problems, and they suggest that genuine progress has been made in this area of philosophy of science.

1. Looking Back: Two Grand Traditions

The classic essay "Studies in the Logic of Explanation" by Carl G. Hempel and Paul Oppenheim ([1948] 1965) constitutes the fountainhead from which almost everything done subsequently on philosophical problems of scientific explanation flows. Strangely enough, it was almost totally ignored for a full decade. Although the crucial parts were reprinted in the famous anthology *Readings in the Philosophy of Science,* edited by Herbert Feigl and May Brodbeck (1953), it is not cited at all in R. B. Braithwaite's well-known book *Scientific Explanation* (1953). During the first decade after publication of the Hempel-Oppenheim paper, very little was published on scientific explanation in general—Braithwaite's book being the main exception. Most of the work on explanation during that period focused either on explanation in history or on teleological/functional explanation.

In the years 1957 and 1958 the situation changed dramatically. At that time a deluge of work on scientific explanation began, much of it highly critical of the Hempel-Oppenheim view. Vigorous attacks came from Michael Scriven (1958, 1959, 1962) and

N. R. Hanson (1959), among others. Sylvain Bromberger (1966) and Israel Scheffler (1957) offered important criticisms, but they were offered more in the spirit of friendly amendments than outright attacks on the Hempel-Oppenheim program (see Salmon, 1990a, pp. 33–46).

When we reflect on what happened, we can see that two grand traditions emerged. Hempel advocated a view of scientific explanation according to which explanation consists in deductive or inductive subsumption of that which is to be explained (the explanandum) under one or more laws of nature. This tradition could find examples that had strong intuitive appeal—for instance, the explanation of the laws of optics by Maxwell's electrodynamics, or the explanation of the ideal gas law by the molecular-kinetic theory. These examples also illustrate what is often called "theoretical reduction" of one theory to another. Another example, if it could be worked out successfully, would be methodological individualism in the social sciences, for it would result in the reduction of the various social sciences to psychology.

Ironically, the very examples that furnish the strongest intuitive appeal for the subsumption approach are of a type that Hempel and Oppenheim found intractable. Although they offered an account of explanations of particular facts, they acknowledged in a notorious footnote (note 33), that they could *not* provide an account of explanations of general laws. To the best of my knowledge, Hempel never returned to this recalcitrant problem. It should be noted that, while Hempel and Oppenheim casually identified their pattern of explanation (later known as the *deductive-nomological* or D-N model) with causal explanation, Hempel later argued emphatically that causality does not play any sort of crucial role in scientific explanation (1965b, § 2.2).

The other major tradition was advanced primarily by Scriven, and it made a strong identification between causality and explanation. Roughly and briefly, to explain an event is to identify its cause. The examples that furnish the strongest intuitive basis for this conception are cases of explanations of particular occurrences—for instance, the sinking of the *Titanic* or the Chernobyl nuclear accident. The most serious problem with this approach has been the lack of any adequate analysis of causality on which to found it. Given Hume's searching critique of that concept, something more was needed. [One of the main aims of this book is to resolve Hume's problems regarding causality.]

As these two traditions developed over the years, there was often conflict, sometimes quite rancorous, between their advocates. At present, I believe, we have searched a stage in which a significant degree of rapprochement is entirely possible.

2. Explanation as Unification

The idea that scientific explanation consists in showing that apparently disparate phenomena can be seen to be fundamentally similar has been around for a long time, long before 1948. However, Michael Friedman, in "Explanation and Scientific Understanding" (1974), seems to have been the first philosopher to articulate this conception clearly and to attempt to spell out the details. His basic thesis is that we increase our scientific understanding of the world to the extent that we can reduce the number of independently acceptable assumptions that are required to explain natural phenomena. By phenomena he means regularities in nature such as Kepler's first law (planets move in elliptical orbits) or Hooke's law (the amount of deformation of an elastic body is proportional to the force applied). It should be noted that Friedman is attempting to furnish an account of the explanation of laws, which is just the sort of explanation Hempel and Oppenheim found themselves unable to handle.

In order for Friedman's program to work, it is obviously necessary to be able to count the number of assumptions involved in any given explanation. In order to facilitate that procedure, Friedman offers a definition of a technical term, "K-atomic statement." This concept is relativized to a knowledge situation K. A statement is K-atomic provided it is not equivalent to two or more generalizations that are independently acceptable in knowledge situation K. A given statement is acceptable independently of another if it is possible to have evidence adequate for the acceptance of the given statement without ipso facto having evidence adequate to accept the other. The problem that arises for Friedman's program is that it seems impossible to have any K-atomic statements—at least, any that could plausibly be taken as fundamental laws of nature. For instance, Newton's law of universal gravitation, which, prior to Einstein, was a good candidate for a fundamental law, can be partitioned into (1) "Between all pairs of masses in which both members are of astronomical dimensions there is a force of attraction proportional to the product of the masses and inversely proportional to the square of the distance between them," (2) "Between all pairs of masses in which one member is of astronomical dimensions and one is smaller there is a force of attraction. . . ," and (3) "Between all pairs of masses in which both are of less than astronomic size there is a force of attraction. . ." Statement (1) is supported by planetary motions and the motion of the moon. Statement (2) is supported by Newton's falling apple and, indeed, by all phenomena to which Galileo's law of falling bodies applies. Statement (3) is supported by the Cavendish torsion-balance experiment. It seems possible to partition virtually any universal statement into two or more independently acceptable generalizations.

If Friedman's program had worked, it would have solved the Hempel-Oppenheim problem of footnote 33. It appears, however, not to be satisfactory in the form originally given. Although Philip Kitcher (1976) offered his own (different) critique of Friedman's paper, he accepted the basic idea of explanation as unification, and he has elaborated it in a different way in a series of papers, of which "Explanatory Unification and the Causal Structure of the World" (1989) is the most recent and most detailed. It is further elaborated in Kitcher (1993).

3. Causality and Mechanism

Around 1970, when I was trying to work out the details of the *statistical relevance* or *S-R* model of scientific explanation, I had hopes that the fundamental causal concepts could be explicated in terms of statistical concepts alone, and that, consequently, the *S-R* model could furnish what was chiefly lacking in the causal approach. By 1980 that no longer seemed possible, and I shifted my focus to an attempt to explicate certain causal mechanisms, in particular, causal interactions and causal processes (see Salmon, 1984b, chaps. 5–6). I took as primitives the notion of a process and that of a spatiotemporal intersection of processes. The aim is to distinguish between processes that are causal and those that are not (causal processes versus pseudo-processes) and to distinguish between those intersections of processes (whether causal or pseudo) that are genuine causal interactions and those that are not.

The basic idea—stated roughly and briefly—is that an intersection of two processes is a *causal interaction* if both processes are modified in the intersection in ways that persist beyond the point of intersection, even in the absence of further intersections. When two billiard balls collide, for instance, the state of motion of each is modified, and those modifications persist beyond the point of collision. A *process* is *causal* if it is capable of transmitting a mark—that is, if it is capable of entering into a causal interaction. For example, a beam of white light becomes and remains red if it passes through a piece of red glass, and the glass absorbs some energy in the same interaction.

However, not all intersections of causal processes are causal interactions. If two light rays intersect, they are superimposed on each other in the locus of intersection, but after they leave that place each of them continues on as if nothing had happened. A process—such as a light beam—is causal if it can be modified or marked in a way that persists beyond the point of intersection as a result of *some* intersection with another process. Causal processes are capable of transmitting energy, information, and causal influence from one part of spacetime to another. I have argued that causal processes are precisely the kinds of causal connections Hume sought but was unable to find. I have also argued that such connections do not violate Hume's strictures against mysterious powers.

It is important to recognize that these causal mechanisms are not necessarily deterministic. In particular, causal processes can interact probabilistically. My favorite example is Compton scattering, in which an energetic photon collides with a virtually stationary electron. The angles at which the photon and electron emerge from the interaction are not strictly determined; there is, instead, a probability distribution over a whole range of pairs of angles. By conservation of momentum and energy, however, there is a strict correlation between the two scattering angles.

The causal mechanisms of interaction and transmission are strongly local; they leave no room for what Einstein called "spooky action-at-a-distance." Interactions occur in a restricted spacetime region, and processes transmit in a spatiotemporally continuous fashion. Regrettably (to me and many others), however, quantum mechanics appears to involve violations of local causality. There seems to be a quantum mechanism, often known as "the collapse of the wave function," which is radically nonlocal, and which is not really understood as yet.

I prefer to think of the conception of explanation that emerges from these considerations as causal/mechanical. The aim of explanations of this sort is to exhibit the ways in which nature operates; it is an effort to lay bare the mechanisms that underlie the phenomena we observe and wish to explain.

4. Some New Perspectives

During the 1960s and 1970s the ideas developed by Hempel constituted a *received view* of scientific explanation. It was based on Hempel-Oppenheim, ([1948] 1965), and was articulated most fully in Hempel's "Aspects of Scientific Explanation" (1965b). As a result of numerous criticisms, it is fair to say, the 'received view' is no longer received. Its natural successor is the unification conception developed chiefly by Friedman and Kitcher.

The causal conception as originally advocated by Scriven and others has also undergone transformation, primarily as a result of more careful and detailed analysis of causality, but also because of the admitted possibility that there are mechanisms of a noncausal type as well. It has involved an explicit recognition of the Humean critique of causality, and an attempt to overcome the Humean difficulties.

Given the history of opposition between the 'received view' and the causal view of scientific explanation, it is not surprising that philosophers continue to find opposition between the successors. Friedman, for example, contrasted *local* and *global* accounts. According to the older views of both Hempel and Scriven, explanation is a local affair, in the sense that one could give a perfectly acceptable explanation of a small and isolated phenomenon without appeal to global theories. One could give a Hempelian explanation of the electrical conductivity of a particular penny by pointing out that it is made of copper, and copper is an electrical conductor. One could give a Scrivenesque explanation of a stain on a carpet by citing the fact that a clumsy professor bumped an open ink bottle off of the desk with his elbow. In contrast to both of the foregoing accounts, Friedman's unification view requires us to look at our entire body of scientific knowledge, to see whether a given attempt at explanation reduces the number of assumptions needed to systematize that body of knowledge. Friedman's conception is patently global.

Kitcher (1985) made a related distinction between conceptions he characterizes as "bottom-up" and "top-down." The Hempelian approach illustrates the bottom-up way. We begin by explaining the conductivity of a penny by appeal to the generalization that copper is a conductor. We can explain why copper is a conductor in terms of the fact that it is a metal. We can explain why metals are conductors in terms of the behavior of their electrons. And so it goes from the particular fact to the more general laws until we finally reach the most comprehensive available theory. The causal/mechanical approach has the same sort of bottom-up quality. From relatively superficial causal explanations of particular facts we appeal to ever more general types of mechanisms until we reach the most ubiquitous mechanisms that operate in the universe. Kitcher's top-down approach, in contrast, looks to the most general explanatory schemes we can find, and works down from there to characterize such items as laws and causal relations.

In a spirit quite different from those of Friedman and Kitcher, Peter Railton advocates an approach that makes the bottom-up and top-down, as well as the local and global, conceptions complementary rather than contrary. In "Probability, Explanation, and Information" (1981) he introduces the concept of an *ideal explanatory text* which is extremely global and detailed. He suggests, however, that we hardly ever seek to articulate fully such an ideal text. Rather, we focus on portions or aspects of the ideal text and try to illuminate these. When we succeed, we have furnished *explanatory information*.

Different investigators, or groups of investigators, have different interests and work on different portions of the ideal text. Pragmatic considerations determine for a given individual or group what portion of the ideal text to look at, and in what depth of detail.

5. Rapprochement?

My main purpose in this essay is to consider the possibility, suggested by Railton's work, that the successors of the 'received view' and its causal opponent are actually compatible and complementary. Let me begin by offering a couple of examples.

(1) A friend recounted the following incident. Awaiting takeoff on a jet airplane, he found himself sitting across the aisle from a young boy who was holding a helium-filled balloon by a string. In order to pique the child's curiosity, he asked the boy what he thought the balloon would do when the airplane accelerated rapidly for takeoff. After considering for a few moments, the boy said he thought it would move toward the back of the cabin. My friend said *he* believed that it would more forward in the cabin. Several other passengers overhead this claim and expressed skepticism. A flight attendant even wagered a miniature bottle of scotch that he was wrong—a wager he was happy to accept. In due course the pilot received clearance for takeoff, the airplane accelerated, and the balloon moved toward the front of the cabin. And my friend enjoyed a free drink courtesy of the flight attendant.

Two explanations of the balloon's strange behavior can be given. First, it can be pointed out that, when the plane accelerates, the rear wall of the cabin exerts a force on the air molecules near the back, which produces a pressure gradient from rear to front. Given that the inertia of the balloon is smaller than that of the air it displaces, the balloon tends to move in the direction of less dense air. This is a straightforward causal explanation in terms of the forces exerted on the various parts of the physical system. Second, one can appeal to Einstein's principle of equivalence, which says that an acceleration is physically equivalent to a gravitation field. The effect of the acceleration of the airplane is the same as that of a gravitational field. Since the helium balloon tends to rise in air in the earth's gravitational field, it will tend to move forward in the air of the cabin in the presence of the aircraft's acceleration. This second explanation is clearly an example of a unification-type explanation, for the principle of equivalence is both fundamental and comprehensive.

(2) A mother leaves her active baby in a carriage in a hall that has a smooth level floor. She carefully locks the brakes on the wheels so that the carriage will not move in her absence. When she returns she finds, however, that by pushing, pulling, rocking, bouncing, etc., the baby has succeeded in moving the carriage some little distance. Another mother, whose education includes some physics, suggests that next time the carriage brakes be left unengaged. Though skeptical, the first mother tries the experiment and finds that the carriage has moved little, if at all, during her absence. She asks the other mother to explain this lack of mobility when the brakes are off.

Two different explanations can be given; each assumes that the rolling friction of the carriage is negligible when the brakes are off. The first (at least in principle) possible explanation would involve an analysis of all of the forces exerted by the baby on the carriage and the carriage on the baby, showing how they cancel out. This would be a detailed causal explanation. The second explanation would appeal to the law of conservation of linear momentum, noting that the system consisting of the baby and the carriage is essentially isolated (with respect to horizontal motion) when the brake is off, but is linked with the floor, the building, and the earth when the brake is on. This is an explanation in the unification sense, for it appeals directly to a fundamental law of nature.

The first point I should like to emphasize in connection with these examples from physics is that both explanations are perfectly legitimate in both cases; neither is intrinsically superior to the other. Pragmatic considerations often determine which of the two types is preferable in any particular situation. Invocation of Einstein's principle of equivalence would be patently inappropriate for the boy with the balloon, and for the other adults in that situation, because it is far too sophisticated. All of them could, however, understand a clear explanation in terms of forces and pressures. The two examples are meant to show that explanations of the two different types are not antithetical but, rather, complementary.

I should like also to consider a famous example from biology, (3) the case of the peppered moth in the vicinity of Liverpool, England. This moth spends much of its life on the trunks of plane trees, which naturally have a light-colored bark. Prior to the industrial revolution the pale form of this moth was prevalent, for its light color matched the bark of the tree, and consequently provided protection against predators. During the industrial revolution in that area, air pollution darkened the color of the tree bark, and the dark (melanic) form of the peppered moth became prevalent, because the darker color then provided better protection. In the post–industrial revolution period, since the pollution has been drastically reduced, the plane trees have again acquired their natural light-colored bark, and the light form of the peppered moth is again becoming dominant.

In this example, like the two preceding, two different explanations are available to account for the changes in color of the moth. The first has already been suggested in the presentation of the example; it involves such evolutionary considerations as natural selection, mutation, and the heritability of traits. This is the unification style of explanation in terms of basic and comprehensive principles of biology. The second kind of explanation is biochemical in nature; it deals with the nitty-gritty details of the causal processes and interactions involved in the behavior of DNA and RNA molecules and the synthesis of proteins leading up to the coloration of the moth. In order to explain the changes in color, it would have to take account also of the births, deaths, and reproductive histories of the individual moths. Although such a causal/mechanical explanation would be brutally complex, it is possible in principle. Again, there is nothing incompatible about the two kids of explanation.

The use of this kind of biological example leads into a more general consideration regarding the status of functional explanations. In the case of the peppered moth, we were clearly concerned with a function of the coloration, namely, its function as camouflage for protection against predators. Although some philosophers have tried to cast doubt upon the legitimacy of functional explanations, I am strongly inclined to consider them scientifically admissible. In my opinion, Larry Wright, a student of Scriven, has given the most convincing theory (1976). Wright makes a distinction between *teleological explanations* and *functional ascriptions,* but his accounts of them are fundamentally similar; they involve what he calls a *consequence-etiology.* It is a *causal* account in which the cause of a feature's presence is the fact that in the past, when it has been present, it has had a certain result or consequence. It is not *just* that it has had such consequences in the past; in addition, the fact that it had such consequences is causally responsible for its coming into being in the present instance. [See Hitchcock (1996) for a technical explication of second-order causation.]

I shall use the term "functional explanation" to cover both teleological explanations and functional ascriptions in Wright's terminology. Although functional explanations in this sense are causal, they do not have a fine-grained causal character—that is to say, they do not go into the small details of the causal processes and interactions involved. They do, of course, appeal to the *mechanisms* of evolution—inheritance and natural selection—but these are coarse-grained mechanisms. Wright is, however, perfectly willing to admit that fine-grained causal explanations are also possible. Just as we can give a straightforwardly mechanistic account of the workings of a thermostat, whose function is to control temperature in a building, so also is it possible, at least in principle, to give a thoroughly physico-chemical account of some item that has a biological function, such as the color of the peppered moth. Although some philosophers have maintained that the mechanistic explanation, when it can be given, supersedes the functional explanation, Wright holds that they are completely compatible, and that the functional explanation need not give way to the mechanistic explanation. I think he is correct in this view.

The philosophical issue of the status of functional explanations is not confined to biology; the problem arises in psychology, anthropology, and the other social or behavioral sciences as well. Whether one regards Freudian psychoanalysis as a science or not, the issue is well illustrated in that discipline. According to Freud, the occurrence and the content of dreams can be explained functionally. The dream preserves sleep by resolving some psychological problem that might otherwise cause the subject to awaken. The content of the dream is determined by the nature of the problem. However, even if it is possible to provide a psychoanalytic explanation of a given dream, it may also be possible to give another explanation in completely neurophysiological terms. This would be a fine-grained causal explanation that incorporates the physical and chemical processes going on in the nervous system of the subject. I am suggesting that the two explanations need not conflict with each other, and I believe that, in this opinion, I am in agreement with Freud.

6. Can Quantum Mechanics Explain?

Ever since the publication of the famous Einstein-Podolsky-Rosen paper (1935), there has been considerable controversy over the explanatory status of the quantum theory. Einstein seems to have taken a negative attitude, while Bohr appears to have adopted an affirmative one. As the discussion has developed, the question of local causality versus action-at-a-distance has become the crucial issue. The paper showed that there could, in principle, be correlations between remote events that seem to defy explanation. Further work by David Bohm, John Bell, and A. Aspect have shown that such correlations actually exist in experimental situations, and that *local hidden-variable* causal explanations are precluded. A clear and engaging account of these issues can be found in N. David Mermin (1985). Because these find-grained causal explanations are not possible, many philosophers, myself included, have concluded that quantum mechanics does not provide explanations of these correlations. As I suggested earlier, there seem to be mechanisms at the quantum level that are noncausal, and that are not well understood.

Other philosophers have taken a different attitude. On the basis of the undeniable claim that quantum mechanics is a highly successful theory in providing precise predictions and descriptions (they are statistical but extremely successful), we need ask for no more. The quantum theory can be formulated on the basis of a small number of highly general principles, and it applies universally.

In terms of the distinct conceptions of scientific explanation I have been discussing, it seems that quantum theory provides explanations of the unification type, but it does not provide those of the causal/mechanical sort. This situation contrasts with that in other scientific disciplines where, as we have seen, explanations of both kinds are possible, at least in principle. The same circumstance may seem to occur in anthropological or sociological explanations of some human institutions, where we can give functional explanations of certain phenomena, but fine-grained causal explanations are far beyond our grasp. In contrast to quantum mechanics, however, there is no solid theoretical basis for claiming that fine-grained causal explanations are impossible in principle in these disciplines.

In answer to the question of this section, "Can quantum mechanics explain?" the answer must be, for the time being at least, "In a sense 'yes,' but in another sense 'no.'" In Salmon, (1984b, pp. 242–59) I had admitted only the negative answer to this question.

7. Two Concepts of Explanation

One of the chief aims and accomplishments of science is to enhance our understanding of the world we live in. In the past it has often been said that this aim is beyond the scope of science—that science can describe, predict, and organize, but that it cannot provide genuine understanding. Among philosophers of science and philosophical scientists at present, there seems to be a fair degree of consensus about the ability of science to furnish explanations, and therefore to contribute to our understanding of the world. As is obvious from the foregoing discussion, however, there is no great consensus on the nature of this understanding. I should like to suggest that it has at least two major aspects, corresponding to the two types of explanation that have been discussed.

On the one hand, understanding of the world involves a general worldview—a *Weltanschauung*. To understand the phenomena in the world requires that they be fitted into the general world-picture. Although it is often psychologically satisfying to achieve this sort of agreement between particular happenings and the worldview, it must be emphasized that psychological satisfaction is not the criterion of success. To have *scientific* understanding, we must adopt the worldview that is best supported by all of our scientific knowledge. The fundamental theories that make up this worldview must have stood up to scientific test; they must be supported by objective evidence. Perhaps we need not ask what makes a scientific world-picture superior to a mythic or religious or poetic worldview. Nevertheless, I would ask, and try to give an answer. The superiority of understanding based on a scientific worldview lies in the fact that we have much better reason to regard that worldview as true—even though some other worldview might have more psychological appeal.

The conception of understanding in terms of fitting phenomena into a comprehensive scientific world-picture is obviously connected closely with the unification conception of scientific explanation. It also corresponds closely to the goal of many contemporary scientists who are trying to find one unified theory of the physical world—for example, those who see in so-called "superstring theory" a TOE (theory of everything). Many scientists seem to believe that it is both feasible and desirable to try to discover some completely unified theory that will explain everything. [This program is discussed in detail in "Dreams of a Famous Physicist" (essay 26).]

Yet there is a different fundamental notion of scientific understanding that is essentially mechanical in nature. It involves achieving a knowledge of how things work. One can look at the world, and the things in it, as black boxes whose internal workings we cannot directly observe. What we want to do is open the black box and expose its inner mechanisms.

This conception of scientific explanation brings us face to face with the problem of realism versus antirealism. Although one can open up a clock to find out how it works by direct observation of its parts, one cannot do so with a container full of a gas. Gases are composed of molecules or atoms (monatomic molecules), and these are too small to be observed by means of the naked eye, a magnifying glass, or a simple optical microscope. The search for mechanistic explanations often takes us into the realm of observables. Although some philosophers, past and present, have adopted a skeptical or agnostic attitude toward unobservables, I think it is possible to argue persuasively that we can have genuine knowledge of such micro-entities as bacteria and viruses, atoms and molecules, electrons and protons, and even quarks and neutrinos. I believe we can have compelling inductive evidence concerning the existence and nature of such entities (Salmon, 1984b, chap. 8). The ideal of this approach is to have the capacity to provide explanations of natural phenomena in terms of the most fundamental mechanisms and processes in the world.

Consideration of these two conceptions of scientific explanation suggests that there may be a kind of explanatory duality corresponding to the two approaches. To invoke Railton's terminology and Kitcher's metaphor, we can think in terms of reading the ideal explanatory text either from the bottom-up or from the top-down. There are, of course, intermediate stages between the two extremes—there are degrees of coarse- or fine-grainedness. The kinds of examples brought up by Wright in his comparison of the course-grained consequence-etiology explanations with the fine-grained mechanical explanations do not usually appeal to either the most general laws of nature or the most fundamental physical mechanisms. Moreover, we often give mechanical explanations of everyday contrivances, such as the hand brake on a bicycle, without any appeal to unobservables.

It is extremely tempting to try to bring a linguistic distinction in English to bear on the explanatory duality I am discussing, but I fear it also holds certain risks. Sometimes we seek explanations by asking "How?" and sometimes by asking "Why?" Consider, for example, "How did the first large mammals get to New Zealand?" and "Why did the first large mammals go to New Zealand?" The answer to the first question is that they were humans, and they went in boats. I do not know the answer to the second question, but it undoubtedly involves human purposes and goals. The danger in making the distinction between how-questions and why-questions in terms of examples of this sort is that it easily leads to anthropomorphism—to the conclusion that 'genuine' explanations always involve an appeal to goals or purposes. That would certainly be a step in the wrong direction. But not all examples have this feature. If one asks *why* a penny conducts electricity, one good answer is that it is made of copper, and copper is a good conductor. If one asks *how* this penny conducts electricity, it would seem that a mechanism is called for. A story about electrons that are free to move through the metal would be an appropriate answer. In this case the why-question elicits an appeal to a general law; the how-question evokes a description of underlying mechanisms.

8. Conclusion

The attempt to gain scientific understanding of the world is a complicated matter. We have succeeded to some extent in reaching this goal, but what we have achieved to date has taken several centuries of effort on the part of many people, some of whom were or are towering geniuses. Many of the explanations that have been found are extraordinarily difficult to understand. When we think seriously about the very concept of scientific understanding, it does not seem plausible to expect a successful characterization of scientific explanation in terms of any simple formal schema or simple linguistic formulation. It is not surprising that there might be the kind of duality I have been discussing.

The situation may be even more extreme. As one of my former graduate students, Kenneth Gemes, has suggested, perhaps it is futile to try to explicate the concept of scientific explanation in a comprehensive manner. It might be better to list various explanatory virtues that scientific theories might possess, and to evaluate scientific theories in terms of them. Some theories might get high scores on some dimensions but low scores on others—recall my brief consideration of quantum mechanics. I have been discussing two virtues, one in terms of unification, the other in terms of exposing underlying mechanisms. Perhaps there are others that I have not considered. The foregoing discussion might serve as motivation to search for additional scientific explanatory qualities.

5

The Importance of Scientific Understanding

As we approach the end of the twentieth century, as well as that of the second millennium, there is an irresistible temptation to look back in order to evaluate the progress or regress that has transpired. At the beginning of the present *millennium* the Western world was in a deplorable state of scientific ignorance; even ancient Greek scientific knowledge had been lost. Scientific understanding was virtually nonexistent. Let us not dwell on that sad situation. At the turn of the present *century* a prettier picture could be seen.

1. Introduction

When the nineteenth century drew to a close, scientists were in possession of an impressive edifice of knowledge. I am thinking primarily of classical physics—which embraces Newtonian mechanics, Maxwellian electrodynamics, and the kinetic-molecular theory of gases—but important achievements had also been accomplished in many other branches of science. The concept of scientific understanding, however, was widely unappreciated. The present century has seen dramatic progress in the sciences, and in philosophy of science as well. In this latter area, it seems to me, one development stands out above the rest. It has to do with scientific explanation. At the turn of the present century many scientists and philosophers—including such eminent philosopher-scientists as Pierre Duhem and Ernst Mach—denied the very existence or possibility of scientific explanation. The realm of science, it was widely held, is confined to the description, systematization, and prediction of *observable* phenomena. Many doubted the reality of unobservable entities such as atoms and molecules; indeed, in some cases it was held that to talk of such things is meaningless. These two ideas are not unconnected; atoms, molecules, electrons, and other micro-entities play indispensable roles in many of our most impressive contemporary scientific explanations.

Duhem and many others did not deny the possibility of explaining natural phenomena; but they held that to do so, one had to go beyond the limitations of science into some other realm such as metaphysics or theology. Other philosophers, rebelling against all forms of supernaturalism, rejected explanation altogether. They saw explanation as a form of anthropomorphism—perhaps a kind of empathic relationship between human beings and inanimate nature as well as other forms of life. We all recognize the desire and need for understanding among humans, but to push the concept of understanding beyond these psychological boundaries was held to violate the inherent limitations of science. In his popular book *Philosophical Foundations of Physics* (1966), later reissued as *An Introduction to the Philosophy of Science* (1974), Rudolf Carnap provides an illuminating discussion of the negative attitude toward scientific explanation that existed in the early decades of the present century. Another interesting and informative account is provided by Mario Bunge ([1959] 1963, pp. 282–286). The dominant attitude at that time can be encapsulated in this slogan: *Science can tell us* what *but not* why. As Karl Pearson wrote in 1911, "Nobody believes now that science *explains* anything; we all look on it as a shorthand description, as an economy of thought" (Pearson, [1911] 1957, p. xi; Pearson's emphasis).

Today the majority of philosophers of science (and scientists too, I suspect) hold an entirely different view of the matter. They maintain that science can and does explain a wide variety of natural phenomena, and that to do so is one of the most basic goals of science. Current scientific journals are filled with explanations. Scientific explanation has both practical and intellectual value. Its practical value is obvious to us now. We want to explain why bridges collapse to discover how to prevent such occurrences in the future. We want to explain why certain diseases occur in order to find out how to cure them. In this practical context, explaining why and explaining how are closely linked. The emphasis is on our manipulative power; understanding involves knowing what will happen if we do or do not do certain things.

This kind of understanding is not, however, my main focus; instead, I want to consider scientific explanation primarily for its intellectual value. My plan is to consider some of the major philosophical and scientific developments that have led from the view that no such thing as scientific explanation can even exist to the view that explanation is a central, if not *the* central, goal of scientific endeavor. A striking example of this latter attitude on the part of an eminent physicist can be found in Steven Weinberg, *Dreams of a Final Theory* ([1992] 1994). The author makes no claim that we already have a "final theory," or even that we know how soon such a theory may be found and established, but he believes that there are strong indications that we are on the way. The key feature is the convergence of what he calls "explanatory arrows," indicating that the kinds of explanations already found strongly suggest that there is one comprehensive theory in terms of which all else can, in principle, be explained. The central argument hinges entirely on explanatory relationships. "Once again I repeat," he says, indicating a recurrent theme, "the aim of physics at its most fundamental level is not just to describe the world but to explain why it is the way it is" (ibid., p. 219). The main text concludes, "Whether or not the final laws of nature are discovered in our lifetimes, it is a great thing for us to carry on the tradition of holding nature up to examination, of asking again and again why it is the way it is" (ibid., p. 275). A critical discussion of Weinberg's book is given in "Dreams of a Famous Physicist" (essay 26).

In the course of this discussion, I shall examine two general forms of scientific understanding, both of which are available to us, and which are neither incompatible with each other nor contrary to the rigor and objectivity of the scientific enterprise. The first of these involves understanding our place in the world and knowing what kind of world it is. This kind of understanding is cosmological. The second involves understanding the basic mechanisms that operate in our world, that is, knowing how things work. This kind of understanding is mechanical. If, however, a "final theory" should be found, encompassing both particle physics and cosmology, then the two kinds of understanding would merge into one at the most fundamental level.

2. The Transition

The change from the attitude that prevailed at the beginning of the century to the view that is generally held today was greatly facilitated in the middle decades of the century by the works of several major philosophers. The first of these was Karl Popper's *Logik der Forschung* (1935), which, because it appeared in German, had little influence on Anglo-American philosophy. At that time, we should recall, Europe was in a state of turmoil because of Hitler's recent rise to power, and many of the most important philosophers of science fled to other parts of the world. Chaos reigned in the German-speaking world. Popper's influence increased dramatically when the subsequent English edition, *The Logic of Scientific Discovery* (1959), including a great deal of new material, was published. Bunge's *Causality* ([1959] 1963) also came out in that same year. It affirms the legitimacy and importance of scientific explanation and offers a useful taxonomy of types (ibid., chap. 11). In the meantime, the classic 1948 Hempel-Oppenheim article "Studies in the Logic of Explanation" appeared, but it had little influence for about a decade. R. B. Braithwaite's *Scientific Explanation* (1953), which made no mention of Hempel-Oppenheim ([1948] 1965), also appeared. During the late 1950s and early to mid-1960s, there was a burst of interest in scientific explanation. Two extremely influential books came out, namely, Ernest Nagel's magnum opus, *The Structure of Science: Problems in the Logic of Scientific Explanation* (1961), and Hempel's *Aspects of Scientific Explanation and Other Essays in the Philosophy of Science* (1965a), containing the magisterial essay "Aspects of Scientific Explanation" (1965b), along with a reprinting of Hempel-Oppenheim ([1948] 1965). By this time the notion that the sciences can provide explanations was strongly consolidated.

Another clear indication lies in the fact that the late 1950s saw the beginning of a rash of critical articles. The criticisms were *not* based on a conviction that scientific explanation does not exist; instead, they attacked specific features of the conceptions of scientific explanation advocated by one or another of the afore-mentioned authors—for example, the thesis that every legitimate scientific explanation must contain, either implicitly or explicitly, a law of nature (or a statement thereof).

It is not my purpose in this essay to give a detailed account of the developments to which I have referred; some of the high points are given in "Scientific Explanation: How We Got from There to Here" (essay 19), and a fuller account can be found in Salmon (1990b). It is worth noting, however, that in all of these discussions surprisingly little attention was devoted to what Carnap called "clarification of the explicandum," that is, to a preliminary informal discussion of the concept to be explicated. Often a few examples were expected to furnish the reader with an adequate idea. Notably lacking, for the most part, was any discussion of the value of scientific explanations or of the reasons for seeking them.

In retrospect this point is brought out forcefully by the view, currently held in some quarters, that science is actually concerned not with providing explanations but rather with the solving of puzzles or problems. Thomas Kuhn is its most influential advocate, and his *Structure of Scientific Revolutions* (1962) is the locus classicus of this view. One is led to wonder why we should devote such enormous human and material resources to the solving of scientific puzzles and problems unless success in that endeavor contributes to our understanding of nature. As I have already mentioned, scientific explanations often do have practical value, but in this essay I want to focus on pure rather than applied science. Important as the practical value of knowing how to prevent airplane crashes may be, my aim will be to characterize the kind of intellectual understanding we can achieve, for example, from knowledge of basic aerodynamic principles.

3. Types of Understanding

In the foregoing paragraphs I have used the term "understanding" several times without trying to clarify its meaning. This is, I believe, the key concept. Figure 5.1 is meant to serve as a sort of road map of the territory it covers. My chief emphasis in this essay will be the region indicated by the fourth column (headed "Natural Phenomena"). As the diagram shows, the concept is extremely broad and extremely ambiguous. For example, Deborah Tannen's book *You Just Don't Understand* (1991) spent about three years on the *New York Times* list of best-selling books. Her general thesis, here and in her more scholarly works, is that women and men speak different languages, and consequently do not understand one another. The so-called "generation gap"—which has been highly publicized in the United States—appears to be a permanent feature of relations between parents and children; parents do not understand their children and children do not understand their parents. "My wife doesn't understand me" is the eternal complaint of husbands, and is the standard 'line' for those who plan to be wayward. Obviously, to understand and be understood is a deep desire for an enormous number of people.

The kind of understanding involved in these situations is *empathy*—the sharing of feelings and emotions. People have often sought a similar kind of understanding with nonhuman parts of the world, leading to various forms of theism, pantheism, and the animistic view, attributed to Thales, that all things are full of gods. Such conceptions often provide great psychological satisfaction, but it was their theological and/or meta-physical character that led many scientists and scientific philosophers to spurn scientific explanation (understanding) altogether.

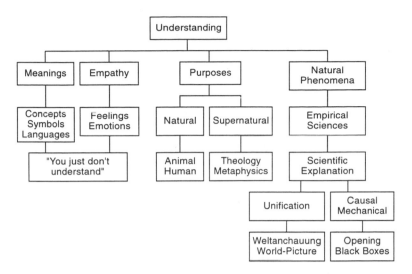

Types of Human Understanding

Figure 5.1

The mention of Tannen's work raises the whole question of meanings. The problem of understanding the *meanings* of various forms of expression arises in many contexts— the understanding of language, symbols, concepts, art objects, and rituals—both within the sciences and outside of them. An outstanding example of revelation of meaning is the deciphering of the Mayan language; in this case archaeologists made intelligible the many inscriptions left by that civilization. Archaeologists also try to interpret pictographs, pottery designs, objects found in burials, and so forth. Anthropologists and sociologists attempt to reveal the meanings of ceremonies and customs practiced in many cultures throughout the world. To understand the meanings of religious artworks of the Renaissance, art historians establish an iconography of standard symbols.

The understanding of meanings over a wide range of contexts is obviously an important aspect of our understanding of the world, in particular the world of human activity past and present. My goal in this essay, however, is to deal with the understanding of events and phenomena that occur in the world. This is not meant to disparage the understanding of meanings; it is rather an attempt to avoid confusion by making an explicit distinction between explanations of meanings and explanations of events and phenomena. We come to understand a meaning when we can say *what* something means; we come to understand a phenomenon when we can explain *why* it occurred.

Where human behavior is concerned, an appeal to *purposes* often provides a suitable explanation. I went to the drugstore yesterday because I had a headache, I had no aspirin, and so I wanted to purchase some. My action obviously involved certain beliefs, namely, that aspirin is an effective medication for headaches and that the drugstore was a convenient place to purchase it. Notice that the actual procurement of the aspirin does not explain my trip to the drugstore; the same explanation would be correct even if I failed to get it because the drugstore happened to be closed or happened to be out of aspirin. The constellation of desires and beliefs that preceded the trip constitute the explanation; there is no element of final causation in terms of the ends that were actually achieved. In some cases the behavior of nonhuman animals can also be explained in terms of explicit purposes, for example, when a dog carries its leash to its mistress or master because it wants to go for a walk.

Explanations of a similar sort are found in the social sciences when we try to understand social institutions or customs. In such cases, however, it is important to distinguish between the explicitly stated aim and the latent function. In a period of drought, for example, a group of people might perform a rain dance. Although the explicit purpose is to bring rain, the ceremony has no causal efficacy with respect to this goal. Even if rain occurs, it cannot be attributed to the performance. However, the performance of the ceremony may achieve an increase in social cohesion, which is valuable in situations that produce social stress.

Because of the psychological immediacy of explanations in terms of conscious purposes, it is tempting to demand that all satisfactory explanations take this form. This is, I think, a primary motive for those who claim that science cannot furnish genuine explanations—that we must appeal to the supernatural to achieve real understanding. For example, creationists explain the existence of species of living things in terms of the will of God. Darwinian evolutionists, in contrast, offer a mechanical account in terms of variation, population pressures, and natural selection.

Although evolutionary explanations exclude appeals to conscious purposes, they often refer to functions. For example, the paloverde tree, which lives in the hot and dry desert of the southwestern United States, has chlorophyll in its bark as well as its leaves. This adaptation has evolved as a water conservation device; it enables the paloverde to survive periods of severe heat and dryness. When moisture is present, the tree is covered with green leaves that perform photosynthesis, but when moisture is no longer present, it readily drops its leaves, thus reducing water loss through transpiration. Photosynthesis continues, though at a reduced rate, because of the chlorophyll in the bark. When moisture returns, new leaves sprout quickly. According to many philosophers such functional explanations are scientifically legitimate; they occur widely in the biological and social sciences and are accepted by competent scientists in these fields. Moreover, as Larry Wright (1976) argues, correctly I believe, they are completely analyzable in causal terms. They do not require appeal to any extrascientific agency.

One manifestation of the desire for understanding of the world is that virtually every culture we have studied has a creation story and a cosmic picture. Such understanding is *cosmological.* In the Judeo-Christian tradition we have the creation story in Genesis, and currently in North America a regrettably large number of people would like to convince us that this story is scientifically accurate. Native North Americans, for example, the Navajo, have other creation stories and cosmologies, as the popular novelist Tony Hillerman has made us aware in a respectful and delightful manner. Steven Weinberg begins his popular book on modern cosmology, *The First Three Minutes* (1977), by remarking, as I have just done, on the irresistable urge to provide creation stories; to illustrate he sketches the Norse myth given in the *Younger Edda* (compiled circa A.D. 1220). Other examples abound. These creation myths arise in response to our desire to comprehend the overall character of our universe and our place within it. A striking feature is their blatant anthropomorphism.

4. Scientific World Pictures

The ancient Greeks also had their myths, but somewhere along the way they began to pursue what we now recognize as a scientific world-picture—that is, a scientific *Weltanschauung*. Because of a variety of historical accidents, the cosmology of Aristotle came to dominate the medieval period. Aristotle knew, for good scientific reasons, that the earth is round, and Eratosthenes made an amazingly accurate determination of its size. The ancient Greeks understood eclipses, solar and lunar, and were able to predict them with some success. Through the development of Aristotelian cosmology and Ptolemaic astronomy, medieval humans had a world-picture, and a clear conception of where they fit into it. Of course, it was scientifically primitive and inadequate, as scientists from Copernicus to Newton taught during the scientific revolution. The transition was psychologically difficult, but ultimately scientific evidence forced the change.

The resulting world-picture is widely known as *the Newtonian Synthesis,* and it is this synthetic feature on which I would focus attention (see fig. 5.2). Newton, with the aid of the famous "giants" on whose shoulders he stood, gave a unified account of a wide variety of phenomena in terms of three simple laws of motion and the law of universal gravitation. Consider the variety. From the laws of motion alone, Newton derived the law of conservation of linear momentum, which we still hold today, even though Newtonian mechanics as a whole was superseded early in the present century by Einstein's special theory of relativity. On the next line in the diagram we see Kepler's three laws of planetary motion, an artificial satellite, Galileo's laws (free fall, the pendulum, and projectile motion), the tides, and comets. This vast diversity of phenomena was known in Newton's day. Although Newton had no rocket fuel or fancy electronics, he understood clearly the principles involved in putting such a satellite into orbit around the earth; in fact, he furnished a diagram. Even today Newtonian mechanics is used to calculate orbits for artificial satellites; the theory of relativity is not required.

Going to the next line in the diagram, we find another diverse group of phenomena that were not manifest until the eighteenth century. We should note that Newton's evidence for universal gravitation included cases in which two bodies of astronomic dimensions (e.g., the earth and the moon, the sun and a planet) or one body of astronomic dimensions and a smaller object (e.g., the earth and an apple) attract each other. In his torsion balance experiment, Henry Cavendish measured the gravitational force between bodies in his own laboratory. An important result was the possibility of determining the mass of the earth. Foucault's pendulum, which seems to change its direction of motion as the hours pass, actually constitutes the first direct evidence for the rotation of the earth. Galileo successfully refuted arguments against the rotation of the earth, but his positive arguments for the earth's motion turned out to be incorrect. During the eighteenth century the oblate shape of the earth was established empirically and was readily explained on Newtonian principles. Application of Newtonian mechanics to the motions of the planets led to the discovery of Neptune, a planet never previously observed.

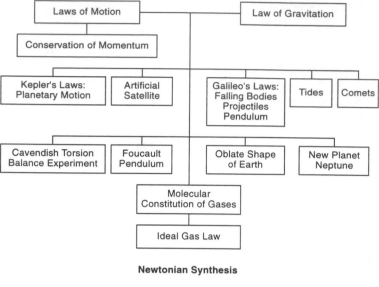

Newtonian Synthesis

Figure 5.2

If, moving on to the nineteenth century, we adopt the hypothesis that gases are composed of molecules in motion, Newtonian physics gives us the molecular-kinetic theory and the ideal gas law. These developments are fundamental to thermodynamics.

Our understanding of the universe was immeasurably increased by this *unification* of a variety of phenomena by means of such a simple and limited basis. The picture was, of course, not perfect. There is the famous "Olbers paradox"—why is the sky dark at night?—which was actually stated by Edmund Halley in 1720, and which could not be resolved within the Newtonian framework. But the Newtonian synthesis was remarkably successful, especially when supplemented by later developments in classical physics such as thermodynamics and electrodynamics.

The twentieth century saw a new *scientific revolution* as classical physics gave way to relativity theory and quantum mechanics, and from this revolution came a new cosmology. There is much merit, I think, in Steven Weinberg's remark that cosmology truly became a science with the discovery of the 3° cosmic background radiation in 1964–1965 (1977, chap. 1). Now we have as a world-picture an expanding universe containing billions of galaxies, each of which contains billions of stars (as well as a lot of other stuff), all of which came into being as a result of the so-called big bang that occurred some 15 billion years ago. We do not know whether the expansion will go on forever, or if it will turn around and contract into a "big crunch." When we find out, we will better understand our universe.

The fact that classical physics broke down at the turn of the twentieth century does not detract from its achievement in providing a comprehensive and unified *scientific world-picture*. Our present world-picture—involving quantum mechanics, relativity, the expansion of the universe, and the "big bang"—departs radically from that of classical physics. With twentieth-century scientific developments we have good reason to believe that we have a high degree of understanding of the universe and our place within it. We obviously have much more to learn, including answers to problems such as the origin of life, the nature of human consciousness, and what the 'missing mass' in the universe consists of. The Copernican revolution and Darwinian evolution may have been psychologically disappointing to many, but they are supported by substantial scientific evidence that tends to enhance our confidence in their partial and approximate accuracy, even if we might prefer that the world were otherwise. In any case, we can say that we have *scientific understanding* of phenomena when we can fit them into the general scheme of things, that is, into the *scientific world-picture.*

5. Understanding of Mechanisms

The second type of understanding I want to discuss also appeals to many people; it is especially prominent in the curiosity of children. We want to know *how things work* and, it should be added, *what they are made of.* This may be characterized as causal-mechanical understanding (but not the nineteenth-century English version satirized by Duhem). It is the kind of understanding we achieve when we take apart an old- fashioned watch, with springs and cogged wheels, and successfully put it back together again, seeing how each part functions in relation to all the others. Before we execute this process, the watch is like a 'black box' whose internal workings are mysterious. What we want to do is open up the black box and see how it works.

Nature presents us with many black boxes whose internal workings are mysterious, but science seeks to open them up to see how they work. A superb example is the epoch-making work of Jean Perrin on Brownian movement in the first dozen years of the present century. The behavior of microscopic particles suspended in a fluid was a mystery from its discovery early in the nineteenth century by the botanist Robert Brown until the first decade of the twentieth century, when Einstein published his famous paper that offered a theoretical explanation, and Perrin's magnificent experimental work confirmed it. Notice how we need to go to the submicroscopic level to explain microscopic phenomena, something that many physical scientists thought impossible in principle at the turn of the present century. Not only did Perrin establish the mechanism of Brownian movement, but also he ascertained Avogadro's number, the number of molecules in a mole (gram molecular weight) of any given substance.

The details of these developments are discussed in a highly illuminating manner in Mary Jo Nye's *Molecular Reality* (1972) and in Perrin's own *Atoms* ([1913] 1916). In summarizing his work on Brownian movement, Perrin emphatically calls attention to the fact that Avogadro's number can be ascertained experimentally in a wide variety of ways. At the end of his book he lists thirteen completely distinct methods, all of which agree quite closely on the value of that constant. Such agreement would be miraculous if matter were not composed of molecules and atoms. Notice what a marvelous epistemological feat has been performed. Avogadro's number is *the link* between the macrocosm and the microcosm: given macro-quantities, it enables us to calculate micro-quantities, and vice versa. Understanding the atomic-molecular constitution of matter enables us to explain wide varieties of phenomena, as Perrin himself points out, including even the blueness of the clear daytime sky. The work of Perrin and Einstein has shown us that it *is possible* to have knowledge of many sorts of entities that are much too small to be observed with the naked eye or any sort of optical microscope, and that such knowledge contributes immeasurably to our understanding of the world. A dramatic statement of this kind and degree of understanding was given by Nobel laureate Richard Feynman at the beginning of his three-volume *Lectures on Physics:*

> If, in some cataclysm, all of scientific knowledge were to be destroyed, and only one sentence passed on to the next generations of creatures, what statement would contain the most information in the fewest words? I believe it is the *atomic hypothesis* (or the atomic *fact,* or whatever you wish to call it) that *all things are made of atoms— little particles that move around in perpetual motion, attracting each other when they are a little distance apart, but repelling upon being squeezed into one another.* In that one sentence, you will see, there is an *enormous* amount of information about the world, if just a little imagination and thinking are applied. (Feynman et al., 1963, vol. 1, § 1.2; Feynman's emphasis)

6. Understanding in the Quantum Domain

Although there can be no doubt about the explanatory value of the atomic theory, we encounter extraordinary difficulties when we consider explanation in quantum mechanics. On the one hand, no theory has had more powerful explanatory success; on the other hand, it presents us with mysteries that at present seem to defy explanation. Suppose we have a black box with a red and a green light on it. At times the red light flashes briefly; at other times the green light flashes briefly. The two lights never flash simultaneously. We have another just like it, situated some distance away, which has no physical connection with the first. On the outside of each box is a dial with a hand that points randomly to one of the three numerals, "1," "2," or "3." Let us call the two black boxes *detectors.* Halfway between the two detectors is a "source," i.e., a device with a button on top. When the button is pressed, either the red light or the green light flashes on each of the detectors; in any given case, lights of the same or different colors may flash (see fig. 5.3). The lights on the detectors do not flash unless the button on the source is pressed. We presume that the source is emitting particles that activate the detectors. If we place a brick between the source and one of the detectors, the lights on that detector will not flash at all when the button on the source is pressed, even though the lights on the other detector continue to flash in the usual fashion. When the brick is removed, the detector resumes its typical flashing. Aside from the particles emitted by the source, there are no physical connections among any of these devices.

We conduct an experiment by pressing the button on the source a large number of times. We record the results on the two detectors, noting in each case which light flashed and which numeral was indicated on the dial. The result of one event might be recorded as 23GR, signifying that for the first detector the pointer indicated "2" and the green light flashed and that for the second detector the pointer indicated "3" and the red light

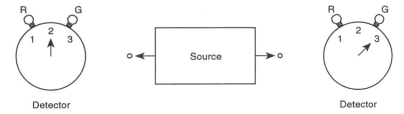

The Aspect Experiment

Figure 5.3

flashed. After an experiment involving a great number of such events, we find two results:

1. Whenever the pointers on the dials of the two detectors indicate the same numeral, lights of the same color flash on the two detectors.
2. Ignoring the indications on the dials of the detectors, we find that R and G occur randomly, each with probability 1/2, and independently of the color that occurs on the other detector.

Notice that the phenomena just described—the pressing of the button on the source and the results on the detectors—are macroscopic.

The example just sketched is offered by N. David Mermin (1985) to illustrate vividly the difficulty posed by certain quantum mechanical situations. When we open up the black boxes—the detectors—we find that each of them contains a set of Stern-Gerlach magnets that can assume any of three different spatial orientations. The orientations are indicated by the pointers on the dials. As Mermin shows, when we try to give a mechanical account of the working of the entire apparatus, extraordinarily difficult problems arise. According to Mermin, those who are not worried about it "have rocks in their heads." The problem presented by this example is discussed in "Indeterminacy, Indeterminism, and Quantum Mechanics" (essay 17). I agree with Mermin's assessment of its gravity.

7. The Values of Scientific Explanations

In this essay I have tried to show that there are at least two intellectual benefits that scientific explanations can confer upon us, namely, (1) a unified world picture and insight into how various phenomena fit into that overall scheme, and (2) knowledge of how things in the world work, that is, of the mechanisms, often hidden, that produce the phenomena we want to understand. The first of these benefits is associated with the unification view of scientific explanation; Philip Kitcher (1989, 1993) is its principal proponent. The second is associated with the causal/mechanical view of scientific explanation that I have advocated (Salmon, 1984b). My current view (sketched in "Scientific Explanation: Causation *and* Unification" [essay 4] and expounded in Salmon [1990b]) is that the two accounts are by no means incompatible. In the process of searching out the hidden mechanisms of nature, we often find that superficially diverse phenomena are produced by the same basic mechanisms. To the extent that we find extremely pervasive basic mechanisms, we are also revealing the unifying principles of nature. Sometimes a certain fact can be explained in either of two equally legitimate ways, that is, by subsumption under highly general principles or by exposure of underlying causal mechanisms. I find no ground for claiming that one is legitimate and the other illegitimate. They complement rather than conflict with each other.

This point is illustrated by the helium-filled balloon example that was brought up in essay 4. As there recounted, my physicist friend was sitting across the aisle from a boy holding a helium-filled balloon on an airplane awaiting departure. Asked what the balloon would do when the airplane began to accelerate, the boy opined that it would move toward the back of the cabin; adult passengers seated nearby agreed. However, when the airplane accelerated for takeoff, the balloon moved forward, and my friend enjoyed a drink of scotch that he won on a bet with a cabin attendant. As we saw, the behavior of the balloon can be explained in either of two ways. The mechanical explanation refers to interactions among the cabin walls, the air molecules, and the balloon. The unification explanation refers to Einstein's principle of equivalence—that an acceleration is equivalent to the presence of a gravitational field—an overarching generalization about the entire universe.

One point that deserves strong emphasis is the absolutely fundamental distinction between "understanding" in the scientific sense and "understanding" in the psychological sense. Understanding in the scientific sense involves the development of a world-picture, including knowledge of the basic mechanisms according to which it operates, that is based on objective evidence—one that we have good reason to suppose actually represents, more or less accurately, the way the world is. In this connection Perrin's work on molecular reality is epoch-making; it demonstrated the possibility of objective knowledge of unobservable reality. This kind of understanding may be psychologically satisfying or psychologically discomforting; regardless, the intellectual value remains. Psychological understanding in the empathic sense may be pleasant and comforting, but it lacks the objective basis furnished by scientific investigation of the world.

8. Conclusion

Let us return to the contrast between the beginning and the end of the twentieth century regarding scientific explanation. Not long ago an anthropologist friend kindly gave me a copy of A. G. Cairns-Smith's *Seven Clues to the Origin of Life* (1985). The author, a distinguished biologist, urges consideration of the hypothesis that life actually originated from clay, and he gives a mechanistic account. Taking into consideration the microstructure of clay, including details of the physical and chemical processes involved, he shows how variation and natural selection might have made possible the evolution of living organisms. He claims not that this is the correct hypothesis, but only that it ought to be considered seriously. Given my meager knowledge of biology, I cannot make any judgment about the scientific adequacy of this explanatory hypothesis. But suppose it is correct. Then I believe that we would have genuine understanding of the origin of life on earth and, by virtue of evolutionary biology, an understanding of how we humans came to be.

According to Genesis 2:7, "And the Lord God formed man of the dust of the ground and breathed into his nostrils the breath of life; and man became a living soul." There are many people who derive spiritual inspiration from the Genesis account, and with this I have no quarrel. But for an understanding of the fact of life on earth, it seems to me that the scientific account is intellectually far more satisfactory because of its mechanical detail and because of the objective basis on which it rests. At the beginning of the present century it was thought that the search for explanation and understanding would necessarily take one outside the domain of science, into the domain of metaphysics or theology. At the end of this century we can seriously argue that, although metaphysics and theology may serve as sources of inspiration or consolation, intellectually illuminating explanations are to be found in the realms of natural science. It is not necessary to depart from science to have genuine understanding of the world and what transpires within it.

Although I have focused attention on the intellectual value of scientific understanding, my conclusion has enormous practical as well as philosophical importance. As we enter the twenty-first century, we realize that humanity faces global problems of staggering proportions involving factors such as population growth, food and water supplies, depletion of atmospheric ozone, greenhouse warming, and atmospheric and oceanic pollution, to name but a few. A necessary prerequisite to finding satisfactory solutions is a sound scientific understanding of the problems, as well as the means available to deal with them. Science cannot set social or political policies, but it can furnish the information needed for responsible policy formation.

Part II

SCIENTIFIC EXPLANATION

The essays in this part present aspects of the evolution of my thought about scientific explanation.

Essay 6, "A Third Dogma of Empiricism" (1977), contains a sustained attack on the idea, almost universally accepted at the time, that explanations are arguments. It concludes that the time has come to put "cause" back into "because." In addition to showing the untenability of the "third dogma," it signals the development of a causal theory of explanation that will supplement the simple statistical-relevance (S-R) model of explanation I had advocated earlier, chiefly in Salmon (1971).

Essay 7, "Causal and Theoretical Explanation" (1975), introduces causal processes and the common cause principle, and it details the strategy for incorporating causal considerations into the theory of scientific explanation. It strongly suggests the thesis of scientific realism, but stops just short of the claiming that this approach establishes that view.

Essay 8, "Why Ask, 'Why?'?" (1977), extends the reasoning concerning causality in scientific explanation. It explicates causal interactions in terms of interactive forks and shows how they differ from Reichenbach's conjunctive forks, in terms of which he formulated his principle of the common cause. This essay shows how reasoning based on the common cause principle provides a basis for scientific realism.

Essay 9, "Deductivism Visited and Revisited" (1988), attacks explanatory deductivism, a view that has strong intuitive appeal to many philosophers. It offers a defense against the claim that there are no statistical explanations of particular facts, i.e., that all statistical explanations are explanations of statistical generalizations—what Hempel designated as the deductive-statistical (D-S) variety. It exposes a glaring conflict between the deductive-nomological (D-N) model of explanation and basic causal considerations relevant to explanation.

Essay 10, "Explanatory Asymmetry" (1993), provides a penetrating analysis of the temporal asymmetry of explanation; it gives reasons why the explanatory facts must precede, rather than follow, the fact to be explained. This is an issue of fundamental importance that has almost always been relegated to declarations based on unanalyzed philosophical or commonsense intuitions.

Essay 11, "Van Fraassen on Explanation" (1987), deals critically with the view—whose most influential proponent is Bas van Fraassen—that the traditional problems of scientific explanation can be resolved by means of pragmatic considerations alone. This approach, elaborated in 1980 in *The Scientific Image*, has found much favor among philosophers of science. As this essay reveals, the traditional problems do not disappear when the resources of pragmatics are brought to bear.

6

A Third Dogma of Empiricism

In 1951 W. V. Quine published his provocative and justly famous article, "Two Dogmas of Empiricism." At about the time Quine was mounting this attack, a number of 'empiricists' were busily establishing what has subsequently become, in my opinion, a third dogma. The thesis can be stated quite succinctly: *scientific explanations are arguments.* This view was elaborated at considerable length by a variety of prominent philosophers, including R. B. Braithwaite (1953), Ernest Nagel (1961), Karl Popper (1959), and most especially Carl G. Hempel.[1] Until the early 1960s, although passing mention was sometimes made of the need for inductive explanation, attention was confined almost exclusively to deductive explanation. In 1962, however, Hempel (1962a) made the first serious attempt to provide a detailed analysis of inductive (or statistical) explanation. In that same year, in a statement referring explicitly to both deductive and inductive explanations, he characterized the "explanatory account" of a particular event as "an *argument* to the effect that the event to be explained . . . was to be expected by reason of certain explanatory facts" (Hempel, 1962b; emphasis added). Shortly thereafter he published an improved and more detailed version of his treatment of inductive-statistical (I-S) explanation (Hempel, 1965a, pp. 381–412). In this newer discussion, as well as in many other places, Hempel has often reiterated the thesis that explanations, both deductive and inductive, are arguments.[2] The purpose of this essay is to raise doubts about the tenability of that general thesis by posing three questions—ones that will, I hope, prove embarrassing to those who hold it.[3]

Question 1.   Why are irrelevancies harmless to arguments but fatal to explanations?

In deductive logic, irrelevant premises are pointless, but they do not undermine the validity of the argument. Even in the relevance logic of Anderson and Belnap, $p \& q \vdash p$ is a valid schema. If one were to offer the argument,

All men are mortal.
Socrates is a man.
Xantippe is a woman.
_____
Socrates is mortal.

it would seem strange, and perhaps mildly amusing, but its logical status would not be impaired by the presence of the third premise. There are more serious examples. When it was discovered that the axioms of the propositional calculus in *Principia Mathematica* were not all mutually independent, there was no thought that the logical system was thereby vitiated. Nor is the validity of Propositions 1–26 of Book I of Euclid called into question as a result of the fact that they all follow from the first four postulates alone, without invoking the famous fifth (parallel) postulate. This fact, which has important bearing upon the relationship between Euclidean and non-Euclidean geometries, does not represent a fault in Euclid's deductive system.

When we turn to deductive explanations, however, the situation is radically different. The rooster who explains the rising of the sun on the basis of his regular crowing is guilty of more than a minor logical inelegancy. So also is the person who explains the dissolving of a piece of sugar by citing the fact that the liquid in which it dissolved is *holy* water. So also is the man who explains *his* failure to become pregnant by noting that he has faithfully consumed birth control pills.[4]

The same lack of parity exists between inductive arguments and explanations. In inductive logic there is a well-known requirement of total evidence.[5] This requirement demands the inclusion of all *relevant* evidence. Since irrelevant 'evidence' has, by definition, no effect on the probability of the hypothesis, inclusion of irrelevant premises in an inductive argument can have no bearing on the degree of strength with which the conclusion is supported by the premises.[6] If facts of unknown relevance turn up, inductive sagacity demands that they be mentioned in the premises, for no harm can come from including them if they are irrelevant, but considerable mischief can accrue if they are relevant and not taken into account.

When we turn our attention from inductive arguments to inductive explanations, the situation changes drastically. If the consumption of massive doses of vitamin C is irrelevant (statistically) to immunity to the common cold, then 'explaining' freedom from colds on the basis of use of that medication is worse than useless.[7] So also would be the 'explanation' of psychological improvement on the basis of psychotherapy if the spontaneous remission rate for neurotic symptoms were equal to the percentage of 'cures' experienced by those who undergo the particular type of treatment.[8]

Hempel recognized from the beginning the need for some sort of requirement of total evidence for inductive-statistical explanation; it took the form of the *requirement of maximal specificity* (Hempel, 1965a, pp. 394–403). This requirement stipulates that the reference class to which an individual is referred in a statistical explanation be narrow enough to preclude, in the given knowledge situation, further relevant subdivision. It does not, however, prohibit irrelevant restriction. I have therefore suggested that this requirement be amended as the *requirement of the maximal class of maximal specificity* (Salmon, 1970b, § 5). This requirement demands that the reference class be determined by taking account of all relevant considerations, but that it not be irrelevantly partitioned.

Inference, whether inductive or deductive, demands a requirement of total evidence—a requirement that *all* relevant evidence be mentioned in the premises. This requirement, which has substantive importance for inductive inferences, is automatically satisfied for deductive inferences. Explanation, in contrast, seems to demand a further requirement—namely, that *only* considerations relevant to the explanandum be contained in the explanans. This, it seems to me, constitutes a deep difference between explanations and arguments.

Questions 2 comes in two distinct forms, which I shall number 2 and 2' respectively. The two forms may actually express different questions, but they are so closely related as to deserve some sort of intimate linkage.

Question 2. Can events whose probabilities are low be explained?

Although they made no attempt to provide an explication of inductive-statistical explanation in their classic 1948 paper, Hempel and Oppenheim did acknowledge the need for explanations of that sort (Hempel, 1965a, pp. 250–251). On Hempel's subsequent account of inductive-statistical explanation, events whose probabilities are high (relative to a suitably specified body of knowledge) are amenable to explanation. A high probability is demanded by the requirement that the explanation be an argument to the effect that the event in question *was to be expected,* if not with certainty, then with high probability, in virtue of the explanatory facts.

If some events are probable, without being certain, others are improbable. If a coin has a strong bias for heads, say 0.9, then tails has a nonvanishing probability, and a small percentage of the tosses will in fact result in tails. It seems strange to say that the results of tosses in which the coin lands heads-up can be explained, while the results of those tosses of the very same coin in which tails show are inexplicable. To be sure, the head-outcomes far outnumber the tail-outcomes, but is it not an eccentric prejudice that leads us to discriminate against the minority, condemning its members to the realm of the inexplicable?

The case need not rest on examples of the foregoing sort. In a number of well-known examples, we seem to be able to offer genuine explanations of events whose nonoccurrence is more probable than not. Michael Scriven has pointed out that the probability of paresis developing in cases of latent untreated syphilis is quite small, but syphilis is accepted as the explanation of paresis in those cases in which it does occur.[9] Similarly, as I understand it, mushroom poisoning may afflict only a small percentage of individuals who eat a particular type of mushroom, but the eating of the mushroom would unhesitatingly be offered as the explanation in instances of the illness in question.[10] Moreover, a uranium nucleus may have a probability as low as 10^{-38} of decaying by spontaneously ejecting an alpha-particle at a particular moment. When decay does occur, we explain it in terms of the tunnel effect, which assigns a low probability to that event.

Imposition of the high probability requirement upon explanations produces a serious malady that Henry Kyburg (1970) has dubbed "conjunctivitis."[11] Because of the basic multiplicative rule of the probability calculus, the joint occurrence of two events is normally less probable than either event occurring individually. This is illustrated by the above-mentioned biased coin. The probability of heads on any given toss is 0.9, while the probability of two heads in a row is 0.81. If 0.9 were the minimal value acceptable in inductive-statistical explanation, we would be able to explain each of the two tosses separately, but their joint occurrence would be unexplainable. The moral to be drawn from examples of this kind is, it seems to me, that there is no reasonable way of answering the question "How high is high enough?"

If conjunctions are the enemies of high probabilities, disjunctions are their indispensable allies. If a fair coin is tossed 10 times, there is a probability of $1/1024$ that it will come up heads on all 10 tosses. This sequence of events constitutes a (complex) low probability event, and as such it is unexplainable. Even if the outcome is 5 heads and 5 tails, however, the probability of that sequence of results, *in the particular order in which they occurred,* is also $1/1024$. It, too, is a low probability event, and as such is unexplainable. If, however, we consider the probability of 5 heads and 5 tails *regardless of order,* we are considering the disjunction of all of the 252 distinct orders in which that outcome can occur. Even this extensive disjunction has a probability of only about 0.246; hence, even it fails to qualify as a high probability event. If, however, we consider the probability of getting almost one-half heads in 10 tosses, i.e., 4 or 5 or 6 heads, this disjunction is ample enough to have a probability somewhat greater than 0.5 (approximately 0.656). The general conclusion would seem to be that, for even moderately complex events, every specific outcome has a low probability, and is consequently incapable of being explained.[12] The only way to achieve high probabilities is to erase the specific character of the complex event by disjunctive dilution.

Richard Jeffrey ([1969] 1971) and James Greeno ([1970] 1971) have both argued, quite correctly I believe, that the degree of probability assigned to an occurrence in virtue of the explanatory facts is not the primary index of the value of the explanation. Suppose, for example, that two individuals, Smith and Jones, both commit suicide. Using our best psychological theories, and summoning all available relevant information about both persons (such as sex, age, race, state of health, marital status, etc.), we find that there is a low probability that Smith would commit suicide, whereas there is a high probability that Jones would do so. This does not mean that the explanation of Jones's suicide is better than that of Smith's, for exactly the same theories and relevant factors have to be taken into account in both.

According to an alternative account of statistical explanation, the statistical-relevance (S-R) model, elaborated in Salmon (1971), an explanation consists not in an argument but in an assemblage of relevant considerations. On this model, high probability is not the desideratum; rather, the amount of relevant information is what counts. According to the S-R model, a statistical explanation consists of a probability distribution over a homogeneous partition of an initial reference class. A homogeneous partition is one that does not admit of further relevant subdivision.

The subclasses in the partition must also be maximal—that is, the partition must not involve any irrelevant subdivisions. The goodness, or epistemic value, of such an explanation is measured by the gain in information provided by the probability distribution over the partition.[13] If one and the same probability distribution over a given partition of a reference class provides the explanations of two separate events, one with a high probability and one with a low probability, the two explanations are equally valuable.

This approach to statistical explanation offers a pleasant dividend. If we insist that the explanation incorporate the probability distribution over the entire partition—not just the probability value associated with the particular cell of the partition into which the event to be explained happens to fall—we are invoking the statistical analogues of both sufficient and necessary conditions, rather than sufficient conditions alone.[14] This feature of the statistical-relevance model overcomes one severe difficulty experienced by Hempel's deductive-nomological and inductive-statistical models in connection with functional explanations, for these models seem always to demand a sufficient condition where the functional explanation itself provides a necessary condition. Although the task has not yet been accomplished, the statistical-relevance model gives promise of providing an adequate account of functional explanations—a type of explanation that has constituted an embarrassment to the standard inferential approach to explanation (Hempel, [1959] 1965a).

There is a strong temptation to respond to examples such as the biased coin, paresis, and mushroom poisoning (as well as functional explanations in general) by relegating them to the status of explanation sketches or incomplete explanations. We are apt to believe—often on good grounds—that further investigation would provide the means to say why it is that this syphilitic develops paresis while that one does not, or why one person has an allergic response to a particular type of mushroom while the vast majority of people do not.[15] Such a tack runs the risk, however, of seducing us into the supposition that *all* inductive-statistical explanations are incomplete. It seems to me that we must ask, however, what to say if *not all* examples of low probability events are amenable to that approach. This problem leads to another question, so closely related to our second question as to be hardly more than a reformulation of it:

Question 2′.   Is genuine scientific explanation possible if indeterminism is true?

The term "determinism" is unquestionably ambiguous. On one plausible construal it can be taken to mean that there are no genuinely homogeneous references classes except in the limiting cases when all A are B or no A are B. Let A be a certain reference class within which the attribute B is present in some but not all cases. According to this version of determinism, there must be a characteristic C in terms of which the class A can be partitioned so that within the subclass $A \cap C$ every element is B and within $A \cap \bar{C}$ every element is $\bar{B}$. Suppose, for example, that an alpha-particle approaches a potential barrier with a certain nonvanishing probability of tunneling through and a certain nonvanishing probability of being reflected back. This form of determinism asserts that there is a characteristic present in some cases and absent in others that 'determines' whether the alpha-particle tunnels through or not. This, I take it, is the thesis of hidden variable theorists.

In order to protect this version of determinism from complete trivialization, it is necessary to place some restrictions on the sort of characteristic C to which we may appeal for purposes of partitioning A. In particular, we must not allow C to be identified with B itself, or any other property whose presence or absence cannot even in principle be ascertained without discovering whether B is present or absent. If, for example, we are discussing the probability of drawing a red ball from an urn, we may partition the class of draws in terms of draws made by males versus draws by females, or draws made in the daytime versus draws made at night; we may not partition the class of draws in terms of draws resulting in a red ball versus draws resulting in other colors, or draws of balls with a color at the opposite end of the visible spectrum from violet versus draws resulting in colors located in other regions of the spectrum.

The problem we confront in attempting to put appropriate restrictions on the attributes permitted in partitioning reference classes is familiar from another context. In order to implement his definition of a "collective," Richard von Mises (1964, chap. 1) introduced the notion of a place selection. Although his original explication of the concept of a place selection was certainly unsatisfactory, he did, I believe, correctly identify the explanandum. Subsequent work has made it possible to supply a serviceable definition of "place selection," and to apply it to the definition of "homogeneous reference class."[16] It then remains an open factual question whether there are nontrivial cases of homogeneous reference classes.[17]

In his published writings Hempel is, I believe, committed to the opposite view, for he categorically asserts that inductive-statistical explanations are *essentially* relativized to knowledge situations; he calls this thesis the "epistemic relativity of inductive-statistical explanation" (Hempel, 1965a, p. 402). He maintains that, although a reference class that satisfies the requirement of maximal specificity is one that we do not know how to partition relevantly, it is in principle capable of further relevant subdivision in the light of additional knowledge.[18] If there were an inductive-statistical explanation whose lawlike statistical premise involved a genuinely homogeneous reference class—one that, even in principle, could not be further relevantly subdivided—then we would have an instance of an inductive-statistical explanation *simpliciter,* not merely an inductive-statistical explanation *relative to a specific knowledge situation.* Since there are no inductive-statistical explanations *simpliciter* on Hempel's view, he must deny the existence of genuinely homogeneous reference classes, except in trivial cases. In the trivial cases we do not have to rest content with inductive-statistical explanations, for universal laws are available by means of which to construct deductive-nomological explanations. In the ideal limit of complete knowledge, inductive-statistical explanation would have no place, for every explanation would be deductive-nomological.[19]

The relationship between inductive-statistical explanations and deductive-nomological explanations closely parallels the relationship between enthymemes and valid deductive arguments. Since an enthymeme is, by definition, an argument with missing premises, there can be no such thing as a valid enthymeme. Enthymemes can be made to approach validity, we might say, by supplying more and more of the missing premises, but the moment a set of premises sufficient for validity is furnished, the argument ceases to be an enthymeme and automatically becomes a valid deductive argument.

Much the same sort of thing can be said about inductive-statistical explanations. The reference class that occurs in a given inductive-statistical explanation and fulfills the requirement of maximal specificity is not genuinely homogeneous; it is still possible in principle to effect a relevant partition, but in our particular knowledge situation we do not happen to know how. As we accumulate further knowledge, we may be able to make further relevant partitions of our reference class, but as long as we fall short of universal laws, we have not exhausted all possible relevant information. Progress in constructing inductive-statistical explanations would thus seem to involve a process of closer and closer approximation to the deductive-nomological ideal. Failure to achieve this ideal would not be a result of the nonexistence of relevant factors sufficient to provide universal laws; failure to achieve deductive-nomological explanations can only result from our ignorance.

As a result of the foregoing considerations, as well as other arguments advanced by J. A. Coffa (1974), I am inclined to conclude that Hempel's concept of *epistemic relativity of statistical explanations,* which demands relativization of *every* such explanation to a knowledge situation (Hempel, 1965a, p. 402), means that Hempel's account of inductive-statistical explanation is completely parasitic upon the concept of deductive-nomological explanation. If, however, indeterminism is true, on any reasonable construal of that doctrine with which I am acquainted, then some reference classes will be actually, objectively, genuinely homogeneous in cases where no universal generalization is possible. In that case, it seems to me, we must have a full-blooded account of inductive-statistical explanation—or statistical explanation, at any rate—that embodies homogeneity of reference classes *not relativized* to any knowledge situation. I do not know whether indeterminism is true; I think we have good physical reasons for supposing it may be true. But regardless of whether indeterminism is true, we need an explication of scientific explanation that is neutral regarding that issue. Otherwise, we face the dilemma of either (1) ruling indeterminism out a priori or (2) holding that events are explainable only to the extent that they are fully determined. Neither alternative seems acceptable: (1) the truth or falsity of indeterminism is a matter of physical fact, not to be settled a priori, and (2) even if the correct interpretation of quantum mechanics is indeterministic, it still must be admitted to provide genuine scientific explanations of a wide variety of phenomena.

In dealing with Question 2', I have said quite a bit about determinism and indeterminism without mentioning causal relations. This omission must be corrected. Consideration of the third question will rectify the situation.

Question 3. Why should requirements of temporal asymmetry be imposed on explanations (while arguments are not subject to the same constraints)?

A particular lunar eclipse can be predicted accurately, using the laws of motion and a suitable set of initial conditions holding prior to the eclipse; the same eclipse can equally well be retrodicted using posterior conditions and the same laws. It is intuitively clear that if explanations are arguments, then only the predictive argument can qualify as an explanation, and not the retrodictive one. The reason is obvious. We explain events on the basis of antecedent causes, not on the basis of subsequent effects (or other subsequent conditions).[20] A similar moral can be drawn from Sylvan Bromberger's flagpole example. Given the elevation of the sun in the sky, we can infer the length of the shadow from the height of the flagpole, but we can just as well infer the height of the flagpole from the length of the shadow. The presence of the flagpole explains the occurrence of the shadow; the occurrence of the shadow does not explain the presence of the flagpole. At first blush we might be inclined to say that this is a case of coexistence: the flagpole and the shadow exist simultaneously. On closer examination, however, we realize that a causal process is involved, and that the light from the sun must either pass or be blocked by the flagpole *before* it reaches the ground where the shadow is cast.

There are, of course, instances in which inference enjoys a preferred temporal direction. As I write, the July Fourth weekend approaches. We can predict with confidence that many people will be killed, and perhaps give a good estimate of the number. We cannot, with any degree of reliability, predict the exact number—much less the identity of each of the victims. By examining next week's newspapers, however, we can obtain an exact account of the number and identities of these victims, as well as a great deal of information about the circumstances of their deaths.[21] By techniques of dendrochronology (tree ring dating), for another example, relative annual rainfall in parts of Arizona is known for some 8000 years into the past. No one could hazard a reasonable guess about relative annual rainfall for even a decade into the future.

Such examples show that the temporal asymmetry reflected by inferences is precisely the opposite to that exhibited in explanation. We have many records, natural and humanly-made, of events that have happened in the past; from these records we can make reliable inferences into the past. We do not have similar records of the future.[22] Prognostication is far more difficult than retrodiction; it has no aid comparable to records. No one would be tempted to 'explain' the accidents of a holiday weekend on the basis of their being reported in the newspaper. No one would be tempted to 'explain' the rainfall of past millennia on the basis of the rings in trees of bristlecone pine. If it is indeed true that being an argument is an essential characteristic of scientific explanations, how are we to account for the total disparity of temporal asymmetry in explanations and in arguments? This is a fundamental question for supporters of the inferential view of explanation.

If one rejects the inferential view of scientific explanation, it seems to me that straightforward answers can be given to the foregoing three questions. On the statistical-relevance model of explanation, an explanation is an assemblage of factors that are statistically relevant to the occurrence of the explanandum-event. To offer an item as relevant when it is, in fact, irrelevant is clearly inadmissible. We thus have an immediate answer to our first question, "Why are irrelevancies harmless to arguments but fatal to explanations?"

The second question, in its first form, receives an equally simple and direct answer. Since additional relevant information may raise or lower probabilities, and since assemblages of relevant information may yield high, middling, or low probabilities for an event of a particular sort, the statistical-relevance model has no problems with low probabilities. It never has to face the question "How high is high enough?" It is absolutely immune to conjunctivitis.

Question 2' poses the problem, mentioned earlier, of characterizing homogeneity in an objective and unrelativized manner. Resolution of this problem must be reserved for another occasion, but there seems no reason to doubt that it can be done.[23]

When we come to the third question, regarding temporal asymmetry, we cannot avoid raising the issue of causation. In the classic 1948 article, Hempel and Oppenheim suggested that deductive-nomological explanations are causal explanations, but in subsequent years Hempel backed away from this position, explicitly dissociating "covering law" from causal explanations.[24] The time has come, it seems to me, to put the "cause" back into "because." Consideration of the temporal asymmetry issue forces reconsideration of causation in explanation. (See "Explanatory Asymmetry" [essay 10] for further elaboration.)

There are two levels of explanation in the statistical-relevance model. At the first level, we invoke statistical regularities to provide a relevant partition of a given reference class into maximal homogeneous subclasses. For example, we place Smith in a subclass of Americans that is defined in terms of characteristics such as age, sex, race, marital status, state of health, etc., which are statistically relevant to suicide. This provides a statistical-relevance explanation of the suicide.[25]

To say that the occurrence of an event of one type is statistically relevant to that of an event of another type is simply to say that the two are not statistically independent. In other words, statistical-relevance explanations on the first level explain individual occurrences on the basis of statistical dependencies. Statistical dependencies are improbable coincidences in the sense that dependent events occur in conjunction with a probability greater (or less in the case of negative relevance) than the product of their separate probabilities. Improbable coincidences demand explanation; hence, the statistical relevance relations invoked at the first level require explanation. The type of explanation required is, I believe, causal. Reichenbach (1956, § 19) formulated this thesis in his *principle of the common cause*. If all of the lights in an entire section of a city go out simultaneously, we explain this coincidence in terms of a power failure, not by the chance burning out of each of the bulbs at the same time. If two term papers are identical, and neither has been copied from the other, we postulate a common source (e.g., a paper in a fraternity or sorority file). Given similar patterns of rings in logs cut from two different trees, we explain the coincidence in terms of the rainfall in the area in which the two trees grew.

Given events of two types A and B that are positively relevant to each other, we hunt for a common cause C that is statistically relevant to both A and B.[26] C absorbs the dependency between A and B in the sense that the probability of A & B given C is equal to the product of the probability of A given C and the probability of B given C. The question naturally arises: Why should we prefer, for explanatory purposes, the relevance of C to A and C to B over the relevance of A to B which we had in the first place? The answer is that we can trace a spatiotemporally continuous causal connection from C to A and from C to B, while the relation between A and B cannot be accounted for by any such direct continuous causal relation. This is especially clear when A and B lie outside each other's light cones.[27]

Improbable coincidences may have common effects as well as common causes, but their common effects do not explain the coincidences. Suppose that the only two ambulances in a town collide as they converge on the scene of a serious automobile accident to which they had been summoned. The coincidence of their meeting is explained in terms of messages sent from a common source calling them to a particular place.[28] Suppose they were called to an accident in which the occupants of an automobile were seriously injured when it crashed into a truck at high speed. Suppose further that the people in the automobile were fleeing from the scene of a dastardly crime, and that, as a result of the collision between the two ambulances, the criminals died because they could not be taken to the hospital for treatment. We would not explain the collision of the ambulances as a case of justice prevailing in the 'punishment' of the criminals. This is not mere prejudice against teleological explanations; it results from the *fact* that the probability of the collision of the ambulances is not affected by the life or death, just or unjust, of any victim of the crash. It seems to be a basic and pervasive feature of the macrocosm that common causes, prior in time, can absorb the statistical relevance relations in improbable coincidences, while common effects (subsequent in time) cannot absorb these relevance relations.[29] Explanations thus exhibit a temporal asymmetry which is quite distinct from that of inferences.

I should like to close by offering a rough, but general, characterization of scientific explanation, followed by a challenge to which it gives rise. It seems to me that the nature of scientific explanation can be summed up as follows:

> To give scientific explanations is to show how events and statistical regularities fit into the causal network of the world.[30]

If this cannot be taken as a thesis supported by example and argument, it can, I believe, be advanced as a reasonable conjecture. It gives rise, however, to one of the most serious problems in current philosophy of science, namely, to provide an explication of causality without violating Hume's strictures against hidden powers and necessary connections.[31] That we need such a characterization of causality, regardless of our attitude toward the role of causality in scientific explanation, is evident from the fundamental role played by causal relations in the basic space-time structure of the physical world. Since we need such an explication anyhow, the fact that our treatment of scientific explanation involves causal relations is no ground for objection to it.

Notes

I should like to express my gratitude to the National Science Foundation for support of research on scientific explanation and related topics.

1. The classic article is Hempel and Oppenheim ([1948] 1965).

2. A serious problem about the nature of inductive inferences or arguments arises because of Carnap's denial of the existence of "rules of acceptance" in his system of inductive logic. I have discussed this issue in some detail in Salmon (1977d). In this essay I am construing "argument" in the usual sense of a logical structure with premises and conclusions, governed by some sort of rule of acceptance. Hempel's writings have conveyed to me, as well as to many others, I believe, the impression that he construes the term in this same way in his discussions of inductive-statistical explanation. In any case, whether Hempel construes explanations as arguments in this straightforward sense or not, there is no shortage of other philosophers who do.

3. Neither Hempel nor anyone else, I suppose, has ever maintained that every sound argument is an explanation, and to attack such a thesis would certainly be to attack a straw man. In order to qualify as explanations, arguments must fulfill a number of conditions, and these have been carefully spelled out. Hempel, as well as many others, have claimed that every scientific explanation is an argument. It is this latter thesis that I am attempting to call into question.

4. These examples, and many others like them, can be schematized so as to fulfill all of Hempel's requirements for deductive-nomological (D-N) explanations. This is shown in detail in Salmon (1970b, § 2).

5. This requirement is explicitly formulated and discussed in Carnap (1950, § 45B).

6. If "*c(h, e)*" designates the inductive probability or degree of confirmation of hypothesis *h* on evidence *e*, then *i* is irrelevant to *h* in the presence of *e* if only if

$$c(h,e.i) = c(h, e).$$

7. If '*P(B|A)*' denotes the statistical probability of attribute *B* in reference class *A*, then *C* is statistically irrelevant to the occurrence of *B* within *A* if and only if

$$P(B|A.C) = P(B|A)$$

Note that this definition of statistical irrelevance is formally identical to the definition of irrelevance of evidence in the preceding note.

8. Examples of this sort are discussed in Salmon (1970b, § 2), where they are shown to conform to Hempel's requirements for inductive-statistical (I-S) explanations.

9. I discuss this example in Salmon (1970b, § 8) and there provide references to a number of other discussions of it.

10. See, for example, the introduction to Smith (1958). The point is illustrated by remarks on the edibility of certain species:

#11 (p. 34), helvella infula, "Poisonous to some, but edible for most people. Not recommended."

#87 (p. 126), cantharellus floccosus, "Edible for some people and NOT for others."

#31 (p. 185), chlorophyllum molybdites, "Poisonous to some but not to others. Those who are not made ill by it consider it a fine mushroom. The others suffer acutely."

11. Hempel was fully aware of this problem, and he discussed it explicitly (1965a, pp. 410–412).

12. It seems to me that Baruch Brody (1975, p. 71) missed this point when he wrote:

It should be noted that there are some cases of statistical explanation where the explanans does provide a high enough degree of probability for the explanandum, so Hempel's requirements laid down in his inductive-statistical model are satisfied, but does not differentiate between the explanandum and some of its alternatives. Thus, one can explain, even according to Hempel, the die coming upon one in 164 of 996 throws by reference to the fact that it was a fair die tossed in an unbiased fashion; such a die has, after all, a reasonably high probability of coming upon one in 164 out of 496 [*sic*] throws. But the same explanans would also explain its coming upon one in 168 out of 996 throws. So it doesn't even follow from the fact than an explanation meets all of Hempel's requirements for statistical explanations that it meets the requirement that an explanans must differentiate between the explanandum and its alternatives.

The fact is that a fair die tossed 996 times in an unbiased fashion has a probability of about 0.0336 of showing side one in 164 throws. By no stretch of the imagination can this be taken as a case in which Hempel's high probability requirement is satisfied. The probability that in the same number of throws with the same die side one will show 168 times is nearly the same, about 0.0333. As Hempel has observed, if two events are incompatible, they cannot both have probabilities that are over 0.5—a minimal value, I should think, for any probability to qualify as 'high'.

13. For a fuller discussion of homogeneity, including a quantitative measure of degree of homogeneity, see Salmon (1970b, § 6). See Greeno ([1970] 1971) for an information-

theoretic treatment of these concepts, especially the definition of "information transmitted." Both are reprinted in Salmon (1971).

14. See Salmon (1970b, § 9) for fuller discussion of this claim.

15. Hempel (1965a, pp. 381–403) offers recovery from a streptococcus infection upon treatment by penicillin as an example of inductive-statistical explanation. I believe it is now possible in principle to provide deductive-nomological explanations of such cures, for the chemistry of bacterial resistance to penicillin seems now to be understood (Cohen, 1975).

16. I have discussed this issue briefly in "Postscript 1971" (Salmon, 1971, p. 106), indicating the difficulties that remain in A. Wald's refinement of von Mises's definition. The basic tool for overcoming these problems was provided by Church (1940). Subsequent work on randomness has not reduced the value of Church's fundamental contribution in the context of the present discussion.

17. By nontrivial I mean homogeneous references classes A in which some, but not all, elements have the attribute B.

18. In personal conversation Hempel expressed what seemed to me to be reservations concerning the *necessity* of relativization of inductive-statistical explanations to knowledge situations in *every* instance, but I have not found such qualifications in his published writings. (It is acknowledged explicitly in [Hempel, 1977, § 3.7].) The question of the *essentiality* of epistemic relativization involves subtle issues whose detailed discussion must be reserved for another occasion. In this essay I confine my efforts to the attempt to draw out the consequences of what I take to be Hempel's published view.

19. This argument is elaborated more fully in Salmon (1974a). In private conversation, I. Niiniluoto pointed out that in infinite reference classes it may be possible to construct infinite sequences of partitions that do not terminate in trivially homogeneous subclasses. It is clear that no such thing can happen in a finite reference class. This is, therefore, one of those important points at which the admitted idealization involved in the use of infinite probability sequences (reference classes) must be handled with care.

20. The issue of temporal asymmetry is discussed at length, including such examples as Bromberger's flagpole, in Salmon (1970b, § 12). For a fuller account, see "Explanatory Asymmetry." (essay 10).

21. Ten people were in fact killed in a tragic head-on collision in Arizona. A pickup truck crossed the center line on a straight stretch of road with clear visibility, striking an oncoming car, with no other traffic present. Prediction of such an accident would have been out of the question.

22. It is often possible to infer the nature of a cause from a partial effect, but it is normally impossible to infer the nature of an effect from knowledge of a partial cause.

23. See Salmon (1984a, chap. 3) for a subsequent treatment of this problem.

24. See Hempel (1965a, p. 250) for the 1948 statement, but see note 6 (added in 1964) on the same page. The later view is more fully elaborated in Hempel (1965a, pp. 347–354).

25. Some philosophers would object to calling such assemblages of probabilities "explanations." Some other term, such as "statistical systematization," might be preferred. I am fairly sympathetic to this view, and have some inclination to believe that explanation in a fuller sense occurs only when we move to the next level.

26. Unless, of course, we can find a *direct* causal dependency, such as one student copying the work of another.

27. This type of causal explanation is discussed in considerable detail in "Causal and Theoretical Explanation" (essay 7).

28. In the normal course of things, they might have collided at a particular location without having received a common call. This presumably, would be even less probable than the type of collision that actually occurred.

29. See Reichenbach (1956, § 19) for fuller discussion, especially his concept of a *conjunctive fork*. See also "Causality: Production and Propagation" (essay 18).

30. Causal relations, as I am conceiving of them in this context, need *not* be deterministic; they are, instead, a species of statistical relevance relations. See "Probabilistic Causality" (essay 14).

31. My solution is offered in "An 'At-At' Theory of Causal Influence" (essay 12).

7

Causal and Theoretical Explanation

In previous discussions of the explanation of particular events (Salmon, 1970b), I have argued—contra Hempel and many others—that such an explanation is not "an *argument* to the effect that the event to be explained . . . *was to be expected* by reason of certain explanatory facts" (Hempel, 1962b, p. 10; emphasis added). Indeed, in the case of inductive or statistical explanation at least, I have maintained that such explanations are not *arguments* of any kind, and that consequently they need not embody the *high* probabilities that would be required to provide reasonable grounds for expectation of the explanandum event. I have argued, instead, that a statistical explanation of a particular event consists of an assemblage of factors relevant to the occurrence or nonoccurrence of the event to be explained, along with the associated probability values. If the probabilities are high, as they will surely be in some cases, the explanation may provide the materials from which an argument can be constructed, but the argument itself is *not* an integral part of the explanation. This model has been called the *statistical-relevance* or *S-R model.*[1]

In addition, I have claimed that the so-called deductive-nomological model of explanation of particular events is incorrect. It is not merely that there are explanandum events that seem explainable only inductively or statistically; Hempel and Oppenheim acknowledged such cases from the very beginning. There are also cases—such as the man who consumes his wife's birth control pills and avoids pregnancy—in which an obviously defective explanation fulfills the conditions for deductive-nomological explanation. All such examples seem to me to exhibit failures of relevance. I have suggested, therefore, that even events that appear amenable to deductive-nomological explanation should also be incorporated, as limiting cases, under the statistical-relevance model.[2]

Arguments by Greeno ([1970] 1971) and others (e.g., Alston, 1971) have convinced me that explanations of particular events seldom, if ever, have genuine scientific import (as opposed to practical value), and that explanations that are scientifically interesting are almost always explanations of classes of events. This leads to the suggestion, elegantly elaborated by Greeno ([1970] 1971), that the goodness or utility of a scientific explanation should be assessed with respect to its ability to account for entire classes of phenomena, rather than by its ability to deal with any particular event in isolation. If, to use Greeno's example, a sociological explanation is offered to account for delinquent behavior in teenage boys, it is to be evaluated in terms of its ability to assign correct probability values to this occurrence among various specifiable classes of boys, not in terms of its ability to predict whether Johnny Jones will turn delinquent. This shift of emphasis is important because it removes any temptation to suppose that we cannot explain Johnny's behavior unless we can cite conditions in relation to which it is highly probable. Perhaps Johnny is a member of a class in which delinquency is very improbable, and no more can be said in the matter. This does not mean that the explanation of *his* delinquency—which is just part of the explanation of delinquency in boys—is defective or weak. As Jeffrey ([1969] 1971) has argued persuasively, the explanation of a low probability event is not necessarily any weaker than the explanation of a high probability event. Even if Billy Smith is a member of a class of boys in which the delinquency rate is very high, the explanation of his delinquency by the afore-mentioned sociological theory is no better or stronger than the explanation of Johnny Jones's delinquency. High probability is not the desideratum, nor is it the standard by which the quality of explanations is to be judged; rather, a correct probability distribution across *relevant* variables is what we should seek.

At the conclusion of my elaboration of the *S-R* model, I expressed certain reservations about it. The two most important problems concerned the involvement of causality in scientific explanation and the nature of theoretical explanation. These two problems are intimately related to each other, and together they form the subject of the present essay. I shall agree from the outset that *causal relevance* (or causal influence) plays an indispensable role in scientific explanation, and I shall attempt to show how this relation can be explicated in terms of the concept of statistical relevance. I shall then argue that the demand for suitable causal relations necessitates reference to theoretical entities, and thus leads to the introduction of theoretical explanations. The theme of the essay will be the centrality of certain kinds of *statistical* relevance relations in the notions of causal explanation and theoretical explanation. The result will be an account of theoretical explanation that differs fundamentally from the received deductive-nomological model.[3]

1. The Common Cause Principle

When all of the lights in a room go off simultaneously, especially if quite a number were on, we infer that a switch has been flipped, a fuse has blown, a power line is down, and so forth, but not that all of the bulbs burned out at once. It is, of course, possible that such a chance coincidence might occur, but so improbable that it is not seriously entertained. The principle is not very different from that by which we conclude that two (or five thousand) identical copies of the same book were produced by a common source. A similar kind of inference is involved when one observes an ordinary bridge deck arranged in perfect order, starting from the ace of spades, and concludes (knowing that cards are packed that way at the factory) that this is a newly opened, unshuffled deck rather than one that arrived at the orderly state by random shuffling. The same principle is involved when two witnesses in court give testimony that is alike in content; if collusion can be ruled out, we have strong grounds for supposing that they are truthfully reporting something they both have observed.

The principle governing these examples has been pointed out by many authors. It is deeply embedded in Russell's famous "postulates of scientific inference" (Russell, 1948, chap. 9), and Reichenbach (1956, § 19) has called it "the principle of the common cause." It may be stated roughly as follows: When apparently unconnected events occur in conjunction more frequently than would be expected if they were independent, then assume that there is a common cause. This principle demands considerable explication, for it involves such obscure concepts as *cause* and *connection*.

Let us take our departure from the standard definition of *statistical independence*. Given two types of events A and B that occur, respectively, with probabilities $P(A)$ and $P(B)$, they are statistically independent if and only if the probability of their joint occurrence $P(A.B)$ is simply the product of their individual occurrences; i.e.,

$$P(A.B) = P(A) \times P(B)$$

If, contrariwise, their joint occurrence is more probable (or less probable) than the product of the probabilities of their individual occurrences, we must say that they are not statistically independent of each other, but rather that they are *statistically relevant* to each other. Statistical independence and statistical relevance, as just defined, are clearly symmetric relations.

It seems fairly clear that events that are statistically independent of each other are completely without explanatory value with regard to one another. If, for example, recovery from neurotic symptoms after psychotherapy occurs with a frequency equal to the spontaneous remission rate, then psychotherapy has no explanatory value concerning the curing of mental illness. (See Salmon, 1970b, for further elaboration.) One reason why independence is of no help whatever in providing explanations is that independent events are inferentially and practically irrelevant; knowing that an event of one type has occurred is of no help in trying to predict the occurrence or nonoccurrence of an event of the other type, or in determining the odds with which to bet on it. Another reason, which will demand close attention, is that statistically independent events are causally irrelevant as well.

If events of the two types are not independent of each other, the occurrence of an event of the one type *may* (but *need* not) help to explain an event of the other type. Suppose, for instance, that the picture on my television receiver occasionally breaks up into a sort of herringbone pattern. At first I may think that this is occurring randomly, but I then discover that there is a nearby police broadcasting station that goes on the air periodically. When I find a strong statistical correlation between the operation of the police transmitter and the breakup of the picture, I conclude that the police broadcast is part of the explanation of the television malfunction. Roughly speaking, the operation of the police transmitter is the cause (or a part of the cause) of the bad TV picture. Obviously, a great deal more has to be filled in to have anything like a complete explanation, but we have identified an important part.

In other cases, however, statistical correlations do not have any such direct explanatory import. The most famous example is the barometer. The rapid dropping of the barometer does not explain the subsequent storm (though, of course, it may enable us to predict it). Likewise, the subsequent storm does not explain the behavior of the barometer. Both are explained by a common cause, namely, the meteorological conditions that cause the storm and are indicated by the barometer. In this case there is a statistical-relevance relation between the barometer reading and the storm, but neither event is invoked to explain the other. Instead, both are explained by a common cause.

The foregoing two examples, the TV interference and the barometer, illustrate respectively cases in which correlated events can and cannot play an explanatory role. The difference is easy to see. The instance in which the event can play an explanatory role is one in which it is cause (or part thereof) of the explanandum event. The case in which the event cannot play an explanatory role is one in which it is not any part of the cause of the explanandum event.

Reichenbach's *basic* principle of explanation seems to be this: *every relation of statistical relevance must be explained by relations of causal relevance.* The various possibilities can be illustrated by a single example. An instructor who receives identical essays from Adams and Brown, two different students in the same class, inevitably infers that something other than a fortuitous coincidence is responsible for their identity. Such an event might, of course, be due to sheer chance (as in the simultaneous burning out of all light bulbs in a room), but that hypothesis is so incredibly improbable that it is not seriously entertained. The instructor may seek evidence that one student copied from the other, i.e., that Adams copied from Brown or that Brown copied from Adams. In either of these cases the identity of the papers can be explained on grounds that one is cause (or part of a cause) of the other. In either of these cases there is a direct causal relation from the one paper to the other, so a causal connection is established. It may be, however, that each student copied from a common source, such as a paper in a sorority or fraternity file. In this case neither of the students' papers is a causal antecedent of the other, but there is a coincidence that has to be explained. The explanation is found in the common cause, the paper in the file, that is a causal antecedent to each.

The case of the common cause, according to Reichenbach's analysis, exhibits an interesting formal property. It is an immediate consequence of our foregoing definition of statistical independence that event A is statistically relevant to event B if and only if $P(B) \neq P(B|A)$. Let us assume positive statistical relevance; then

$$P(B|A) > P(B) \text{ and } P(A|B) > P(A).$$

From this it follows that

$$P(A.B) > P(A) \times P(B).$$

To explain this improbable coincidence, we attempt to find a common cause C such that

$$P(A.B|C) = P(A|C) \times P(B|C),$$

which is to say that, in the presence of the common cause C, A and B are once more rendered statistically independent of each other. The statistical dependency is, so to speak, swallowed up in the relation of causal relevance of C to A and C to B. Under these circumstances C must, of course, be statistically relevant to both A and B; that is,

$$P(A|C) > P(A) \text{ and } P(B|C) > P(B).$$

These *statistical*-relevance relations must be explained in terms of two causal processes in which C is *causally* relevant to A and C is *causally* relevant to B.

A further indirect causal relation between two correlated events may obtain, namely, both may serve as partial causes for a common effect. Perhaps Adams and Brown are basketball stars on a championship team that can beat its chief rival if and only if either Adams or Brown plays. Caught at plagiarism, however, both are disqualified and the team loses. As Reichenbach points our, a common effect that follows a combination of partial causes cannot be used to explain the coincidence in the absence of a common cause. In the absence of any common source, and in the absence of copying one from the other, we cannot attribute the identity of the two papers to a conspiracy of events to produce the team's defeat.[4] Thus, there is no 'principle of the common effect' to parallel the principle of the common cause. This fact provides a basic temporal asymmetry of explanation that is difficult to incorporate into the standard deductive-nomological account of explanation (see "Explanatory Asymmetry" [essay 10]).

2. Causal Explanation of Statistical Relevance

To provide an explanation of a particular event, we may make reference to a statistically relevant event, but the statistical relevance relation itself is a statistical generalization. I agree with the standard nomological account of explanation, which demands that an explanation have at least one general statement in the explanans. As indicated in the preceding section, however, we are adopting a principle that says that relations of statistical relevance must be explained in terms of relations of causal relevance. This brings us to the problem of explanations of general relations.

Most of the time (though I *am* prepared to admit exceptions) we do not try to explain statistical independencies or irrelevancies. If the incidence of sunny days in Bristol is independent of the occurrence of multiple human births in Patagonia, no explanation seems called for.[5] Statistical dependencies often do demand explanation, however, and causal relations constitute the explanatory device. Plagiarism, unfortunately, is not a unique occurrence; identical papers turn up with a frequency that cannot be attributed to chance. In such cases it is possible to trace observable chains of events from the essays back to a *causal* antecedent. In these instances nothing of a theoretical nature has to be introduced, for the explanation can be given in terms of observable events and processes.[6] In other cases, such as the breakup of the television picture, it is necessary to invoke theoretical considerations if we want to give a causal explanation of the statistical dependency. The statistical relevance between the events of the two types may help to explain the breakup of the picture, and this correlation is essentially observable—for example, by telephoning the station to ask if they have just been on the air. The statistical dependency itself, however, cannot be explained without reference to theoretical entities such as electromagnetic waves.

Spatiotemporal continuity obviously makes the critical difference in the two examples just mentioned. In the instance of cheating on the essay, we can provide spatiotemporally continuous processes from the common cause to the two events whose coincidence was to be explained. Having provided the continuous *causal* connections, we have furnished the explanation. In the case of trouble with the TV picture, a statistical correlation is discovered between events that are remote from one another spatially, and this correlation itself requires explanation in terms of processes such as the propagation of electromagnetic waves in space. We invoke a theoretic process that exhibits the desired continuity requirements. When we have provided spatiotemporally continuous connections between correlated events, we have fulfilled a major part of the demand for an explanation of the correlation. We shall return in a subsequent section to a more thorough discussion of the introduction of theoretic entities into explanatory contexts.

The propagation of electromagnetic radiation is generally taken to be a continuous causal process. In characterizing it as continuous we mean, I suppose, that given any two spatiotemporally distinct events in such a process, we can interpolate other events between them in the process.[7] But, over and above continuity, what do we mean by characterizing a process as causal? At the very least it would seem reasonable to insist that events that are causally related exhibit statistical dependencies. This suggests that we require, as a necessary but not sufficient condition, that explanation of statistical dependencies between events that are not contiguous be given by means of statistical relevance between neighboring or contiguous events.[8]

I have been talking about *causes* and *causal relations;* these seem to figure essentially in the concept of explanation. The principle I am considering (as enunciated by Reichenbach) is the principle of the common *cause;* Russell's treatment of scientific knowledge relies heavily and explicitly on *causal* relations. It seems to be a serious shortcoming of the received doctrine of scientific explanation that it does not incorporate any full-blooded requirement of causality.[9] But we must not forget the lessons Hume taught us. The question is whether we can explicate the concept of causality in terms that do not surreptitiously introduce any 'occult' concepts of 'power' or 'necessary connection'. Statistical relevance relations represent the type of constant conjunction Hume relied upon, and spatiotemporal contiguity is also consonant with his strictures. Hume's attempt to explicate causal relations in terms of constant conjunction was admittedly inadequate because it was an oversimplification; Russell's was also inadequate for the same reason, as I shall show in the next section. Our problem is to see whether we can provide a more satisfactory account of causal processes using only such notions as statistical relevance. We shall see in a moment that processes that satisfy the conditions of continuity and mutual statistical relevance are not necessarily causal processes. We shall, however, remain true to the Humean spirit if we can show that more complicated patterns of statistical relevance relations will suffice to do the job.

3. Causal Processes and Pseudo-Processes

Reichenbach tried, in various ways, to show how the concept of causal relevance could be explicated in terms of statistical relevance. He believed, essentially, that causal relevance is a special case of statistical relevance. One of his most fruitful suggestions, in my opinion, employs the concept of a *mark*.[10] Since we are not, in this context, attempting to deal with the problem of 'time's arrow', and correlatively, with the nature and existence of irreversible processes, let us assume that we have provided an adequate physical basis for identifying irreversible processes and ascertaining their temporal direction. Thus, to use one of Reichenbach's favorite examples, we can mark a beam of light by placing a red filter in its path. A beam of white light, encountering such a filter, will lose all of its frequencies except those in the red range, and the red color of the beam will thus be a mark transmitted onward from the point at which the filter is placed in its path. Such marking procedures can obviously be used for the transmission of information along causal processes.

In the context of relativity theory, it is essential to distinguish causal processes, such as the propagation of light ray, from various pseudo-processes, such as the motion of a spot of light cast on a wall by a rotating beacon. The light ray itself can be marked by the use of a filter, or it can be modulated to transmit a message. The same is not true of the spot of light. If it is made red at one place because the light beam creating it passes through a red filter at the wall, that red mark is not passed on to the successive positions of the spot. The motion of the spot is a well-defined process of some sort, but it is not a causal process. The causal processes involved are the passages of light rays from the beacon to the wall, and these can be marked to transmit a message. But the direction of message transmission is from the beacon to the wall, not across the wall. This fact has great moment for special relativity, for the light beam can travel no faster than the universal constant c, while the spot can move across the wall at arbitrarily high velocities. Causal processes can be used to synchronize clocks; pseudo-processes cannot. The arbitrarily high velocities of pseudo-processes cannot be exploited to undermine the relativity of simultaneity.[11]

Consider a car traveling along a road on a sunny day. The car moves along in a straight line at 60 m.p.h., and its shadow moves along the shoulder at the same speed. If the shadow encounters another car parked on the shoulder, it will be distorted, but will continue on unaffected thereafter. If the car collides with another car and continues on, it will bear the marks of the collision. If the car passes a building tall enough to cut off the sunlight, the shadow will be destroyed, but it will exist again immediately when the car passes out of the shadow of the building. If the car is totally destroyed, say by an atomic explosion, it will not automatically pop back into existence after the blast and continue on its journey as if nothing had happened.

There are many causal processes in this physical world; among the most important are the transmission of electromagnetic waves, the propagation of sound waves and other deformations in various material media, and the motion of physical objects. Such processes transpire at finite speeds no greater than that of light; they involve the transportation of energy from one place to another,[12] and they can carry messages. Assuming, as we are, that a temporal direction has been established, we can say that the earlier members of such causal processes are *causally relevant* to the later ones, but not conversely.[13] Causal relevance thus becomes an asymmetric relation, one that we might also call "causal influence." We can test for the relation of causal relevance by making marks in the processes we suspect of being causal and seeing whether the marks are, indeed, transmitted. Radioactive "tagging" can, for example, be used to trace physiological causal processes. The notion of causal relevance has been aptly characterized by saying, "You wiggle something over here and see if anything wiggles over there." This formulation suggests, of course, some form of human intervention, but that is obviously no essential part of the definition. It does not matter what agency is responsible for the marking of the process. At the same time, experimental science is built on the idea that *we* can do the wiggling.[14] There is an obvious similarity between this approach and Mill's methods of difference and concomitant variation.

Just as it is necessary to distinguish causal processes from pseudo-processes, so also is it important to distinguish the relation of causal relevance from the relation of statistical relevance, especially in view of the fact that pseudo-processes exhibit striking instances of statistical relevance. Given the moving spot of light on a wall produced by our rotating beacon, the occurrence of the spot at one point makes it highly probable that the spot will appear at a nearby point (in the well-established path) at some time very soon thereafter. This is not a certainty, of course, for the light may burn out, an opaque object may block the beam or the beacon may stop rotating in its accustomed fashion. The same is true of causal processes. Given an occurrence at some point in the process, there is a high probability of another occurrence at a nearby point in the well-established path. Again, however, there is no certainty, for the process may be disturbed or stopped by some other agency. These considerations show that pseudo-processes may exhibit both continuity and statistical relevance among members; this establishes my earlier contention that these two properties, though perhaps necessary, are not sufficient to characterize causal processes.

Pseudo-processes exhibit the same basic characteristics as correlated events or improbable coincidences that require explanation in terms of a common cause. There is a strong correlation between the sudden drop of the barometer and the occurrence of a storm; however, fiddling with a barometer will have no effect upon the storm, and marking or modifying the storm (assuming we had power to do so) would not be transmitted to the (earlier) barometer reading. The pseudo-process is, in fact, just a fairly elaborate pattern of highly correlated events produced by a common cause (the rotating beacon). Pseudo-processes, like other cases of noncausal statistical relevance, require explanation; they do not provide it, even when they possess the sought-after property of spatiotemporal continuity.

One very basic and important principle concerning causal relevance—i.e., the transmission of marks—is, nevertheless, embedded in continuous processes. Marks (or information) are transmitted continuously in space and time. Spatiotemporal continuity, I shall argue, plays a vital role in theoretical explanation. The fact that it seems to break down in quantum mechanics—that quantum mechanics seems unavoidably to engender causal anomalies—is a source of great distress. It is far more severe, to my mind, than the discomfort we should experience on account of the apparent breakdown of determinism in that domain. The failure of determinism is one thing, the violation of causality quite another. As I understand it, determinism is the thesis that (loosely speaking) the occurrence of an event has probability zero or one in the presence of a complete set of statistically relevant conditions. Indeterminism, by contrast, obtains if there are complete sets of statistically relevant conditions (i.e., homogeneous reference classes) with respect to which the event may either happen or not—that is, the probability of its occurrence has some intermediate value other than zero or one.[15] The breakdown of causality lies in the fact that (in the quantum domain) causal influence is not transmitted with spatiotemporal continuity. This, I take it, formulates a fundamental aspect of Bohr's principle of complementarity as well as Reichenbach's principle of anomaly (see Reichenbach, 1956, p. 216; also 1946). Causal influence need not be deterministic to exhibit continuity; I am construing causal relevance as a species of statistical relevance. Causality, in this sense, is entirely compatible with indeterminism, but quantum mechanics goes beyond indeterminism in its admission of familiar spatiotemporal discontinuities.[16] In classical physics and relativity theory, however, we retain the principle that all causal influence is via action by contact. It is doubtful, to say the least, that action by contact can be maintained in quantum mechanics. Even in the macrocosm, however, pseudo-processes may display obvious discontinuities, as, for example, when the spot of light from the rotating beacon must "jump" from the edge of a wall to a cloud far in the background.

Another fundamental characteristic of causal influence is its asymmetric character; in this respect it differs from the relation of statistical relevance. It is an immediate consequence of the foregoing definition of statistical relevance that A is relevant to B if and only if B is relevant to A. This has the consequence that effects are statistically relevant to causes if (as must be the case) causes are statistically relevant to their effects. As we shall see, Reichenbach defines the *screening-off relation* in terms of statistical relevance; it is a nonsymmetric relation from which the relation of causal relevance inherits its asymmetry. The property of asymmetry is crucial, for the common cause that explains a coincidence always precedes it.

4. Theoretical Explanation

In our world the principle of the common cause works rather nicely. We can explain the identical student essays by tracing them back to a common cause via two continuous causal processes. These causal processes are constituted, roughly speaking, of events that are in principle observable, and that were in fact observed by the two plagiarists. Many authors, including Hume very conspicuously, have explained how we may endow our world of everyday physical objects with a high degree of spatiotemporal continuity by suitably interpolating observable objects and events between observed objects and events. Russell has discussed at length the way in which similar structures grouped around a center could be explained in terms of the propagation of continuous causal influence from the common center; indeed, this principle became one of Russell's postulates of scientific inference.[17] In many of his examples, if not all, the continuous process is in principle observable at any point in its propagation from the center to more distant points at later times.

Although we can endow our world with lots of continuity by reference to observable (though unobserved) entities, we cannot do a very complete job of it. In order to carry through the task, we must introduce some entities that are unobservable, at least for ordinary human capabilities of perception. If, for example, we notice that the kitchen windows tend to get foggy on cold days when water is boiling on the stove, we connect the boiling on the stove with the fogging of the windows by hypothesizing the existence of water molecules that are too small to be seen by the naked eye, and by asserting that they travel continuous trajectories from the pan to the window. Similar considerations lead to the postulation of microbes, viruses, and genes for the explanation of such phenomena as the spread of disease and the inheritance of biological characteristics. Note, incidentally, how fundamental a role the transmission of a mark or information plays in modern molecular biology. Electromagnetic waves are invoked to fulfill the same kind of function; in the explanation of the TV picture disturbance, the propagation of electromagnetic waves provided the continuous connection. These unobservable entities are not fictions—not simple-minded fictions at any rate—for we maintain that it is possible to detect them at intermediate positions in the causal process. Hertz detected electromagnetic waves; he could have positioned his detector (or additional detectors) at intermediate places. The high correlation between a spark in the detecting loop and a discharge at the emitter had to be explained by a causal process traveling continuously in space and time. Moreover, the water molecules from the boiling pan will condense on a chilled tumbler anywhere in the kitchen. Microbes and viruses, chromosomes and genes, can all be detected with suitable microscopes; even heavy atoms can now be observed with the electron scanning microscope. The claim that there are continuous causal processes involving unobservable objects and events is one that we are willing to test; along with this claim goes some sort of theory about how these intermediate parts of the process can be detected. The existence of causal-relevance relations is also subject to test, of course, by the use of marking processes.

Many philosophers, most especially Berkeley, have presented detailed arguments against the view that there are unobserved physical objects. Berkeley did, nevertheless, tacitly admit the common cause principle, and consequently invoked God as a theoretical entity to explain statistical correlations among observed objects. Many other philosophers, among them Mach, presented detailed arguments against the view that there are unobservable objects. Such arguments lead either to phenomenalism (as espoused, for example, by C. I. Lewis) or instrumentalism (as espoused by many early logical positivists). Reichenbach strenuously opposed both of these views, and in the course of his argument he offers a strange analogy, namely, his cubical world (1938, esp. § 14).

Reichenbach invites us to consider an observer who is confined to the interior of a cube in which a bunch of shadows appear on the various walls. Careful observation reveals a close correspondence between the shadows on the ceiling and those on one of the walls; there is a high statistical correlation between the shadow events on the ceiling and those on the wall. For example, when one notices what appears to be the shadow of one bird pecking at another on the ceiling, one finds the same sort of shadow pattern on the wall. Reichenbach argues that these correlations should be explained as shadows of the same birds cast on the ceiling and the wall; that is, birds outside of the cube should be postulated. It is further postulated that they are illuminated by an exterior source, which makes the shadows of the same birds appear on the translucent material of both the ceiling and the wall. He stipulates that the inhabitant of the cube cannot get to the ceiling or walls to poke holes in them or any such thing, so that it is physically impossible for the inhabitant to observe the birds directly. Nevertheless, according to Reichenbach, one should infer their existence.[18] Reichenbach is doing precisely what he advocated explicitly in his later work: he is explaining a relation of statistical relevance in terms of relations of causal relevance, invoking a common cause to explain the observed noncontiguous coincidences. The causal processes he postulates are, of course, spatiotemporally continuous.

In *Experience and Prediction* Reichenbach claims that the theory of probability enables us to infer, with a reasonable degree of probability, the existence of entities of unobservable types. This claim seems problematic, to say the least, and I was never quite clear how he thought it could be done. One could argue that all we can observe in the cubical world are constant conjunctions between patterns on the ceiling and patterns on the wall. If constant (temporal) conjunction were the whole story as far as causality is concerned, then we could say that the patterns on the ceiling cause the patterns on the wall, or vice versa. There would be no reason to postulate anything beyond the shadows, for the constant conjunctions are given observationally, and they are all we need. The fact that they are not connected to one another by continuous causal lines would be no ground for worry; there would be no reason to postulate a common cause to link the observed coincidences via *continuous* causal processes. This, a very narrow Humean might say, is the entire empirical content of the situation; we cannot infer even with probability that the common cause exists. Such counterarguments might be offered by phenomenalists or instrumentalists.

Reichenbach is evidently invoking (though not explicitly in 1938) his principle that statistical relevance must be explained by causal relevance, where causal relevance is defined by continuity and the ability to transmit a mark. In the light of this principle, we may say that there is a certain probability $P(A)$ that a particular pattern (the shadow of one bird pecking at another) will appear on the ceiling, and a certain probability $P(B)$ that a similar pattern will appear on the wall. There is another probability $P(A.B)$ that this pattern will appear both on the ceiling and on the wall at the same time. This latter probability seems to be much larger than it would be if the events were independent, i.e.,

$$P(A.B) \gg P(A) \times P(B).$$

Reichenbach's principle asserts that this sort of statistical dependency demands causal explanation if, as in this example, A and B are not spatiotemporally contiguous. Using this principle, Reichenbach can certainly claim that the existence of the common cause can be inferred with a probability; otherwise we would have to say that the probability of $A.B$ *is equal to* the product of the two individual probabilities, and that we were misled into thinking that an inequality holds because the observed frequency of $A.B$ is much larger than the actual probability. In other words, the choice is between a common cause and an exceedingly improbable coincidence. This makes the common cause the less improbable hypothesis. But the high frequency of the joint occurrence is statistically miraculous only if there are no alternatives except fortuitous coincidence or a continuous connection to a common cause. If we could have causal relevance without spatiotemporal contiguity, no explanation would be required, and hence there would be no probabilistic evidence for the existence of the common cause. If, however, we can find an adequate basis for adopting the principle that statistical relevancies must be explained by continuous causal processes, then it seems we have sufficient ground for postulating or inferring the existence of theoretical entities.

In rejecting the notion that we have an impression of necessary connection, Hume analyzed the causal relation in terms of constant conjunction. As he realized explicitly, his analysis of causation leaves open the possibility of filling the spatiotemporal gaps in the causal chain by interpolating events between observed causes and observed effects. In so doing, he maintained, we simply discover a larger number of relations of constant conjunction with higher degrees of spatiotemporal contiguity. In recognition of the fact that causal relations often serve as a basis for inference, Hume attempts to provide this basis in the "habit" or "custom" to which observed constant conjunction naturally gives rise.

Russell has characterized causal lines as continuous series of events in which it is possible to infer the nature of some members of the series from the characteristics of other events in the same series. This means, in our terms, that there are relations of statistical relevance among the members of such series. Although causal series have enormous epistemological significance for Russell, providing a basis for our knowledge of the physical world, his characterization of causal series is by no means subjective. It is by virtue of factual relations among the members of causal series that we are enabled to make the inferences by which causal processes are characterized.

Statistical-relevance relations do provide a basis for making certain kinds of inferences, but they do not have all of the characteristics of causal relevance as defined by Reichenbach; in particular, they do not always have the ability to transmit a mark. Although Russell did not make explicit use of mark transmission in his definitions, his approach would seem hospitable to the addition of this property as a further criterion of causal processes. Russell emphasizes repeatedly the idea that perception is a causal process by which structure can be transmitted. He frequently cites processes such as radio transmission as physical analogues of perception, and he obviously considers such examples extremely important. The transmission of messages by the modulations of radio waves is a paradigm of a mark. In similar ways, the absorption of all frequencies but those in the green range from white light falling on a leaf is a suggestive case of the marking of a causal process involved in human perception. The transmitted mark conveys information about the interaction that is responsible for the mark. The mark principle thus seems to me to be a desirable addition to Russell's definition of causal processes, and one that can be fruitfully incorporated into his postulates of scientific knowledge.

I do not wish to create the impression that ability to transmit a mark is any mysterious kind of necessary connection or "power" of the sort Hume criticized in Locke. Ability to transmit a mark is simply a species of constant conjunction. We observe that certain kinds of events tend to be followed by others in certain kinds of processes. Rays of white light are series of undulatory events that are spatiotemporally distributed in well-defined patterns. Events that we would describe as passage of light through a red filter are followed by undulations with frequencies confined to the red range; undulations characterized by other frequencies do not normally follow thereupon. It is a fact about this world (at least as long as we stay out of the quantum realm) that there are many continuous causal processes that do transmit marks. This is fortunate for us, for such processes are highly informative. Russell was probably right in saying that without them we would not have anything like the kind of knowledge of the physical world we actually do have. It is not too surprising that causal processes capable of carrying information figure significantly in our notion of scientific explanation. To maintain that such processes are continuous, we must invoke theoretical entities. Let us then turn to the motivation for the continuity requirement.

5. Spatiotemporal Continuity

Throughout this essay I have been discussing the continuity requirement on causal processes; it is now time to see why such processes figure so importantly in the discussion. If special relativity is correct, there are no essential spatiotemporal limitations on relations of statistical relevance, but there are serious limitations on relations of causal relevance. Any event A that we choose can be placed at the apex of a Minkowski light cone, and this cone establishes *a cone of causal relevance*. The backward section of the cone, containing all of the events in the absolute past of A, contains all events that can bear the relation of causal relevance to A. The forward part of the light cone, which contains all events in the absolute future of A, contains all events to which A may bear the relation of causal relevance. In contrast, an event B that is in neither the backward nor the forward section of the cone cannot bear the relation of causal relevance to A, nor can A bear that relation to B. Nevertheless, B can sustain a relation of *statistical* relevance to A. When this occurs, according to Reichenbach's principle, there must be a common cause C somewhere in the region of overlap of the backward sections of the light cones of A and B. The relation of statistical relevance is *not* explainable, as mentioned earlier, by a common effect in the region of overlap of the forward sections of the two light cones.[19]

If our claims are correct, any *statistical* relevance relation between two events can be explained in terms of *causal*-relevance relations. Causal-relevance relations are embedded in continuous causal processes. If, therefore, an event C is causally relevant to A, then we can, so to speak, mark off a boundary in the backward part of the light cone (i.e., the causal relevance cone) and be sure either that C is within that part of the cone or else that it is connected with A by a continuous causal process that crosses that boundary. Hence, to investigate the question of what events are causally relevant to A, we have only to examine the interior and boundary of some spatial neighborhood of A for a certain time in the immediate past of A. We can thus ascertain whether such an event lies within that neighborhood, or whether a connecting causal process crosses the boundary. We have been assuming, let us recall, that a continuous causal process can be detected anywhere along its path. This means that we do not have to search the whole universe to find out what events bear relations of *causal* relevance to A.[20]

If we make it our task to find out what events are *statistically* relevant to A, all of the events in the universe are potential candidates. There are, in principle, no spatiotemporal limitations on statistical relevance. But, it might be objected, statistical-relevance relations can serve as a basis for inductive inference, or at least for inductive behavior (for example, betting). How are we therefore justified, if knowledge is our aim, in restricting our considerations to events that are causally relevant? The answer lies in the *screening-off* relation (Reichenbach, 1956, p. 189; Salmon, 1970b, § 7).

If A and B are two events that are statistically relevant to each other, but neither is causally relevant to the other, then there must be a common cause C in the region of overlap of the past light cones of A and B. It is possible to demonstrate the causal relevance of C to A by showing that a mark can be transmitted along the causal process from C to A, and the causal relevance of C to B can be demonstrated in a similar fashion. There is, however, no way of transmitting a mark from B to A or from A to B. When we have that kind of situation, which can be unambiguously defined by the use of marking techniques, we find that the statistical relevance of B to A is absorbed in the *statistical* relevance of C to A. That is just what the screening-off relation amounts to. Given that B is statistically relevant to A, and C is statistically relevant to A, we have

$$P(A|B) > P(A) \text{ and } P(A|C) > P(A).$$

To say that C screens off B from A means that, given C, B become statistically irrelevant to A; i.e.,

$$P(A|B.C) = P(A|C).$$

Thus, for example, though the barometer drop indicates a storm and is statistically relevant to the occurrence of the storm, the barometer becomes statistically irrelevant to the occurrence of the storm, given the meteorological conditions that led to the storm and that are indicated by the barometer reading. The claim that statistical-relevance relations can always be explained in terms of causal-relevance relations therefore means that causal-relevance relations screen off other kinds of statistical-relevance relations.

The screening-off relation can be used, moreover, to deal with questions of causal proximity. We can say in general that more remote causal-relevance relations are screened off by more immediate causal-relevance relations. Part of what we mean by saying that causation operates via action by contact is that the more proximate causes absorb the entire influence of more remote causes. Thus, we do not even have to search the entire backward section of the light cone to find *all* factors relevant to the occurrence of A. A *complete* set of factors statistically relevant to the occurrence of a given event can be found by examining the interior and boundary of an appropriate neighboring section of its past light cone. Any factor outside of that portion of the cone that is, by itself, statistically relevant to the occurrence of the event in question is screened off by events within that neighboring portion of the light cone. These are strong factual claims; if correct, they have an enormous bearing on our conception of explanation.

6. Conclusions

In this essay I have been trying to elaborate the view of scientific explanation that is present, at least implicitly I think, in the works of Russell and Reichenbach. Such explanation is causal in a very deep and pervasive sense; yet I believe is does not contain causal notions that have been proscribed by Hume's penetrating critique. This causal treatment accounts in a natural way for the invocation of theoretical entities in scientific explanation. It is therefore, I hope, an approach to scientific explanation that fits especially well with scientific realism (as opposed to instrumentalism). Still, I do not wish to claim that this account of explanation establishes the realistic thesis regarding theoretical entities. An instrumentalist might well ask: Is the world understandable because it contains continuous causal processes, or do we make it understandable by imputing continuous causal processes? This is a difficult and far-reaching query.

It is tempting to try to argue for the realist alternative by saying that it would be a statistical miracle of overwhelming proportions if there were statistical dependencies between remote events that reflect precisely the kinds of dependencies we should expect if there were continuous causal connections between them. At the same time, the instrumentalist might retort: What makes remote statistical dependencies any more miraculous than contiguous ones? Unless one is willing to countenance (as I am not) some sort of pre-Humean concept of power or necessary connection, I do not know quite what answer to give.[21] We may have reached a point at which a pragmatic vindication, a posit, or a postulate is called for. It may be possible to argue that scientific understanding can be achieved most efficiently (if such understanding is possible at all) by searching for spatiotemporally continuous processes capable of transmitting marks. This may be the situation with which Russell attempted to cope by offering his postulates of scientific inference.[22] The preceding section was an attempt to spell out the methodological advantages we gain if the world is amenable to explanations of this sort, but I do not intend to suggest that the world is otherwise totally unintelligible. After all, we still have to cope with quantum mechanics, and that does not make scientific understanding seem hopeless.

Regardless of the merits of the foregoing account of explanation, and regardless of the stand one decides to take on the realism-instrumentalism issue, it is worthwhile, I think, to contrast this account with the standard deductive-nomological account. According to the received view, empirical laws, whether universal or statistical, are explained by deducing them from more general laws or theories. Deductive subsumption is the key to theoretical explanation. According to the present account, statistical dependencies are explained by, so to speak, filling in the causal connections in terms of spatiotemporally continuous causal processes. I do not mean to deny, of course, that there are physical laws or theories that characterize the causal processes to which we are referring—laws of mechanics which govern the motions of material bodies, laws of optics and electromagnetism which govern the propagation of electromagnetic waves, etc. The point is, rather, that explanations are not arguments on this view. Causal or theoretical explanation of a statistical correlation between distinct types of events is an exhibition of the way in which those regularities fit into the causal structure of the world—an exhibition of the causal connections between them that give rise to the statistical-relevance relations.

Notes

The author wishes to express his gratitude to the National Science Foundation (USA) for support of research on scientific explanation and other related topics.

1. See especially the introduction in Salmon (1971).

2. This approach has been elaborated in some detail in the three essays by Richard C. Jeffrey, James G. Greeno, and myself in Salmon (1971).

3. Richard Beven Braithwaite (1953); Carl G. Hempel (1965a); and Ernest Nagel (1961) are among the major proponents of the received view.

4. The temporal asymmetry of explanation is discussed at length in connection with the common cause principle (and lack of a parallel common effect principle) in Salmon (1970b, § 11–12).

5. In a situation in which we expect to find a statistical correlation and none is found, we may demand an explanation. Why, for example, is the presence of a certain insecticide irrelevant to the survival of a given species of insect? Because of an adaptation of the species? An unnoticed difference between that species and another that finds the substance lethal? And so on.

6. I realize that a full and complete explanation would require references to the theoretical aspects of perception and other psycho-physiological mechanisms, but for the moment the example is being taken in commonsense terms.

7. For present purposes I ignore the distinction between denseness and genuine continuity in the Cantorean sense. For a detailed discussion of this distinction and its relevance to physics, see my anthology *Zeno's Paradoxes* (Salmon, 1970a), especially my introduction and the selections by Adolf Grünbaum.

8. In the present context I am ignoring the perplexities about discontinuities and causal anomalies in quantum mechanics.

9. In Hempel's account of deductive-nomological explanation. there is some mention of nomological relations constituting causal relations, but this passing mention of causality is too superficial to capture the features of causal processes with which we are concerned, and which seems ineradicably present in our intuitive notions about explanation. [In Hempel (1965b) causality is unambiguously excluded from explanatory constraints.]

10. Although Reichenbach often discussed the "mark method" of dealing with causal relevance, the following discussion is based chiefly on Reichenbach (1956, § 23).

11. See Reichenbach ([1928] 1957, § 23) for a discussion of "unreal sequences," which I have chosen to call pseudo-processes.

12. See "Causality without Counterfactuals" (essay 16) for further discussion of transmission of energy and other conserved quantities.

13. Although Reichenbach ([1928] 1957) seemed to maintain that the mark method could be taken as an independent criterion of temporal direction (without any other basis for distinguishing irreversible processes), he rejected that view in his (1956).

14. We must, however, resist the strong temptation to use intervention as a criterion of temporal direction; see Reichenbach (1956, § 6).

15. This conception of determinism, which seems to me especially suitable in the context of discussions of explanation, is elaborated in Salmon (1970b, § 4). Note also that it is technically illegitimate to identify probability one with invariable occurrence and probability zero with universal absence, but that technicality need not detain us. I ignore it in the context of this essay.

16. It would be completely compatible with indeterminism and causality to suppose that a "two-slit experiment" performed with macroscopic bullets would yield a two-slit statistical distribution that is just the superposition of two one-slit patterns when large numbers of bullets are involved. At the same time, it might be that no trajectory of any individual bullet is precisely determined by the physical conditions. This imaginary situation differs sharply, of course, from the familiar two-slit experiment of quantum mechanics. See "Indeterminacy,

Indeterminism, and Quantum Mechanics" (essay 17) for further discussion of the situation in quantum mechanics.

17. In Russell (1948, pt. 6, chap. 6) it is called "the structural postulate."

18. Reichenbach does not say whether there are any birds *inside* the cube, so that the inference is to entities outside the cube quite like those on the inside, or no birds on the inside to give a clue to the nature of the inferred exterior birds. If his analogy is to be interesting, we must adopt the latter interpretation and demand that the observer postulate theoretical entities quite unlike those observed. See Salmon (1994).

19. These statements obviously represent factual claims about this world. We believe they are true, and if true they are very important. But we have no reason to think they are true in all possible worlds.

20. In this connection it is suggestive to remember Popper's distinction between falsifiable and unfalsifiable existential statements.

21. [I now believe that an adequate response to this question is given by the "at-at" theory of causal transmission discussed in essays 1, 12, 16, and 18.]

22. I have discussed Russell's views on his postulates in some detail in Salmon (1974b, pp. 183–208). In the same essay I have discussed aspects of Popper's methodological approach that are relevant to this context.

8

Why Ask, "Why?"?

An Inquiry Concerning Scientific Explanation

Concerning the first order question "Why?" I have raised the second order question "Why ask, 'Why?'?" to which you might naturally respond with the third order question "Why ask, 'Why ask, "Why?"?'?" But this way lies madness, to say nothing of an infinite regress. While an infinite sequence of nested intervals may converge upon a point, the infinite series of nested questions just initiated has no point to it, and so we had better cut it off without delay. The answer to the very natural third order question is this: the question "Why ask, 'Why?'?" expresses a deep philosophical perplexity which I believe to be both significant in its own right and highly relevant to certain current philosophical discussions. I want to share it with you.

The problems I shall be discussing pertain mainly to scientific explanation, but before turning to them, I should remark that I am fully aware that many—perhaps most—why-questions are requests for some sort of *justification* (Why did not employee receive a larger raise than another? Because she had been paid less than a male colleague for doing the same kind of job) or *consolation* (Why, asked Job, was I singled out for such extraordinary misfortune and suffering?). Since I have neither the time nor the talent to deal with questions of this sort, I shall not pursue them further, except to remark that the seeds of endless philosophical confusion can be sown by failing carefully to distinguish them from requests for scientific explanation (see "The Importance of Scientific Understanding" [essay 5]).

Let me put the question I do want to discuss to you this way. Suppose you had achieved the epistemic status of Laplace's demon—the hypothetical superintelligence who knows all of nature's regularities and the precise state of the universe in full detail at some particular moment (say now, according to some suitable simultaneity slice of the universe). Possessing the requisite logical and mathematical skill, you would be able to predict any future occurrence, and you would be able to retrodict any past event. Given this sort of apparent omniscience, would your scientific knowledge be complete, or would it still leave something to be desired? Laplace asked no more of his demon; should we place further demands upon ourselves? And if so, what should be the nature of the additional demands?

If we look at most contemporary philosophy of science texts, we find an immediate *affirmative* answer to this question. Science, the majority say, has at least two principal aims—prediction (construed broadly enough to include inference from the observed to the unobserved, regardless of temporal relations) and explanation. The first of these provides knowledge of *what* happens; the second is supposed to furnish knowledge of *why* things happen as they do. This is not a new idea. In the *Posterior Analytics* (bk. I.2, 71b) Aristotle distinguishes syllogisms that provide scientific understanding from those that do not. In the *Port Royal Logic,* Arnauld distinguishes demonstrations that merely *convince* the mind from those that also *enlighten* the mind.[1]

This view has not been universally adopted. It was not long ago that we often heard statements to the effect that the business of science is to predict, not to explain. Scientific knowledge is descriptive—it tells us *what* and *how.* If we seek explanations—if we want to know *why*—we must go outside of science, perhaps to metaphysics or theology. In his preface to the third edition of *The Grammar of Science,* Karl Pearson wrote, "Nobody believes now that science *explains* anything; we all look upon it as a shorthand description, as an economy of thought" (Pearson, [1911] 1957, p. xi).[2] This doctrine is not very popular nowadays. It is now fashionable to say that science aims not merely at describing the world; it also provides *understanding, comprehension,* and *enlightenment.* Science presumably accomplishes such high-sounding goals by supplying scientific explanations.

The current attitude leaves us with a deep and perplexing question, namely, if explanation does involve something over and above mere description, just what sort of thing is it? The use of such honorific near-synonyms as "understanding," "comprehension," and "enlightenment" makes it sound important and desirable, but helps not at all in the philosophical analysis of explanation—scientific or other. What, over and above its complete descriptive knowledge of the world, would Laplace's demon require in order to achieve understanding? I hope you can see that this is a real problem, especially for those who hold what I shall call "the inferential view" of scientific explanation, because Laplace's demon can infer every fact about the universe, past, present, and future. If you were to say that the problem does not seem acute, I would make the same remark Russell made about Zeno's paradox of the arrow: "The more the difficulty is meditated, the more real it becomes" (1922b, p. 179).

It is not my intention to discuss the details of the various formal models of scientific explanation that have been advanced in the last five decades.[3] Instead, I want to consider the general conceptions that lie beneath the most influential theories of scientific explanation. Two powerful intuitions seem to have guided much of the discussion. Although they have given rise to disparate basic conceptions and considerable controversy, both are, in my opinion, quite sound. Moreover, it seems to me, both can be incorporated into a single overall theory of scientific explanation.

(1) The first of these intuitions is the notion that the explanation of a phenomenon essentially involves *locating and identifying its cause or causes.* This intuition seems to arise rather directly from common sense, and from various contexts in which scientific knowledge is applied to concrete situations. It is strongly supported by a number of

paradigms, the most convincing of which are explanations of particular occurrences. To explain a given airplane crash, for example, we seek "the cause"—a mechanical failure, perhaps, or pilot error. To explain a person's death, again we seek the cause—strangulation or drowning, for instance. I shall call the general view of scientific explanation that comes more or less directly from this fundamental intuition *the causal conception;* Michael Scriven (e.g., 1975) has been one of its chief advocates.

(2) The second of these basic intuitions is the notion that all scientific explanation involves *subsumption under laws.* This intuition seems to arise from consideration of developments in theoretical science. It has led to the general covering law conception of explanation, as well as to several formal models, including the well-known deductive-nomological and inductive-statistical models. According to this view, a fact is subsumed under one or more general laws if the assertion of its occurrence follows, either deductively or inductively, from statements of the laws (in conjunction, in some cases, with other premises). Since this view takes explanations to be arguments, I shall call it *the inferential conception;* Carl G. Hempel has been one of its ablest champions.[4]

Although the proponents of this inferential conception have often chosen to illustrate it with explanations of particular occurrences—e.g., why did the Bunsen flame turn yellow on this particular occasion?—the paradigms that give it strongest support are explanations of general regularities. When we look to the history of science for the most outstanding cases of scientific explanations, examples such as Newton's explanation of Kepler's laws of planetary motion or Maxwell's electromagnetic explanation of optical phenomena come immediately to mind.

It is easy to guess how Laplace might have reacted to my question about his demon, and to the two basic intuitions I have just mentioned. The superintelligence would have everything needed to provide scientific explanations. When, to mention one of Laplace's favorite examples, ([1820] 1951, pp. 3–6), a seemingly haphazard phenomenon, such as the appearance of a comet, occurs, it can be explained by showing that it actually conforms to natural laws. On Laplace's assumption of determinism, the demon possesses explanations of all happenings in the entire history of the world—past, present, and future. Explanation, for Laplace, seemed to consist in showing how events conform to the laws of nature, and these very laws provide the causal connections among the various states of the world. The Laplacian version of explanation thus seems to conform both to the causal conception and to the inferential conception.

Why, you might well ask, is not the Laplacian view of scientific explanation basically sound? Why do twentieth-century philosophers find it necessary to engage in lengthy disputes over this matter? There are, I think, three fundamental reasons: (1) the causal conception faces the difficulty that no adequate treatment of causation has yet been offered; (2) the inferential conception suffers from the fact that it seriously misconstrues the nature of subsumption under laws; and (3) both conceptions have overlooked a central explanatory principle.

The inferential view, as elaborated in detail by Hempel and others, has been the dominant theory of scientific explanation in recent years—indeed, it has become virtually "the received view." From that standpoint, anyone who had attained the epistemic status of Laplace's demon could use the known laws and initial conditions to predict a future event, and when the event comes to pass, the argument that enabled us to predict it would ipso facto constitute an explanation of it. If, as Laplace believed, determinism is true, then every *future* event would thus be amenable to deductive-nomological explanation.

When, however, we consider the explanation of past events—events that occurred earlier than our initial conditions—we find a strange disparity. Although, by applying known laws, we can reliable *retrodict* any past occurrence on the basis of facts subsequent to the event, our intuitions rebel at the idea that we can *explain* events in terms of subsequent conditions. (But see "Explanatory Asymmetry" [essay 10] for reasons that go beyond mere appeals to intuition.) Thus, although our inferences to future events qualify as explanations according to the inferential conception, our inferences to the past do not. Laplace's demon can, of course, construct explanations of past events by inferring the existence of still earlier conditions and, with the aid of the known laws, deducing the occurrence of the events to be explained from these conditions that held in the more remote past. But if, as the inferential conception maintains, explanations are essentially inferences, such an approach to past events seems strangely roundabout. Explanations demand an asymmetry not present in inferences.

When we drop the fiction of Laplace's demon and relinquish the assumption of determinism, the asymmetry becomes even more striking. The demon can predict the future and retrodict the past with complete precision and reliability. We cannot. When we consider the comparative difficulty of prediction versus retrodiction, it turns out that retrodiction enjoys a tremendous advantage. We have records of the past—tree rings, diaries, fossils—but none of the future. As a result, we can have extensive and detailed knowledge of the past that has no counterpart in knowledge about the future. From a newspaper account of an accident, we can retrodict all sorts of details that could not have been predicted an hour before the collision. But the newspaper story—even though it may *report* the explanation of the accident—surely does not *constitute* the explanation. We see that *inference* has a preferred temporal direction, and that *explanation* also has a preferred temporal direction. The fact that these two are opposite to each other is one thing that makes me seriously doubt that explanations are essentially arguments.[5] As we shall see, however, denying that explanations are arguments does not mean that we must give up the *covering law* conception. Subsumption under laws can take a different form.

Although the Laplacian conception bears strong similarities to the received view, a fundamental difference must be noted. Laplace apparently believed that the explanations provided by his demon would be *casual explanations,* and the laws invoked would be *casual laws.* Hempel's deductive-nomological explanations are often casually called "causal explanations," but this is not accurate. Hempel (1965a, pp. 352–354) explicitly notes that some laws, such as the ideal gas law,

$$PV = nRT,$$

are noncausal. This law states a mathematical functional relationship among several quantities—pressure P, volume V, temperature T, number of moles of gas n, universal gas constant R—but gives no hint as to how a change in one of the values would lead causally to changes in others. As far as I know, Laplace did not make any distinction between causal and noncausal laws. Hempel has recognized the difference, but he allows noncausal as well as causal laws to function as covering laws in scientific explanations.

This attitude toward noncausal laws is surely too tolerant. If someone inflates an air mattress of a given size to a certain pressure under conditions that determine the temperature, we can deduce the value of *n,* the amount of air blown into it. The *subsequent* values of pressure, temperature, and volume are thus taken to explain the quantity of air *previously* introduced. Failure to require covering laws to be causal laws leads to a violation of the temporal requirement on explanations. This is not surprising. The asymmetry of explanation is inherited from the asymmetry of causation, namely, that causes precede their effects. At this point, it seems to me, we experience vividly the force of the intuitions underlying the causal conception of scientific explanation.

There is another reason for maintaining that noncausal laws cannot bear the burden of covering laws in scientific explanations. Noncausal regularities, instead of having explanatory force that enables them to provide understanding of events in the world, cry out to be explained. Mariners, long before Newton, were fully aware of the correlation between the behavior of the tides and the position and phase of the moon. But inasmuch as they were totally ignorant of the causal relations involved, they rightly made no claim to any understanding of why the tides ebb and flow. When Newton provided the gravitational links, understanding was achieved. Similarly, I should say, the ideal gas law had little or no explanatory power until its causal underpinnings were furnished by the molecular-kinetic theory of gases. Keeping this consideration in mind, we realize that we must give at least as much attention to the explanations of regularities as we do to explanations of particular facts. I will argue, moreover, that these regularities demand causal explanation. Again, we must give the causal conception its due.

Having considered a number of preliminaries, I should now like to turn to an attempt to outline a general theory of causal explanation. I shall not be trying to articulate a formal model; I shall be focusing on general conceptions and fundamental principles rather than technical details. I am not suggesting, of course, that the technical details are dispensable—merely that this is not the time or place to try to go into them.

Developments in twentieth-century science should prepare us for the eventuality that some of our scientific explanations will have to be statistical—not merely because our knowledge is incomplete (as Laplace would have maintained), but rather because nature itself is inherently statistical. Some of the laws used in explaining particular events will be statistical, and some of the regularities we wish to explain will also be statistical. I have been urging that causal considerations play a crucial role in explanation; indeed, I have just said that regularities—and this certainly includes statistical regularities— require causal explanation. I do not believe there is any conflict here. It seems to me that, by employing a statistical conception of causation along the lines developed by Patrick Suppes (1970) and Hans Reichenbach (1956, chap. 4), it is possible to fit together harmoniously the causal and statistical factors in explanatory contexts. Let me attempt to illustrate this point by discussing a concrete example.

A good deal of attention has been given in the press to cases of leukemia in military personnel who witnessed an atomic bomb test (code name "Smoky") at close range in 1957.[6] Statistical studies of the survivors of the bombings of Hiroshima and Nagasaki have established the fact that exposure to high levels of radiation, such as occur in an atomic blast, is statistically relevant to the occurrence of leukemia—indeed, that the probability of leukemia is closely correlated with the distance from the explosion.[7] A clear pattern of statistical relevance relations is exhibited here. If somebody contracts leukemia, this fact may be explained by citing the fact that they were, say, 2 kilometers from the hypocenter at the time of the explosion. This relationship is further explained by the fact that individuals located at specific distances from atomic blasts of specified magnitude receive certain high doses of radiation.

This tragic example has several features to which I should like to call special attention:

(1) The location of the individual at the time of the blast is statistically relevant to the occurrence of leukemia; the probability of leukemia for a person located 2 kilometers from the hypocenter of an atomic blast is radically different from the probability of the disease in the population at large. Notice that the probability of such an individual contracting leukemia is not high; it is much smaller than one-half—indeed, in the case of Smoky it is much less than 1/100. But it is markedly higher than for a random member of the entire human population. It is the *statistical relevance* of exposure to an atomic blast, not a *high probability,* that has explanatory force.[8] Such examples defy explanation according to an inferential view that requires high inductive probability for statistical explanation.[9] The case of leukemia is subsumed under a statistical regularity, but it does not "follow inductively" from the explanatory facts.

(2) There is a *causal process* that connects the occurrence of the bomb blast with the physiological harm done to people at some distance from the explosion. High energy radiation, released in the nuclear reactions, traverses the space between the blast and the individual. Although some of the details may not yet be known, it is a well-established fact that such radiation does interact with cells in a way that makes them susceptible to leukemia at some later time.

(3) At each end of the causal process—i.e., the transmission of radiation from the bomb to the person—there is a *causal interaction.* The radiation is emitted as a result of a nuclear interaction when the bomb explodes, and it is absorbed by cells in the body of the victim. Each of these interactions may well be irreducibly statistical and indeterministic, but that is no reason to deny that they are causal.

(4) The causal processes begin at a central place, and travel outward at a finite velocity. A rather complex set of statistical relevance relations is explained by the propagation of a process, or set of processes, from a common central event.

In undertaking a general characterization of causal explanation, we must begin by carefully distinguishing between causal processes and causal interactions. The transmission of light from one place to another, and the motion of a material particle, are obvious examples of causal processes. The collision of two billiard balls, and the emission or absorption of a photon, are standard examples of causal interactions. Interactions are the sorts of things we are inclined to identify as events. Relative to a particular context, an event is comparatively small in its spatial and temporal dimensions; processes typically have much larger durations, and they may be more extended in space as well. A light ray, traveling to earth from a distant star, is a process that covers a large distance and lasts for a long time. What I am calling a causal process is similar to what Russell called a "causal line" (1948, p. 459).

When we attempt to identify causal processes, it is of crucial importance to distinguish them from pseudo-processes such as a shadow moving across the landscape. This can best be done, I believe, by invoking Reichenbach's *mark criterion*.[10] Causal processes are capable of propagating marks or modifications imposed on them; pseudo-processes are not. An automobile traveling along a road is an example of a causal process. If a fender is scraped as a result of a collision with a stone wall, the mark of that collision will be carried on by the car long after the interaction with the wall occurred. The shadow of a car moving along the shoulder is a pseudo-process. If it is deformed as it encounters a stone wall, it will immediately resume its former shape as soon as it passes by the wall. It will not transmit a mark or modification. For this reason, we say that a causal process can transmit information or causal influence; a pseudo-process cannot.[11]

When I say that a causal process has the capability of transmitting a causal influence, it might be supposed that I am introducing precisely the sort of mysterious power Hume warned us against. It seems to me that this danger can be circumvented by employing an adaptation of the 'at-at' theory of motion, which Russell used so effectively in dealing with Zeno's paradox of the arrow. (See "An 'At-At' Theory of Causal Influence" [essay 12].) The arrow—which is, by the way, a causal process—gets from one place to another by being *at* the appropriate intermediate points of space *at* the appropriate instants of time. Nothing more is involved in getting *from* one point *to* another. A mark, analogously, can be said to be propagated from the point of interaction at which it is imposed to later stages in the process if it appears *at* the appropriate intermediate stages in the process *at* the appropriate times without additional interactions that regenerate the mark. The precise formulation of this condition is a bit tricky, but I believe that the basic idea is simple, and the details can be worked out. (See "Causality without Counterfactuals" [essay 16].)

If this analysis of causal processes is satisfactory, we have an answer to the question, raised by Hume, concerning the connection between cause and effect. If we think of a cause as one event and of an effect as a distinct event, then the connection between them is simply a spatiotemporally continuous causal process. This sort of answer did not occur to Hume because he did not distinguish between causal processes and causal interactions. When he tried to analyze the connections between distinct events, he treated them as if they were chains of events with discrete links rather than processes analogous to continuous filaments. I am inclined to attribute considerable philosophical significance to the fact that each link in a chain has adjacent links, while the points in a continuum do not have next-door neighbors. This consideration played an important role in Russell's discussion of Zeno's paradoxes.[12]

After distinguishing between causal interactions and causal processes, and after introducing a criterion by means of which to discriminate the pseudo-processes from the genuine causal processes, we must consider certain configurations of processes that have special explanatory import. Russell noted that we often find similar structures grouped symmetrically about a center, for example, concentric waves moving across an otherwise smooth surface of a pond, or sound waves moving out from a central region, or perceptions of many people viewing a stage from different seats in a theater. In such cases, Russell (1948, pp. 460–475) postulates the existence of a central event—a pebble dropped into the pond, a starter's gun going off at a racetrack, or a play being performed upon the stage—from which the complex array emanates. It is noteworthy that Russell never suggests that the central event is to be explained on the basis of convergence of influences from remote regions upon that locale.

Reichenbach (1956, § 19) articulated a closely related idea in his *principle of the common cause*. If two or more events of certain types occur at different places, but occur at the same time more frequently than would be expected if they occurred independently, then this apparent coincidence is to be explained in terms of a common causal anteced-ent. If, for example, all of the electric lights in a particular area go out simultaneously, we do not believe that all of the bulbs just happened by chance to burn out at the same time. We attribute the coincidence to a common cause such as a blown fuse, a downed transmission line, or trouble at the generating station. If all of the students in a dormitory fall ill on the same night, it is attributed to spoiled food in the meal which all of them ate. Russell's similar structures arranged symmetrically about a center obviously qualify as the sorts of coincidences that require common causes for their explanations.

In order to formulate his common cause principle more precisely, Reichenbach de-fined what he called a *conjunctive fork*. Suppose we have events of two types A and B that happen in conjunction more often than they would if they were statistically indepen-dent of each other. For example, let A and B stand for color blindness in two brothers. There is a certain probability that a male, selected from the population at random, will have that affliction, but since it is often hereditary, occurrences in male siblings are not independent. The probability that both will have it is greater than the product of the two respective probabilities. In cases of such statistical dependencies, we invoke a common cause C that accounts for them; in this case it is a genetic factor carried by the mother. In order to satisfy the conditions for a conjunctive fork, events of the types A and B must occur independently in the absence of the common cause C—that is, for two unrelated males, the probability of both being color-blind is equal to the product of the two separate probabilities. Furthermore, the probabilities of A and B must each be increased above their overall values if C is present. Clearly the probability of color blindness is greater in sons of mothers carrying the genetic factor than it is among all male children regardless of the genetic makeup of their mothers. Finally, Reichenbach stipulates, the dependency between A and B is absorbed into the occurrence of the common cause C, in the sense that the probability of A and B given C equals the product of the probability of A given C and the probability of B given C. This is true in the color blindness case. Excluding pairs of identical twins, the question of whether a male child inherits color blindness from the mother, who carries the genetic trait, depends only on the genetic relationship between that child and his mother, not on whether other sons happened to inherit the trait.[13] Note that screening-off occurs here.[14] While the color blindness of a brother is statistically relevant to color blindness in a boy, it becomes irrelevant if the genetic factor is known to be present in the mother.

Reichenbach obviously was not the first philosopher to notice that we explain coinci-dences in terms of common causal antecedents. Leibniz postulated a preestablished harmony for his windowless monads which mirror the same world, and the occasional-ists postulated God as the coordinator of mind and body. Reichenbach (1956, pp. 162–163) was, to the best of my knowledge, the first to give a precise characterization of the conjunctive fork, and to formulate the general principle that conjunctive forks are open only to the future, not to the past. The result is that we cannot explain coincidences on the basis of future effects, but only on the basis of antecedent causes. A widespread blackout is explained by a power failure, not by the looting that occurs as a consequence. (*A* common effect *E* may form a conjunctive fork with *A* and *B*, but only if there is also a common cause *C*.) The principle that conjunctive forks are not open to the past accounts for Russell's principle that symmetrical patterns emanate from a central source; they do not converge from afar upon the central point. It is also closely related to the operation of the second law of thermodynamics and the increase of entropy in the physical world.

The common cause principle has, I believe, deep explanatory significance. Bas van Fraassen (1977) has subjected it to careful scrutiny, and he has convinced me that Reichenbach's formulation in terms of the conjunctive fork, as he defined it, is faulty. (We do not, however, agree about the nature of the flaw.) There are, its seems, certain sorts of causal *interactions* in which the resulting effects are more strongly correlated with one another than is allowed in Reichenbach's conjunctive forks. If, for example, an energetic photon collides with an electron in a Compton scattering experiment, there is a certain probability that a photon with a given smaller energy will emerge, and there is a certain probability that the electron will be kicked out with a given kinetic energy (see fig. 8.1). However, because of the law of conservation of energy, there is strong correspondence between the two energies: their sum must be close to the energy of the incident photon. Thus, the probability of getting a photon with energy E_1 and an electron with energy E_2, where $E_1 + E_2$ is approximately equal to E (the energy of the incident photon), is much greater than the product of the probabilities of each energy occurring separately. Assume, for example, that there is a probability of 0.1 that a photon of energy E_1 will emerge if a photon of energy E impinges on a given target, and assume that there is a probability of 0.1 that an electron with kinetic energy E_2 will emerge under the same circumstances (where E, E_1, and E_2 are related as the law of conservation of energy demands). In this case the probability of the joint result is not 0.01, the product of the separate probabilities, but 0.1, for each result will occur if and only if the other does.[15] The same relationships could be illustrated by such macroscopic events as collisions of billiard balls, but I have chosen Compton scattering because there is good reason to

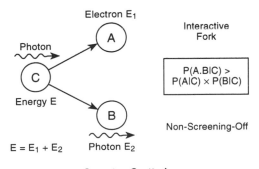

Compton Scattering

Figure 8.1

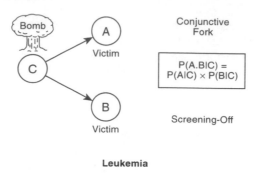

Leukemia

Figure 8.2

believe that events of that type are irreducibly statistical. Given a high energy photon impinging on the electron in a given atom, there is no way, even in principle, of predicting with certainty the energies of the photon and electron that result from the interaction.

This sort of interaction stands in sharp contrast with the sort of statistical dependency we have in the leukemia example (see fig. 8.2, which also represents the relationships in the color blindness case). In the absence of a strong source of radiation, such as the atomic blast, we may assume that the probability of next-door neighbors contracting the disease equals the product of the probabilities for each of them separately. If, however, we consider two next-door neighbors who lived at a distance of 2 kilometers from the hypocenter of the atomic explosion, the probability of both of them contracting leukemia is much greater than it would be for any two randomly selected members of the population at large. This apparent dependency between the two leukemia cases is not a direct physical dependency between them; it is merely a statistical result of the fact that the probability for each of them has been enhanced independently of the other by being located in close proximity to the atomic explosion. But the individual photons of radiation that impinge on the two victims are emitted independently, travel independently, and damage living tissues independently.

It thus appears that there are two kinds of causal forks: (1) Reichenbach's *conjunctive forks*, in which the common cause screens off the one effect from the other, which are exemplified by the color blindness and leukemia cases, and (2) *interactive forks*, exemplified by the Compton scattering of a photon and an electron. In forks of the interactive sort, the common cause does not screen off the one effect from the other. The probability that the electron will be ejected with kinetic energy E_2 given an incident photon of energy E is *not equal to* the probability that the electron will emerge with energy E_2 given an incident photon of energy E *and* a scattered photon of energy E_1. In the conjunctive fork, the common cause C absorbs the dependency between the effects A and B, for the probability of A and B given C is *equal to* the product of the probability A given C and the probability of B given C. In the interactive fork, the common cause C does not absorb the dependency between the effects A and B, for the probability of A and B given C is *greater than* the product of the two separate conditional probabilities.[16]

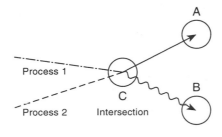

Causal Interaction

Figure 8.3

Recognition and characterization of the interactive fork enables us to fill a serious lacuna in the treatment up to this point. I have discussed causal processes, indicating roughly how they are to be characterized, and I have mentioned causal interactions, but have said nothing about their characterization. Indeed, the criterion by which I distinguished causal processes from pseudo-processes involved the use of marks, and marks are obviously results of causal interactions. Thus, my account stands in serious need of a characterization of causal interactions, and the interactive fork enables us, I believe, to furnish it.

There is a strong temptation to think of events as basic types of entities, and to construe processes—real or pseudo-—as collections of events. This viewpoint may be due, at least in part, to the fact that the space-time interval between events is a fundamental invariant of the special theory of relativity, and that events thus enjoy an especially fundamental status. I suggest, nevertheless, that we reverse the approach. Let us begin with processes (which have not yet been sorted out into causal and pseudo-) and look at their intersections. We can be reassured about the legitimacy of this new orientation by the fact that the basic space-time structure of both special relativity and general relativity can be built on processes without direct recourse to events.[17] An electron traveling through space is a process, and so is a photon; if they collide, that is an intersection. A light pulse traveling from a beacon to a screen is a process, and a piece of red glass standing in the path is another; the light passing through the glass is an intersection. Both of these intersections constitute interactions. If two light beams cross each other, we have an intersection without an interaction—except in the extremely unlikely event of a particle-like collision between photons. What we want to say, very roughly, is that when two processes intersect, and both are modified in such ways that the changes in one are correlated with changes in the other—in the manner of an interactive fork (see fig. 8.3)—we have a causal interaction. There are technical details to be worked out before we can claim to have a satisfactory account, but the general idea seems clear enough.[18]

I should like to commend the principle of the common cause—so construed as to make reference to both conjunctive forks and interactive forks—to your serious consideration.[19] Several of its uses have already been mentioned and illustrated. *First,* it supplies a schema for the straightforward explanations of everyday sorts of otherwise improbable coincidences. *Second,* by means of the conjunctive fork, it is the source of the fundamental temporal asymmetry of causality, and it accounts for the temporal asymmetry we impose on scientific explanations. *Third,* by means of the interactive fork, it provides the key to the explication of the concept of causal interaction.[20] These considerations certainly testify to its philosophical importance.

There are, however, two additional applications to which I should like to call attention. *Fourth,* as Russell (1948, pp. 491–492) showed, the principle plays a fundamental role in the causal theory of perception. When various observers (including cameras as well as human beings) arranged around a central region (such as a stage in theater-in-the-round) have perceptions that correspond systematically with one another in the customary way, we may infer, with reasonable reliability, that they have a common cause, namely, a drama being performed on the stage. This fact has considerable epistemological import.

Fifth, the principle of the common cause can be invoked to support scientific realism.[21] Suppose, going back to a previous example, we have postulated the existence of molecules to provide a causal explanation of the phenomena governed by the ideal gas law. We will naturally be curious about their properties—how large they are, how massive they are, how many there are. An appeal to Brownian motion enables us to infer such things. By microscopic examination of smoke particles suspended in a gas, we can ascertain their average kinetic energies, and since the observed system can be assumed to be in a state of thermal equilibrium, we can immediately infer the average kinetic energies of the molecules of the gas in which the particles are suspended. Since average velocities of the molecules are straightforwardly ascertainable by experiment, we can easily find the masses of the individual molecules, and hence, the number of molecules in a given sample of gas. If the sample consists of precisely one mole (gram molecular weight) of the particular gas, the number of molecules in the sample is Avogadro's number—a fundamental physical constant. Thus, the causal explanation of Brownian motion yields detailed quantitative information about the micro-entities of which the gas is composed.[22]

Now, consider another phenomenon which appears to be of an altogether different sort. If an electric current is passed through an electrolytic solution—for example, one containing a silver salt—a certain amount of metallic silver is deposited on the cathode. The amount deposited is proportional to the amount of electric charge that passes through the solution. In constructing a causal explanation of this phenomenon (known as electrolysis), we postulate that charged ions travel through the solution, and that the amount of charge required to deposit a singly charged ion is equal to the charge on the electron. The magnitude of the electron charge was empirically determined through the work of J. J. Thomson and Robert Millikan. The amount of electric charge required to deposit one mole of a monovalent metal is known as the Faraday, and by experimental determination it is equal to 96,485 coulombs. When this number is divided by the charge on the electron (-1.602×10^{-19} coulombs), the result is Avogadro's number. Indeed, the Faraday is simply Avogadro's number of electron charges.

The fundamental fact to which I wish to call attention is that the value of Avogadro's number ascertained from the analysis of Brownian motion agrees, within the limits of experimental error, with the value obtained by electrolytic measurement. Without a common causal antecedent, such agreement would constitute a remarkable coincidence.

The point may be put in this way. From the molecular-kinetic theory of gases we can derive the statement form, "The number of molecules in a mole of gas is _____." From the electrochemical theory of electrolysis, we can derive the statement form, "The number of electron charges in a Faraday is _____." The astonishing fact is that the same number fills both blanks. In my opinion the instrumentalist cannot with impunity ignore what must be an amazing correspondence between what happens when one scientist is watching smoke particles dancing in a container of gas while another scientist in a different laboratory is observing the electroplating of silver. Without an underlying causal mechanism—of the sort involved in the postulation of atoms, molecules, and ions—the coincidence would be as miraculous as if the number of grapes harvested in California in any given year were equal, up to the limits of observational error, to the number of coffee beans produced in Brazil in the same year. Avogadro's number, I must add, can be ascertained in a variety of other ways as well—e.g., X-ray diffraction from crystals—which also appear to be entirely different unless we postulate the existence of atoms, molecules, and ions. The principle of the common cause thus seems to apply directly to the explanation of observable regularities by appeal to unobservable entities. In this instance, to be sure, the common cause is not some sort of event; it is rather a common constant underlying structure that manifests itself in a variety of different situations.

Let me now summarize the picture of scientific explanation I have tried to outline. If we wish to explain a particular event, such as death by leukemia of GI Joe, we begin by assembling the factors statistically relevant to the occurrence—for example, his distance from the atomic explosion, the magnitude of the blast, and the type of shelter he was in. There will be many others, no doubt, but these will do for purposes of illustration. We must also obtain the probability values associated with the relevancy relations. The statistical relevance relations are statistical regularities, and we proceed to explain them. Although this differs substantially from things I have said previously, I no longer believe that the assemblage of relevant factors provides a complete explanation—or much of anything in the way of an explanation.[23] We do, I believe, have a bona fide explanation of an event if we have a complete set of statistically relevant factors, the pertinent probability values, *and* causal explanations of the relevance relations. Subsumption of a particular occurrence under statistical regularities—which, we recall, does not imply anything about the construction of deductive or inductive arguments—is a necessary part of any adequate explanation of its occurrence, but it is not the whole story. The causal explanation of the regularity is also needed. This claim, it should be noted, is in direct conflict with the received view, according to which the mere subsumption—deductive or inductive—of an event under a lawful regularity constitutes a complete explanation. One can, according to the received view, go on to ask for an explanation of any law used to explain a given event, but that is a different explanation. I am suggesting, on the contrary, that if the regularity invoked is not a causal regularity, then a causal explanation of that very regularity must be made part of the explanation of the event.

If we have events of two types, A and B, whose respective members are not spatiotemporally contiguous, but whose occurrences are correlated with one another, the causal explanation of this regularity may take either of two forms. Either there is a direct causal connection from A to B or from B to A, or there is a common cause C that accounts for the statistical dependency. In either case, those events that stand in the cause-effect relation to one another are joined by a causal process.[24] The distinct events *A, B,* and *C* that are thus related constitute interactions—as defined in terms of an interactive fork— at the appropriate places in the respective causal processes. The interactions *produce* modifications in the causal processes, and the causal processes *transmit* the modification. Statistical dependency relations arise out of local interactions—there is no action-at- a-distance (as far as macro-phenomena are concerned, at least)—and they are propa- gated through the world by causal processes. In our leukemia example, a slow neutron, impinging upon a uranium atom, has a certain probability of inducing nuclear fission, and if fission occurs, gamma radiation is emitted. The gamma ray travels through space, and it may interact with a human cell, producing a modification that may leave the cell open to attack by the virus associated with leukemia. The fact that many such interac- tions of neutrons with fissionable nuclei are occurring in close spatiotemporal proximity, giving rise to processes that radiate in all directions, produces a pattern of statistical dependency relations. After initiation, these processes go on independently of one an- other, but they do produce relationships that can be described by means of the conjunc- tive fork.

Causal processes and causal interactions are, of course, governed by various laws— e.g., conservation of energy and momentum. In a causal process, such as the propagation of a light wave or the free motion of a material particle, energy is being transmitted. The distinction between causal processes and pseudo-processes lies in the distinction be- tween the transmission of energy from one space-time locale to another and the mere appearance of energy at various space-time locations. When causal interactions occur— not merely intersections of processes—we have energy and/or momentum transfer. Such laws as conservation of energy and momentum are causal laws in the sense that they are regularities exhibited by causal processes and interactions. [This paragraph strongly anticipates the theory of causal processes advocated in "Causality without Counterfactuals" (essay 16).]

Near the beginning I suggested that deduction of a restricted law from a more general law constitutes a paradigm of a certain type of explanation. No theory of scientific explanation can hope to be successful unless it can handle cases of this sort.[25] Lenz's law, for example, which governs the direction of flow of an electric current generated by a changing magnetic field, can be deduced from the law of conservation of energy. But this deductive relation shows that the more restricted regularity is simply part of a more comprehensive physical pattern expressed by the law of conservation of energy. Sim- ilarly, Kepler's laws of planetary motion describe a restricted subclass of the class of all motions governed by Newtonian mechanics. The deductive relations *exhibit* what amounts to part-whole relationship, but it is, in my opinion, the physical relationship between the more comprehensive physical regularity and the less comprehensive physi- cal regularity that has explanatory significance. I should like to put it this way. An explanation may sometimes provide the materials out of which an argument, deductive or inductive, can be constructed; an argument may sometimes exhibit explanatory rela- tions. It does not follow, however, that explanations are arguments.

Earlier in this discussion I mentioned three shortcomings in the most widely held theories of scientific explanation. I should now like to indicate the ways in which the theory I have been outlining attempts to cope with these problems. First, the causal conception, I claimed, has lacked an adequate analysis of causation. The foregoing explications of causal processes and causal interactions were intended to fill that gap. Second, the inferential conception, I claimed, had misconstrued the relation of subsumption under law. When we see how statistical relevance relations can be brought to bear upon facts-to-be-explained, we discover that it is possible to have a *covering law conception* of scientific explanation without regarding explanations as arguments. The recognition that subsumption of narrower regularities under broader regularities can be viewed as a part-whole relation reinforces that point. At the same time, the fact that deductive entailment relations mirror these inclusion relations suggests a reason for the tremendous appeal of the inferential conception in the first place. Third, both of the popular conceptions, I claimed, overlooked a fundamental explanatory principle. That principle, obviously, is the principle of the common cause. I have tried to display its enormous explanatory significance. The theory I have outlined is designed to overcome all three of these difficulties.

On the basis of the foregoing characterization of scientific explanation, how should we answer the question posed at the outset? What does Laplace's demon lack, if anything, with respect to the explanatory aim of science? Several items may be mentioned. The demon *may* lack an adequate recognition of the distinction between causal laws and noncausal regularities; it *may* lack adequate knowledge of causal processes and of their ability to *propagate* causal influence; and it *may* lack adequate appreciation of the role of causal interactions in *producing* changes and regularities in the world. None of these capabilities was explicitly demanded by Laplace, for his analysis of causal relations—if he actually had one—was at best rather superficial.

What does scientific explanation offer, over and above the inferential capacity of prediction and retrodiction, at which the Laplacian demon excelled? It provides knowledge of the mechanisms of *production* and *propagation* of structure in the world. That goes some distance beyond mere recognition of regularities, and of the possibility of subsuming particular phenomena thereunder. It is my view that knowledge of the mechanisms of production and propagation of structure in the world yields scientific understanding, and that this is what we seek when we pose explanation-seeking why-questions. The answers are well worth having. That is why we ask not only "What?" but "Why?"

Notes

The author wishes to express his gratitude to the National Science Foundation for support of research on scientific explanation.

1. Such demonstrations may convince the mind, but they do not enlighten it; and enlightenment ought to be the principal fruit of true knowledge. Our minds are unsatisfied unless they know not only *that* a thing is but *why* it is" (Arnauld, [1662] 1964, p. 330).

2. The first edition appeared in 1892, the second in 1899, and the third was first published in 1911. In the preface to the third edition, Pearson remarked, just before the statement quoted in the text, "Reading the book again after many years, it was surprising to find how the

heterodoxy of the 'eighties had become the commonplace and accepted doctrine of to-day." Since the "commonplace and accepted doctrine" of 1911 has again become heterodox, one wonders to what extent such changes in philosophic doctrine are mere matters of changing fashion.

3. The classic paper by Carl G. Hempel and Paul Oppenheim, "Studies in the Logic of Explanation," which has served as the point of departure for almost all subsequent discussion, was first published in 1948.

4. Hempel's conceptions have been most thoroughly elaborated in his monographic essay, "Aspects of Scientific Explanation" (Hempel, 1965b).

5. In "A Third Dogma of Empiricism" (essay 6) I have given an extended systematic critique of the thesis (dogma?) that scientific explanations are arguments.

6. See *Nature*, 2 February 1978, p. 399.

7. Copi (1972, pp. 396–397) cites this example from Pauling (1959, pp. 85–91).

8. According to the article in *Nature* (note 6), "the eight reported cases of leukaemia among 2235 [soldiers] was 'out of the normal range.'" Dr. Karl Z. Morgan "had 'no doubt whatever' that [the] radiation had caused the leukaemia now found in those who had taken part in the manoeuvers."

9. Hempel's inductive-statistical model, as formulated in his (1965b), embodied such a high probability requirement, but in "Nachwort 1976," inserted into a German translation of this article (Hempel, 1977), this requirement is retracted.

10. Reichenbach ([1928] 1957, § 21) offers the mark criterion as a criterion for temporal direction, but as he realized in his (1956), it is not adequate for this purpose. I am using it as a criterion for a symmetric relation of causal connection.

11. See "Causal and Theoretical Explanation" (essay 7, § 3), for a more detailed discussion of this distinction. It is an unfortunate lacuna in Russell's discussion of causal lines—though one which can easily be repaired—that he does not notice the distinction between causal processes and pseudo-processes.

12. Russell (1922b, lecture 6). The relevant portions are reprinted in my anthology (1970a).

13. Reichenbach (1956, p. 159) offers the following formal definition of a conjunctive fork ACB:

$$P(A.B|C) = P(A|C) \times P(B|C)$$
$$P(A.B|\bar{C}) = P(A|\bar{C}) \times P(B|\bar{C})$$
$$P(A|C) > P(A|\bar{C})$$
$$P(B|C) > P(B|\bar{C})$$

14. C screens off A from B if

$$P(A|C.B) = P(A|C) \neq P(A|B)$$

15. The relation between $E_1 + E_2$ and E is an approximate rather than a precise equality because the ejected electron has some energy of its own before scattering, but this energy is so small compared with the energy of the incident X-ray or γ-ray photon that it can be neglected. When I refer to the probability that the scattered photon and electron will have energies E_1 and E_2 respectively, this should be taken to mean that these energies fall within some specified interval, not that they have exact values.

16. As the boxed formulas in figures 8.1 and 8.2 indicate, the difference between a conjunctive fork and an interactive fork lies in the difference between

$$P(A.B|C) = P(A|C) \times P(B|C)$$

and

$$P(A.B|C) > P(A|C) \times P(B|C)$$

One reason why Reichenbach may have failed to notice the interactive fork is that, in the special case in which

$$P(A|C) = P(B|C) = 1$$

the conjunctive fork shares a fundamental property of the interactive fork, namely, a perfect correlation between A and B given C. Many of his illustrative examples are instances of this special case.

17. For the special theory of relativity, this has been shown by John Winnie (1977), who utilizes much earlier results of A. A. Robb. For general relativity, the approach is discussed under the heading "The Geodesic Method" in Grünbaum (1973, pp. 735–750).

18. The whole idea of characterizing causal interactions in terms of forks was suggested by Philip von Bretzel (1977).

19. It strikes me as an unfortunate fact that this important principle seems to have gone largely unnoticed by philosophers ever since its publication in Reichenbach's *Direction of Time* (1956).

20. The interactive fork, unlike the conjunctive fork, does not seem to embody a temporal asymmetry. Thus, as seen in figure 8.3, the intersection C along with two *previous* stages in the two processes, constitute an interactive fork. This fact is, I believe, closely related to Reichenbach's analysis of intervention in The *Direction of Time* (1956, § 6), where he shows that this concept does not furnish a relation of temporal asymmetry.

21. Scientific realism is a popular doctrine, and most contemporary philosophers of science probably do not feel any pressing need for additional arguments to support this view. Although I am thoroughly convinced (in my heart) that scientific realism is correct, I am largely dissatisfied with the arguments usually brought in support of it. The argument I am about to outline seems to me more satisfactory than others.

22. [The situation regarding Brownian motion is not as simple as this. See Salmon (1984b, chap. 8) for a more detailed and accurate account.]

23. Compare Salmon (1971, p. 78), where I ask, "What more could one ask of an explanation?" The present essay attempts to provide at least part of the answer.

24. Reichenbach believed that various causal relations, including conjunctive forks, could be explicated entirely in terms of the statistical relations among the events involved. I do not believe this is possible; it seems to me that we must also establish the appropriate connections via causal processes. See "Causal Propensities: Statistical Causality versus Aleatory Causality" (essay 13) for further details.

25. [Note that in Hempel and Oppenheim ([1948] 1965, n. 33) the authors acknowledge their inability to account for explanations of this sort. To the best of my knowledge Hempel never returned to this issue.]

9

Deductivism Visited and Revisited

D eductive chauvinism—an apt term coined by J. Alberto Coffa—comes in two forms, inferential and explanatory. Inferential chauvinism is associated mainly with Karl R. Popper and his followers; surprisingly, perhaps, it was also espoused, though in a rather different form, by Bertrand Russell (see Salmon, 1974b). In this essay I shall focus on explanatory deductive chauvinism. I shall pay it a brief visit in the context of Laplacian determinism, and then revisit it in the context of modern science, where there is a strong presumption that inderterminism holds sway. In the indeterministic setting, I shall argue, explanatory deductive chauvinism cannot prevail.

1. Deductivism and Determinism

Explanatory deductivism flourishes in the fertile soil of Laplacian determinism. Consider Laplace's demon—the imaginary being that knows all of the laws of nature and the precise state S_1 of the universe at just one moment, and is capable of solving any mathematical problem that is in principle amenable to solution. This being can provide a deductive-nomological (D-N) explanation of any particular event whatever. If the event to be explained comes after the special momentary state S_1, known in complete detail by the demon, then the demon can summon laws that, in conjunction with facts drawn from S_1, entail that the explanandum event occurs. This deduction constitutes a D-N explanation. If the event to be explained precedes S_1, the demon can make a retrodictive inference to facts preceding the explanandum event. These preceding facts constitute initial conditions that, in conjunction with the pertinent laws, entail the occurrence of the event to be explained. The statements describing the preceding facts, in conjunction with applicable law statements, constitute the explanans for an acceptable D-N explanation. The same general strategy will work if the demon wants to explain some event contained within S_1. Earlier facts can be inferred, and these can be used to provide the desired explanations. Moreover, the demon can construct a D-N explanation of any regularity, provided it is not one of the basic regularities that constitute fundamental laws of nature. These most general regularities—for example, for Laplace, Newton's laws of motion—have enormous explanatory power, but they cannot be explained, because there are no laws of still greater generality under which they can be subsumed.

Laplace realized, of course, that human beings never achieve the capacities of the demon. At any given stage of human knowledge, there will be many facts for which our best scientific knowledge will not provide any D-N explanation. In such situations we may see fit to resort to probabilistic inferences or probabilistic explanations or both. Under these circumstances the probabilities that are invoked reflect our ignorance. Explanations that do not qualify as D-N are simply incomplete; they do not represent bona fide explanations of a different type, such as inductive-statistical (I-S). When confronted by a putative explanation that is not D-N, the natural and legitimate response is to ask what is missing—not to seek a different model to characterize acceptable scientific explanations.

Although Carl G. Hempel gave the first precise, detail formulation of the D-N model of scientific explanation, and defended it ably against many objections, he was not a partisan of explanatory deductivism. The classic paper, Hempel and Oppenheim ([1948] 1965), contains an explicit statement to the effect that not all legitimate scientific explanations fit the D-N model; instead, some are probabilistic or statistical (Hempel 1965a, pp. 250–251). The task of analyzing such explanations and providing a precise model was, to be sure, not attempted in that paper; it was postponed until another time. The I-S model was first presented (though not under that name) in Hempel (1962a); an improved version was offered in his *Aspects of Scientific Explanation* (1965a). In spite of these facts, I have often (even after 1962) encountered the belief that Hempel was committed to explanatory deductivism.

According to Hempel's theory of D-N explanation, the explanandum may be either a particular event or a general regularity. In either case the same model applies. The explanation is a valid deduction; at least one statement of a law occurs essentially in the explanans; and the conclusion states that the event (or fact) to be explained actually occurs (or obtains). Similarly, Hempel (1962a) maintained that in the case of probabilistic or statistical explanation the explanandum may be either a particular event or a general statistical regularity—but there is a fundamental difference.

When the explanandum is a statistical generalization, the explanation may be deductive. For example, it might be wondered why a player who tosses a pair of standard dice twenty-four times has less than a fifty-fifty chance of getting double six at least once. It is reported that this problem bothered the Chevalier de Méré in the seventeenth century, and that Pascal was able to solve it by proving, with statistical generalizations about standard dice, that twenty-five throws are required to have better than a fifty-fifty chance for double six (Salmon, 1967, p. 68). Although Hempel called attention to statistical explanations of this sort in (1962a, p. 122), the formal model—known as deductive-statistical (D-S) explanation—was first presented three years later (Hempel, 1965b).

When, however, we explain some particular event or fact on the basis of statistical laws, the explanation cannot have the form of a deductive argument but must rather assume an inductive form. One of Hempel's most famous examples (1965a, pp. 381–382 and 394–398) accounts for the quick recovery of John Jones from a streptococcus infection on the basis of the fact that penicillin was administered and the statistical regularity that almost all (but not all) such infections clear up promptly after treatment with penicillin. Explanations of this type have the inductive form often called "statistical syllogism"; they conform to the I-S model of explanation.

The explanatory deductivist can comfortably admit three kinds of explanation (at least those who are willing to overlook the problem stated in footnote 33 of Hempel and Oppenheim [[1948] 1965a]: see "Scientific Explanation: How We Got from There to Here" [essay 19]): explanation of individual events by subsumption under universal laws, explanation of universal regularities by subsumption under more general universal laws, and explanation of statistical regularities by subsumption under more general statistical laws. Indeed, to handle these three types, there is no need for any model beyond the D-N, for in all of them some fact (particular or general) is explained by deduction from premises that include essentially at least one statement of a law. The D-S model as a separate entity is not needed.

The point at which the explanatory deductivist must take umbrage is when models of explanation—such as the I-S model—that characterize explanations as nondeductive arguments are introduced.[1] If there are bona fide scientific explanations that conform to the I-S model, that fact reveals a crucial limitation on deductivism.

Let us reconsider the streptococcus example. When Hempel first presented this case, it was known that most streptococcus infections could be handled effectively by the administration of penicillin, but that some strains of these bacteria are penicillin resistant. If Jones is infected by a non–penicillin-resistant strain, then his prompt recovery after treatment is practically certain, though not absolutely certain. In response to this example, the deductivist can say that the 'explanation' of the quick recovery on the basis of treatment with penicillin has explanatory value, though it is not a complete explanation. The 'explanation' in terms of a non–penicillin-resistant strain of bacteria and treatment by penicillin is more complete, and hence possesses more explanatory value than the 'explanation' in terms of treatment by penicillin alone. But it falls short of being a bona fide explanation. Most of us would agree that it is very probably incomplete, for there is good reason to suppose that further research will reveal additional factors that help to determine which victims of streptococcus infection by non–penicillin-resistant strains will recover quickly after treatment with penicillin and which will not.

Cases of this sort need not cause the deductivist any discomfort. The inferential deductivist may readily grant that many proffered arguments—ones that seem compelling to common sense—are actually enthymemes. Enthymemes are incomplete deductive arguments; it is possible to complete them by furnishing missing premises.[2] Similarly, the explanatory deductivist can, with impunity, accept explanations that conform to the I-S model, as long as they are regarded as incomplete D-N explanations. As such, these I-S explanations are literally enthymemes. In many such cases we do not yet have access to the additional true premises needed to transform the enthymeme into an acceptable D-N explanation, but we can be reasonably confident that, in principle, the required premises can be established. As J. Alberto Coffa argued brilliantly (1974), Hempel achieved essentially the same result through his doctrine of *essential ambiguity* of I-S explanation and his consequent doctrine of *essential epistemic relativization* of I-S explanation. Although Hempel is not a determinist, his treatment of the status of I-S explanations fits harmoniously with determinism; in fact, it is tempting to suspect that Hempel belongs to the class of philosophers—characterized by Peter Railton—who, though they do not hold the doctrine of determinism, are nevertheless held by it (1980, p. 241). Determinism assures us that there are enough unknown factors to complete any I-S explanation, and thus to transform it into a D-N explanation. So, for the determinist, there are no complete I-S explanations, because whenever any such explanation is completed it is automatically transformed into a D-N explanation. This makes I-S explanation completely parasitic on D-N explanation—a move that is entirely congenial to the deductivist.

2. Sufficient versus Necessary Conditions

Let us now abandon the Laplacian *Weltanschauung* and begin our discussion of deductivism in the more modern, indeterministic context. It seems obvious that indeterminism is compatible with the existence of some universal laws, for there are cases in which an event has a sufficient cause but no necessary cause or a necessary cause but no sufficient cause. In discussing the nature of causality, J. L. Mackie (1974, pp. 40–41) invites us to consider three machines, two of which are indeterministic. These two constitute special cases that have a profound bearing on explanatory deductivism.

> [L]et us consider three different shilling-in-the-slot machines, *K, L,* and *M.* Each of them professes to supply bars of chocolate; each of them also has a glass front, so that its internal mechanism is visible. But in other respects, the three are different. *K* is deterministic, and conforms to our ordinary expectations about slot-machines. It does not always produce a bar of chocolate when a shilling is put in the slot, but if it does not there is some in principle discoverable fault or interference with the mechanism. Again, it can be induced to emit a bar of chocolate without a shilling's being inserted, for example by the use of some object which sufficiently resembles a shilling. . . . Inserting a shilling is neither absolutely necessary nor absolutely sufficient for the appearance of a bar of chocolate, but in normal circumstances it is both necessary and sufficient for this. . . . *L,* on the other hand, is an indeterministic machine. It will not, indeed, in normal circumstances produce a bar of chocolate unless a shilling is inserted, but it may fail to produce a bar even when this is done. And such failure is a matter of pure chance. *L*'s failures, unlike *K*'s, are not open to individual explanation even in principle, though they may be open to statistical explanation. With *L,* in normal circumstances, putting a shilling in the slot is necessary, but not sufficient, for the appearance of a bar of chocolate. *M* is another indeterministic machine, but its vagaries are opposite to *L*'s. *M* will, in ordinary circumstances, produce a bar of chocolate whenever a shilling is inserted; but occasionally, for no reason that is discoverable even in principle, the mechanism begins to operate even though nothing has been inserted, and a bar of chocolate comes out. With *M,* in normal circumstances, putting a shilling in the slot is sufficient, but not necessary, for the appearance of a bar of chocolate.

Suppose, now, that a shilling is inserted in one of these machines and a chocolate bar is forthcoming. Concerning machine *K,* Mackie says, *without hesitation,* that putting a shilling in the slot under normal circumstances causes the candy bar to be emitted, for insertion of the coin is necessary and sufficient for that result under those conditions. Concerning machine *L,* Mackie concludes, *after deliberation,* that insertion of the shilling causes the candy bar to come out, for without the shilling no chocolate bar would have appeared. Concerning machine *M,* Mackie concludes, *again after deliberation,* that we are not entitled to maintain that the insertion of the coin is the cause, for we have no way of knowing whether a candy bar would have been forthcoming at that time even if no coin had been inserted. I am inclined to agree with Mackie's assessment of these cases, as long as we are not entertaining a probabilistic concept of causality.

If we shift our attention from causality to explanation, an interesting wrinkle appears. The deductivist can maintain that, with the deterministic machine *K,* we can explain why the chocolate bar is forthcoming in terms of the insertion of the coin in a machine of that sort. Causation and explanation coincide in this case. With the other two machines this correspondence does not obtain. Applying the D-N schema, we find that we can explain the appearance of the candy bar in terms of the insertion of the shilling if putting in the coin is sufficient for the result. This is the situation for machine *M.* Moreover, if putting in the coin is necessary but not sufficient, it cannot provide a D-N explanation of the emission of the candy bar. This characterizes machine *L.* To those of us who see a close relationship between causation and explanation, this outcome seems wrong. If one were to accept Mackie's account of causality and the deductivist's account of explanation, it would be necessary to conclude that putting the coin in machine *L causes* the candy bar to come out but does not *explain* its appearance, whereas putting the coin in machine *M explains* the appearance of the chocolate bar but does not *cause* it to emerge. This result is quite paradoxical.

The difficulty that arises in connection with machine *M,* it should be noted, strongly resembles a well-known problem for D-N explanation, namely, the problem of preemption. Consider a California ticky-tacky house built near the San Andreas Fault. If an earthquake measuring 7.0 or greater on the Richter scale is centered nearby, the house will collapse. Likewise, if a tornado touches down right there, the house will also collapse. One day a major earthquake does occur in that area and the house collapses. We have all the makings of a D-N explanation. However, the collapse of the house is not a result of the earthquake, for a tornado knocks it down just before the earthquake occurs. In the case of machine *M,* it may be that the candy bar would have been delivered quite by chance, and thus that the insertion of the shilling had nothing to do with its appearance.

Since Mackie's machines—especially *L* and *M*—may seem rather artificial, let us consider a more realistic scientific example, one in which it is plausible to suppose that genuine quantum indeterminacy is involved. In 1972–1973, Coffa and I signed up informally as lab partners in a course in which some of the landmark experiments in early twentieth-century physics were repeated (but not necessarily in the original way). As our first project we did a Compton scattering experiment, using a monochromatic gamma ray beam. It was necessary, of course, to detect scattered gamma ray photons. We constructed a detector consisting of a sodium iodide crystal, which scintillates when the scattered photons impinge upon it, and a photomultiplier tube, which detects the scintillations and sends a signal to a counter. Such detectors are not perfect.

Suppose, for the sake of argument, that our detector has an efficiency of 0.95 (a realistic value for commercially available detectors)—that is, it responds, on the average, to ninety-five out of a hundred photons impinging upon it. If it fails to detect some photons but never responds when no such photon is present, it seems that we would have little hesitation in explaining a detection event in terms of the incidence of a photon. If someone asked why the counter clicked, it would be appropriate to answer that a photon interacted with the sodium iodide crystal, which produced a flash of light that passed through the window of the photomultiplier tube and struck a metal surface, which in turn ejected a photoelectron. This electron initiated a cascade of secondary electrons, thus producing a brief pulse of electricity that activated the counter and made it click. Even if a small percentage of photons fail to activate the detector, we explain each response on the basis of an incident photon. It should be carefully noted that this explanation does not fit the D-N pattern, because the impinging of the photon on the device is a necessary but not sufficient condition for the occurrence of the click. This case corresponds to Mackie's machine *L.*

Suppose instead—again, for the sake of argument—that the detector sometimes produces a click when no scattered photon is present—for example, on account of a stray thermal electron in the photomultiplier tube—but never fails to detect an entering photon that was Compton-scattered into the detector. Again, it is realistic to suppose that five counts out of a hundred are spurious. It seems highly dubious that the deductivist, if asked to explain a given click, would be right in insisting that the story told in the preceding paragraph constitutes a bona fide explanation, even though it now fits the D-N pattern by providing a sufficient, but not necessary, condition for the click. In this case, if someone asked why the click occurred, it would seem far more appropriate to say that it might have been because a scattered photon entered the detector, producing a scintillation that then produced a cascade of electrons in the photomultiplier tube, or it might have been because of a thermal electron in the photomultiplier tube, though most probably it was on account of a scattered photon. This explanation obviously fails to fit the D-N pattern. It seems irrelevant to the explanation that every photon entering the tube produces a click. This case corresponds to Mackie's machine *M.*

Mackie's two indeterministic candy machines and the gamma ray detectors resurrect—rather surprisingly, I think—the old problem about the explanatory value of necessary conditions. Consider Michael Scriven's syphilis/paresis example (1959). Paresis is one form of tertiary syphilis, and about one victim of latent syphilis out of four (untreated by penicillin) develops paresis. No one else ever contracts paresis. There is no known way to predict which victims of latent untreated syphilis will develop paresis and which will not. Scriven maintained that latent untreated syphilis explains paresis in those cases in which it occurs. Deductivists steadfastly claimed that we cannot explain paresis unless we can discover other factors that will serve to distinguish those victims of latent untreated syphilis who will develop paresis from those who will not. If the world is deterministic—at least with respect to syphilis and paresis—then there will be characteristics that make just this distinction; but if paresis occurs randomly and intrinsically unpredictably among victims of latent untreated syphilis, no such in-principle-discoverable characteristics exist. In this latter case, latent untreated syphilis is a necessary cause of paresis, and there are no sufficient causes. Under these circumstances, I would suggest, untreated latent syphilis sometimes causes paresis and sometimes does not; but in all cases in which paresis occurs, it is caused by latent untreated syphilis, and latent un- treated syphilis explains the paresis. But for the latent untreated syphilis, the paresis would not have occurred.

The same sort of issue has arisen over the years in connection with the possibility of functional explanations. It might be asked, for example, why the jackrabbit—an animal that inhabits hot desert regions—has such large ears. The answer is that they act as devices for the dissipation of excess body heat. When the jackrabbit becomes over- heated, it seeks the shade and dilates the many blood vessels in its ears. Heat is carried by the blood from the interior parts of the body to the ears and radiated into the environ- ment. This provides a way for the animal to regulate the temperature of its body. I gather that elephants' ears function in a similar manner.

The trouble with functional explanations, from the standpoint of the deductivist, is that they furnish necessary conditions where the D-N model requires sufficient conditions. If we claim that regulation of body temperature explains the size of the jackrabbit's ears, then we can say that having big ears is a sufficient condition for regulation of body temperature, but we cannot say that regulation of body temperature is a sufficient condition for big ears, because there are other mechanisms of body temperature control—for example, panting, perspiring, or nocturnal habits.

One response that is often given by deductivists consists of the claim that functional explanations of the foregoing sort are incomplete explanations. When we know more about the development of body temperature control mechanisms we will find additional antecedent factors that determine that humans perspire, dogs pant, and mice avoid the heat of the day by staying in their shelters. It may turn out that, in the evolutionary context in which the large ears of the jackrabbit emerged, other mechanisms for the control of body temperature were not genetically available.

While I think that there is considerable hope in dealing with the syphilis/paresis example in terms of as-yet-unknown factors, I am far more skeptical about the feasibility of dealing with functional explanations in that way. One reason for this skepticism lies in the fact that some functional explanations involve characteristics that result from biolog- ical evolution. Since mutations play a vital role in such developments, we must not neglect that fact that these mutations arise by chance. This appeal to chance may sometimes be merely a euphemism for our ignorance. However, it is known that muta- tions can be produced by cosmic rays, and in such cases the fundamental quantum indeterminacy may well prevail.

There is another—far more basic—reason. Functional explanations are found in many scientific contexts, in social and behavioral sciences as well as evolutionary biology. Ethnologists, for example, explain social practices on the basis of their roles in fostering social cohesiveness. Freudian psychologists explain dreams in terms of wish fulfillments. In most, if not all, such cases, functional equivalents—alternative mecha- nisms that would achieve the same result—appear to be possible. I find no reason to suppose that such explanations are acceptable only on the supposition that it is possible in principle to show that one rather than another of these alternatives had to occur in the circumstances. Totemic worship of the wolf can be explained in terms of its social function even if totemic worship of the bear would have worked as well. Functional explanations do not cease to be explanations just because we cannot rule out functional equivalents.

We have looked at several examples in which there is a necessary cause but no sufficient cause, known or unknown, that invariably produces the event to be explained, and some in which there is a sufficient cause but no necessary cause, known or unknown, without which the event to be explained could not occur. Such examples necessarily involve some sort of indeterminism. In these indeterministic settings, it appears that necessary causes have at least some degree of explanatory force, but sufficient causes do not. This raises a serious question about the apparent plausibility of the deductivist's demand for sufficient causes in D-N explanations.

Consider the following example in contrast to the foregoing ones. On a stormy summer evening Smith's barn catches fire and burns to the ground. An insurance investigator wants to find out why. There are many possible causes of such a fire: lightning striking the barn, deliberate arson, a lighted cigarette discarded by a careless smoker, spontaneous combustion of green hay stored in the barn, and many others. The investigator establishes the fact that, even though there was a thunderstorm, lightning did not strike in the vicinity of the barn at the time the fire started; moreover, the barn was protected by adequate lightning rods. It is further established that no smoker was in the barn at the time. Careful examination of the remains establishes the absence of any incendiary substance such as gasoline. The burn pattern fits with the supposition that the cause was spontaneous combustion. This explanation satisfies the insurance investigator.

It would be natural at this point to ask what difference between the photon detector example and the burning barn example accounts for the fact (as I see it, anyhow) that a necessary condition has explanatory force in the former while a sufficient condition has explanatory force in the latter. One important difference is this: in the case of the barn it is plausible to assume that there always is a sufficient cause, and that different causes of burned barns leave different traces in the effect. The photon detector does not have this characteristic. The click that results from a genuine photon detection is utterly indistinguishable from the click that results from a spurious count.

One outstanding feature of examples such as the burned barn is that they are assumed to be fully deterministic. Several possible sufficient conditions have to be considered, but (assuming no overdetermination or preemption) it is supposed that one and only one must have been present. Indeed, the presumption is that conditions surrounding this particular occurrence can be specified in enough detail to establish the existence of a unique necessary and sufficient cause. This situation is a deductivist's dream come true; the case fits the D-N model to a tee. When, however, we allow for the possibility of ineluctable indeterminacy, the D-N model loses a good deal of its earlier appeal.

In this section I have discussed a very special aspect of indeterminism—one in which I did not need to make reference to any genuinely statistical laws. I found a significant limitation of explanatory deductivism, namely, its demand for sufficient causes in cases in which not they but rather necessary causes seem manifestly to have explanatory import. In the next section I shall deal with issues that arise directly from the statistical character of basic physical laws. Additional limits of deductivism will appear.

3. Scientific Explanation and Irreducibly Statistical Laws

If Laplacian determinism is fertile soil for explanatory deductivism, it might be supposed that the indeterministic context of twentieth-century physics would prove quite barren. On the standard Born–Pauli statistical interpretation of quantum mechanics, at least some of its basic laws are *ineluctably* statistical. Moreover, it has achieved important explanatory successes. Consider some of the most pressing problems that led to the discovery of quantum mechanics. Classical physics could not explain the energy distribution in the spectrum of blackbody radiation; quantum mechanics could. Classical physics could not explain the stability of the hydrogen atom or the discrete character of the hydrogen spectrum; quantum mechanics could. Similar remarks can be made about specific heats, the photoelectric effect, and radioactivity. Quantum mechanics thus seems to provide explanatory power of an irreducibly statistical kind. Modern physics therefore appears to mandate a theory of statistical explanation.

Even if it should turn out that quantum mechanics in the standard interpretation is incorrect and has to be replaced by a deeper, deterministic theory, it is surely advisable at this juncture to leave the door open to indeterminism and not to close off that possibility by some a priori fiat. If we have a model of statistical explanation, yet determinism is true after all, then at worst we have a model of scientific explanation that is not fundamental. If we have no model of statistical explanation, yet indeterminism is true, then we will have failed to comprehend a fundamental type of scientific explanation. It is obvious, in my opinion, that the second error would be much more serious than the first.

In an earlier section I discussed Hempel's example of the streptococcus infection and considered ways in which the deductivist might handle it. This example did not appear to present any insuperable difficulties. Not all of Hempel's classic examples can be handled as readily by the deductivist. Consider a particular 10 milligram sample of radon, a radioactive element with a half-life of 3.82 days (Hempel, 1965a, p. 392). The statistical law concerning the half-life of radon could be invoked to explain why, after 7.64 days, this sample contains about 2.5 milligrams of radon. Radioactive decay is one of the phenomena in which the ineluctably statistical character of quantum mechanics is displayed. To the best of our current knowledge, there is no strict law that determines just which nuclei will decay within a given period of time and which will remain intact. Indeed, the best current theory gives deep reasons for maintaining that there cannot be any deterministic explanation. Each atom has a fifty-fifty chance of decaying within any period of 3.82 days; that is the whole story. It is, consequently, impossible to deduce the statement that approximately 2.5 milligrams remain after 7.64 days from the given laws and initial conditions.

As I mentioned earlier, Hempel offered two models of statistical explanation, D-S and I-S. The radon example does not qualify as a D-S explanation, for it offers an explanation of the fact that this particular sample contains approximately 2.5 milligrams of radon at the end of a period of 7.64 days. Using the information given in the example, it would be possible to construct a D-S explanation of the fact that, in general, 10 milligram samples of radon almost always contain about 2.5 milligrams of radon 7.64 days later. However, it is impossible to deduce from the information given anything about the radon content of this particular sample (beyond, perhaps, the conclusion that the value must be somewhere between 0 and 10 milligrams). Moreover, even if we agree with Hempel's analysis of this example as an instance of I-S explanation, it seems plausible to deny that it is epistemically relativized and, hence, an incomplete D-N explanation, for we have good reasons for believing that there are no additional factors, as yet unknown, that determine precisely which radon atoms will decay in 7.64 days, or precisely how many. Instead, it is a strong candidate to qualify as a bona fide complete I-S explanation.

Even philosophers who are uncommitted to determinism find basic difficulties with the I-S model. In his revised presentation of that model Hempel (1965a) required that the explanandum be highly probable in relation to the explanans. One fundamental problem concerns the question how high is high enough. Take, for example, the explanation of the melting of an ice cube on the basis of the fact that it was placed and left in a tumbler of tepid water. Even in classical statistical mechanics we must admit that the melting is not a necessary consequence of the temperature of the surrounding water, for within the class of states of lukewarm water there is a minute fraction that will not eventuate in the melting of the ice cube within a given span of time. This is a consequence of the fact that the entropy in the system consisting of the water and the ice cube is overwhelmingly likely, but not absolutely certain, to increase. Because of the incredibly small difference between unity and the probability of melting, most philosophers will, I am confident, have little if any uneasiness in accepting this explanation as legitimate, even though strictly speaking it does not qualify as D-N. Hempel's example of radon decay would count as legitimate in the same fashion. These examples are in the same category as Richard Jeffrey's "beautiful cases," in which "the probability of the phenomenon to be explained is so high, given the stochastic law governing the process that produces it, as to make no odds in any gamble or deliberation" ([1969] 1971, p. 27). As Jeffrey remarks, such cases surely provide "practical certainty"—though I am not sure what role that concept can play in the deductivist's scheme of things. A serious attempt to deal with this question has been made by J. W. N. Watkins (1984, pp. 242–246).

Let us consider the opposite extreme. It seems unlikely that anyone would admit a probability value in the closed interval from zero to one-half as a high probability; a minimal demand would be that the explanandum be more likely than not relative to the explanans. However, most people would find it hard to accept an I-S explanation of some event if the explanans renders it just a little more likely than not to occur. Few would be willing to agree that the sex of a newborn baby boy is explained in terms of the slight superiority of the statistical probability for boys over girls. Most people would have similar reluctance to admit an I-S explanation of the outcome heads on a toss of a coin, given that the coin has only a very slight bias toward heads—say, fifty-one to forty-nine.

The problem we confront involves a basic slippery slope. A probability virtually indistinguishable from one seems adequate for an I-S explanation; a probability scarcely above one-half seems clearly inadequate. If I-S explanations (in their full-blooded, unrelativized sense) are admissible at all, it is hard to see where to draw the line between probability values that are high enough and those that are not.

At various times I have invited consideration of examples in which there are two alternative outcomes having statistical probabilities of three-fourths and one-fourth, respectively. One of these examples—taken from Hempel (1965a, pp. 391–392)— involves a Mendelian genetic experiment on the color of blossoms of pea plants in a population having a certain genetic makeup, in which the probability of a red blossom is three-fourths and the probability of a white blossom is one-fourth. With admitted over-simplification, the occurrence of brown or blue eye color in children of brown-eyed, heterozygous parents has the same probability distribution (Salmon, 1985b). Similarly, the probability of radioactive disintegration of an unstable nucleus—for example, one of Hempel's radon nuclei—within a period of two half-lives, or of its survival for at least that length of time, also exhibits the same probability distribution (Hempel 1965a, p. 392; see also "Scientific Explanation: Three Basic Conceptions" [essay 20]).

It should be clearly noted that Hempel's use of these examples is quite different from mine. He does not discuss the explanation of a single instance of a red blossom or the decay of an individual nucleus. In the radon example, he specifies the size of the sample (10 milligrams); it contains about 3×10^{19} atoms. Obviously, as previously remarked, in a sample of this size the probability of getting approximately seventy-five percent decays in two half-lives is extremely close to one.[3]

Hempel's discussion of the distribution of red and white blossoms involves samples of unspecified size taken from a finite population of pea plants (1965a, pp. 391–392). It therefore nicely illustrates the problem of the slippery slope. In a sample containing only one member, there is a probability of three-fourths that the distribution of red flowers will approximate the probability of red in the population, for a frequency of one is closer to three-fourths than is a frequency of zero. Is three-fourths large enough to qualify as a high probability? I shall return to that question shortly. In the meantime it should be noted that, in a sample of ten, there is a probability of 0.42 that the frequency of red will be in the range 0.675 to 0.825—that is, within ten percent of 0.75. In a sample of fifty, the probability is 0.78 that the frequency will lie within that range; in a sample of one hundred, the probability is 0.92; in a sample of five hundred, the probability is virtually indistinguishable from one. The general situation is clear. Choose any degree of approximation you wish. By taking larger and larger samples, the probability that the frequency matches three-fourths within that degree of approximation may be made as close to unity as you wish. Therefore, if there is a probability, less than one, that qualifies as a high probability, we can find a finite sample size in which we will have a high probability of getting approximately three-fourths red blossoms. If your chosen degree of approximation is ten percent, and if any probability exceeding 0.9 is high enough for I-S explanation, when we can give an I-S explanation of the fact that the frequency of red in a sample of one hundred lies between 0.675 and 0.825, but we cannot give that kind of explanation of the fact that the frequency of red in a sample of fifty lies between thirty-four and forty-one. The question is how to draw the line between probabilities that are high enough for I-S explanation and those that are not in any nonarbitrary way.

It is interesting to note that almost no one seems to accept three-fourths as a sufficiently high value; nearly everyone seems to maintain that, if there is any explanation of the color of a single blossom (or the blossoms on a single plant), it must be in terms of the details of the chromosomal processes that causally determine the color. The reluctance to allow explanation of the statistical makeup of larger samples in terms of the probability distribution is not nearly as great.

There is another way to look at these cases in which the two outcomes have probabilities of three-fourths and one-fourth. It involves an important symmetry consideration.

Given our knowledge of Mendelian genetics and the ancestors of the pea plants, it seems clear to me that we understand the occurrence of a white blossom just as well or just as poorly as we understand the occurrence of a red blossom. We can explain either red or white—whichever happens to occur—or we can explain neither.[4] Similarly, our knowledge of radioactive disintegration provides us with equally good understanding either of disintegration of a radon nucleus within two half-lives or of survival intact for the same period.

The conclusion I have drawn from this symmetry argument (as well as many other considerations) is that the high probability requirement is not a suitable constraint to impose on statistical explanations. My suggestion has been to adopt something akin to the statistical-relevance (S-R) model. Deductivists have been understandably reluctant to accept this suggestion, for not only does it reject the thesis that explanations are deductive arguments, but also it rejects the thesis that explanations are arguments of any variety.

A frequent response to this symmetry argument is to deny that we can give explanations of any of these particular cases. What can be explained, many people say, is a statistical distribution in a class of individuals (see Watkins, 1984, chap. 6, for a very clear account). As I have noted, Hempel introduced the genetic example to illustrate the explanation of the proportion of red blossoms in a limited sample of pea plants. He introduced the nuclear disintegration example to illustrate the explanation of the proportion of radon atoms in a particular sample that undergo transmutation. He fully realized that no deductive explanation can be given of the statistical distribution in any sample that is smaller than the population from which it is drawn. These examples were explicitly offered as instances of I-S explanation in which the probability of the explanandum relative to the explanans falls short of unity.

The symmetry consideration is rather far-reaching. For any event E that occurs with probability p, no matter how high—provided $p < 1$—there is a probability $(1 - p) > 0$ that E does not occur. For any unstable nucleus, the probability that it will disintegrate before n half-lives have transpired is $1 - \frac{1}{2^n}$, which can be made arbitrarily close to one by choosing an n large enough. If this symmetry argument is cogent, it shows that there is no probability high enough for I-S explanations, because the larger p is, the smaller is $1 - p$. The symmetry argument says that we can explain the improbable outcome whenever it occurs if we can explain the highly probable outcome. This argument strikes me as sound.

A natural response of those who appreciate the force of the symmetry consideration is to deny the existence of genuine I-S explanations altogether. Instead of attempting to find refuge in sufficiently high probabilities, they maintain that there are no statistical explanations of individual events but only D-S explanations of statistical regularities. This means, of course, that there are no statistical explanations of frequency distributions in limited samples, no matter how large, provided they fall short of the total population. Since the D-S species of D-N explanations must appeal to statistical laws, and since such laws typically apply to potentially infinite classes, we can say, in principle, that there can be no D-S explanations of frequency distributions in finite samples.

The deductivist who takes this tack can readily agree that quantum mechanics does provide deductive explanations of all such phenomena as the energy distribution in the spectrum of blackbody radiation, the stability of the hydrogen atom, the discreteness of the hydrogen spectrum, the photoelectric effect, radioactivity, and any others that quantum mechanics has treated successfully. In each of these examples the explanandum is a general regularity, and it is explained by deductive subsumption under more general laws. Following this line the deductivist maintains that we can explain why $_{86}Rn^{222}$ has a half-life of 3.82 days, but we cannot explain the decay of a single atom, the survival of a single atom, or the decay of a certain proportion of radon atoms in a particular sample of that gas. We can explain why, in the overwhelming majority of 10 milligram samples of radon, approximately 2.5 milligrams of radon remain at the end of 7.64 days. We can remark *that* the particular sample described by Hempel behaved in the typical way, but we cannot explain *why* it did so.

In the preceding section I discussed two types of hypothetical gamma ray detectors, each of which was subject to a particular kind of inaccuracy. In the real world we should expect inaccuracies of both kinds to occur within any such detecting devices. A gamma ray detector generally fails to respond to every photon that impinges upon it, and it may also give spurious counts. Thus, realistically, the impinging of a photon is neither a necessary nor a sufficient condition for the click of the counter. In addition, the probabilities in this example are modest—they are *not* for all practical purposes indistinguishable from one or zero. The discrepancies from the extreme values do make a significant difference. Under these circumstances we must conclude, I believe, that it is impossible in principle to provide a deductive explanation of such a simple phenomenon as the click of a counter. If any explanation of such an individual event is possible, it will have to fit a nondeductive pattern such as Hempel's I-S model, my S-R model, or Railton's deductive-nomological-probabilistic (D-N-P) model.[5]

4. The Profit and the Price

If indeterminism is true, there will be types of circumstances C and types of events E such that E sometimes occurs in the presence of C and sometimes does not. We can know the probabilities with which E happens or fails to happen given C, but we cannot, even in principle, know of any circumstances in addition to C that fully determine whether E happens or not. Even if the probability of E given C is high, many philosophers will reject the notion that an explanation of a particular case of E can be given. It may, of course, be possible to give a D-N explanation of the statistical regularity that E follows C with a certain probability p.

Georg Henrik von Wright (1971, p. 13) has argued that, in such cases, we cannot have an explanation of E, given that E occurs, for we can still always ask why E occurred in this case but fails to occur in others; Watkins (1984, p. 246) expresses a similar view. Circumstances C, von Wright holds, may explain why it is reasonable to expect E, but they cannot explain why E occurred. Wolfgang Stegmüller (1973, p. 284) responds to the same situation by observing that, if we claim that E is explained by C when E occurs, then we must admit that C sometimes explains E and sometimes does not. Since, by hypothesis, given C, it is a matter of chance whether E occurs or does not, it becomes a matter of chance whether C explains anything or not. Stegmüller finds this conclusion highly counterintuitive. While he admits the value of what he calls statistical deep analysis—something that is closely akin to the I-S and S-R models—he denies that it qualifies as any type of scientific explanation.

The counterintuitive character of statistical explanation becomes more dramatic if we invoke the symmetry consideration discussed in the preceding section. Given that E follows C in the vast majority (say ninety-five percent) of cases, E fails to happen a small minority (five percent) of the time. If, for example, we have an electron with a certain amount of energy approaching a potential barrier of a certain height, there will be a probability of 0.95 that it will tunnel through and a probability of 0.05 that it will be reflected back. If we claim to understand why the electron got past the barrier in one case, then we must understand just as well why in another case with the same conditions it was turned back. The explanatory theory and the initial conditions are the same in both cases. Thus, it must be admitted, circumstances C sometimes explain why the electron is on one side of the barrier and sometimes why it is on the other side. Circumstances C are called upon to explain whatever happens.

A fundamental principle is often invoked in discussions of scientific explanation. Watkins (1984, pp. 227–228) explicitly adopts it; Stegmüller calls it "Leibniz's principle" (1973, pp. 311–317); D. H. Mellor refers to it as "adequacy condition S" (1976, p. 237). I shall call it principle 1 (Salmon, 1984b, p. 113). It might be formulated as follows:

> It is impossible that, on one occasion, circumstances of type C adequately explain an outcome of type E and, on another occasion, adequately explain an outcome of type E' that is incompatible with E.

It is clear that D-N explanations never violate this condition; from a given consistent set of premises it is impossible to derive contradictory conclusions. It is clear, in addition, that I-S explanations complying with the high probability requirement satisfy it, for on any consistent set of conditions, the sum of the probabilities of two inconsistent outcomes is, at most, one. Since any high probability must be greater than one-half, at most one of the outcomes can qualify. As I noted in the preceding section, however, it is difficult to see how the high probability requirement can be maintained without an extreme degree of arbitrariness. If that result is correct, then it appears that Principle 1 draws the line constituting the limit of explanatory deductivism. In making this statement, I am claiming that anyone who rejects explanatory deductivism must be prepared to violate Principle 1. The question then becomes, Is it worth the price?

In an earlier work (Salmon, 1984b, pp. 112–120), I argued at some length and in some detail that violation of Principle 1 does not have the dire consequences that are generally assumed to ensue. If we pay sufficient attention to avoidance of ad hoc and vacuous explanatory 'laws', we can disqualify the undesirable candidates for scientific explanations without invoking Principle 1. Abandonment of that principle need not be an open invitation to saddle science with pseudo-explanations.

In the same place I argued, on the basis of several striking examples, that twentieth-century science contains statistical explanations of the nondeductive sort. One of these was a Mendelian genetic experiment on eye color in fruit flies conducted by Yuichiro Hiraizumi in 1956. Although it looked in advance just about the same as the genetic experiments mentioned earlier, it turned out that in a small percentage of matings, the statistical distribution of eye color was wildly non-Mendelian, while in the vast majority of matings the distribution was just what would be expected on Mendelian principles. A new theory was needed to provide an explanation of the exceptional outcomes in the particular matings that were observed. A possible explanation would attribute the exceptional distributions to chance fluctuations under standard Mendelian rules; that explanation could not, however, be seriously maintained. The preferred explanation postulates "cheating genes" that violate Mendel's rules (see Cohen, 1975).

Another example involved the spatial distribution of electrons bouncing off of a nickel crystal in the Davisson–Germer experiment. The periodic character of the distribution, revealing the wave aspect of electrons, was totally unanticipated in the original experiment and demanded theoretical explanation. The observed pattern involved a finite number of electrons constituting a limited sample from the class of all electrons diffracted by some sort of crystal.

The deductivist can reply that in both of these cases what is sought is a statistical theory that will explain, in general, the occurrence of statistical distributions in limited samples of the types observed. When the mechanism of "cheating genes" is understood, we can explain why occasional matings will produce results of the sort first observed by Hiraizumi. Similarly, quantum mechanics explains why, in general, electron diffraction experiments will very probably yield periodic distributions. In both cases, the deductivist might say, theoretical science shows us how such occurrences are possible on some ground other than the supposition that they are incredibly improbable chance fluctuations. The deductivist can maintain, in short, that all such statistical explanations in pure science are of the D-S type.

Consider another example. Suppose that an archaeologist, studying a particular site, comes across a piece of charcoal that has a concentration of C^{14} in its carbon content that is about half the concentration found in trees growing there at present. Since the half-life of C^{14} is 5,715 years, the archaeologist explains the difference in C^{14} concentration by supposing that the tree from which this charcoal came was felled, and consequently ceased absorbing CO_2 from the atmosphere, about 5,715 years ago. Although there are many potential sources of error in radiocarbon dating, it is not seriously supposed that the tree was felled 2,857 years ago, and that by chance its C^{14} decayed at twice the normal rate. This example—like the melting ice cube, the radon decay, and the Davison–Germer experiment—qualifies as a Jeffrey-type "beautiful" case.

The deductivist might, it seems to me, reply to the "beautiful" cases that, strictly speaking, we can furnish only D-S explanations of the statistical distributions of melting of ice cubes, the behavior of diffracted electrons, and rates of decay of radioactive isotopes. However, in the "beautiful" cases the statistical distributions show that a different outcome is so improbable that, though it is not physically impossible, we can be confident that neither we nor our ancestors nor our foreseeable descendants have ever seen or will ever see anything like it—anything, that is, as egregiously exceptional as an ice cube that does not melt in tepid water or a large collection of radioactive atoms whose rate of disintegration differs markedly from the theoretical distribution. We may therefore be practically justified in treating the theoretical statistical relationship as if it were a universal law.[6]

Archaeology, unlike many areas of physics, perhaps, is usually concerned with particulars—particular sites, particular populations, particular artifacts, and particular pieces of charcoal. In the preceding example the archaeologist could invoke a precise physical law concerning radioactive decay to explain the C^{14} content of a particular piece of charcoal. In other cases no such exact general laws are available; at best there may be vague statistical relationships to which an appeal can be made. For instance, archaeologists working in the southwestern United States would like to find out why one particular habitation site (Grasshopper Pueblo) was abandoned at the end of the fourteenth century, and more generally why the same thing happened all over the Colorado Plateau within a relatively short period of time (see Martin and Plog, 1973, pp. 318–333). Various factors such as overpopulation followed by drought can be adduced by way of explanation, but there is no real prospect of a D-N explanation in either the more restricted or the more general case. At best, any explanation will be probabilistic.

To examples of this sort the deductivist can readily respond that they involve obviously incomplete explanations. This point can hardly be denied. Certainly there are as yet undiscovered factors that contribute to the explanation of the phenomenon in question. The same kind of response is often made to examples of another type, namely, controlled experiments. Such cases occur very frequently, especially in the biological and behavioral sciences. Consider one well-known instance.

In the late 1970s some Canadian researchers studied the relationship between bladder cancer and saccharin in laboratory rats (see Giere, 1984, pp. 274–276). The experiment involved two stages. First, large quantities of saccharin were added to the diet of one group ($n = 78$), while a similar group ($n = 74$) were fed the same diet except for the saccharin. Seven members of the experimental group developed bladder cancer; one member of the control group did so. The null hypothesis—that there is no genuine association between saccharin and bladder cancer—could be rejected at the 0.075 level. This result was not considered statistically significant.

The second stage of the experiment involved the offspring of the original experimental and control groups. In each case the offspring were fed the same diets as their parents; consequently, the members of the second-generation experimental group were exposed to saccharin from the time of conception. Fourteen members of this group ($n = 94$) developed bladder cancer, while none of the members of the second-generation control group did. This result is significant at the 0.003 level. It was taken to show that there is a genuine positive association between ingestion of saccharin and bladder cancer among rats. The difference between the incidence of bladder cancer in the experimental and control groups is explained not as a chance occurrence but by this positive correlation, which is presumed to be indicative of some sort of causal relationship.

What can we say about this explanation? First, if no such positive correlation exists, the proffered explanation is not a genuine explanation. Second, if the positive correlation does exist, it is clear that, by chance, on some occasions, no statistically significant difference will appear when we conduct an experiment of the type described. The deductivist must, consequently, reject the foregoing explanation because the explanandum—the difference in frequency between the two groups—cannot be deduced from the explanans. The deductivist can, in this case, reiterate the response to the pueblo abandonment example, namely, the statistical character of the explanation arises from its incompleteness. If we possessed complete information, it would be possible to deduce, for each rat, whether it contracts bladder cancer or not.

Nevertheless, it seems to me, the claim that all such explanations are necessarily incomplete insofar as they fall short of the D-N model is extreme. If it arises from the assumption that all such cases are absolutely deterministic—and that all appeals to probability or statistics simply reflect our ignorance—then it is based on a gratuitous metaphysics that appears to be incompatible with contemporary physical science. If we do not have an a priori commitment to determinism, there is no reason to deny that indeterminacy arises in the domains of the biological or behavioral sciences. At the same time, if indeterminacy does occur, why should we withhold the appellation "complete explanation" from an explanation that cites all factors that are statistically or causally relevant to the occurrence of the explanandum event?[7]

In an earlier work (Salmon, 1984b), I maintained that pure theoretical science includes nondeductive statistical explanations. In the present context, *for the sake of argument,* I am prepared to relinquish that claim. Let us therefore agree—for now—that all of the statistical explanations that occur in theoretical science are either D-S explanations of statistical regularities or incomplete D-N explanations of particular facts. For purposes of the present discussion, I want to give deductivism the benefit of the doubt. I will therefore grant that the deductivist can—by treating every statistical explanation as either D-S or incomplete D-N—avoid admitting nondeductive statistical explanations as long as the discussion is confined to the realm of pure science. With respect to these examples, I know of no knockdown argument with which to refute the deductivist claim. But I shall try to show that the price the deductivist must pay is still exorbitant. It requires relinquishing the capacity to account for explanations in the more practical context of applied science.

5. Explanation in Applied Science

I am not prepared to concede, even for the sake of argument, that applied science can dispense with nondeductive statistical explanations. Granted that many explanations encountered in practical situations may reasonably be regarded as I-S explanations that are incomplete D-N explanations, there are others that defy such classification. Let us look at a couple of examples.

When Legionnaires' disease was first diagnosed in 1976, it was found that every victim had attended an American Legion convention in Philadelphia, and that all of them had stayed at one particular hotel. In the population of individuals attending that convention, residence at that hotel was a necessary but by no means sufficient condition of contracting the disease. Later, after the bacillus responsible for the disease had been isolated and identified, it was found that cooling towers for air-conditioning systems in large buildings sometimes provide both a favorable environment for their growth and a mechanism to distribute them inside the building. In this case, as well as in subsequent outbreaks in other places, only a small percentage of the occupants of the building contracted the disease. Since quantum fluctuations may lead to large uncertainties in the future trajectories of molecules in the air, and to those of small particles suspended in the atmosphere, I believe it quite possible that there is, even in principle, no strictly deterministic explanation of which bacteria entered which rooms and no strictly deterministic explanation of which people occupying rooms infested with the bacteria contracted the disease. Nevertheless, for purposes of assigning responsibility and taking preventive steps in the future, we have an adequate explanation of the disease in this very limited sample of the population of Americans in the summer of 1976. It is a nondeductive statistical explanation that, admittedly, may be incomplete. There is, however, no good reason to suppose that it can, even in principle, be transformed by the addition of further relevant information into a D-N explanation of the phenomenon with which we are concerned (see Salmon, 1984b, p. 212).

Eight soldiers, out of a group of 2,235 who participated in Operation Smoky in 1957, witnessing the detonation of an atomic bomb at close range, subsequently developed leukemia. The incidence—which is much greater in this group than it is in the population at large—is explained by the high levels of radiation to which they were exposed (see "Why Ask, 'Why?'?" [essay 8]). Because leukemia occurs with nonzero frequency in the population at large, it is possible, but not likely, that the high incidence of leukemia in this sample of the population was due to a chance fluctuation rather than an increased probability of leukemia as a result of exposure to a high level of radiation. From a practical standpoint, the fact to be explained is the high incidence in this particular sample, not the incidence among all people who ever have been or will be exposed to that amount of radiation; thus, it would be a mistake to construe the explanation as D-S. It is the occurrence of leukemia in this particular sample that has obvious importance in deciding such questions as whether these soldiers should receive extra compensation from the federal government. I am by no means certain that there is, in principle, no deterministic explanation of the onset of leukemia; however, because of the crucial involvement of radiation, it is not implausible to suppose that certain aspects are irreducibly statistical. In that case it would be impossible in principle to provide a D-N explanation of this phenomenon.

When it comes to the question of explaining the individual cases of leukemia, we must admit that we know of no factors that are either necessary or sufficient. Any given member of the group might have contracted leukemia even if he had not participated in Operation Smoky, and the vast majority of those who were involved did not contract leukemia. It is quite possible that other relevant factors bearing on the occurrence of leukemia were operative, but there is no guarantee that they add up to either sufficient or necessary conditions.

When we try to explain some occurrence, we may have any of several purposes. First, we may be seeking purely intellectual understanding of the phenomenon in question. Depending on one's philosophical biases, such understanding may result from finding the causes that produced the phenomenon or from subsuming it under a universal law. When there are no strict deterministic causes by which to account for it, or when there are no universal laws, we may be willing to settle for knowledge of the frequency with which *events of that type* are produced under specific conditions. The deductivist can accept this kind of understanding as the product of D-N explanation (recalling that D-S explanation is one of its subtypes). Such explanations can be attributed to pure science.

Second, when the occurrence to be explained is undesirable, we may wish to understand why it happened in order to take steps to prevent such occurrences in the future. Our practical purposes will be served if there is a necessary cause that is within our power to eliminate—for example, paresis would be prevented if syphilis were eradicated or treated with penicillin. Failing a strict necessary condition, our practical purposes will be served if we can find conditions—again within our power to control—whose elimination will reduce the frequency with which the undesirable occurrence takes place. Finding that automobile accidents on snow-covered roads occur probabilistically as a result of inadequate traction, we see that accidents of this type can be reduced—though not completely eliminated—through the use of adequate snow tires.

Third, if the occurrence in question is one we consider desirable, we may seek to understand why it happened in terms of sufficient causes. If sufficient causes that are under our control exist, they may be brought about in order to produce the desired result. For example, we can explain why a satellite remains at a fixed location above the earth in terms of the radius of the geosynchronous orbit (on the basis of Kepler's third law). A satellite can be placed in a geosynchronous orbit by boosting it, via rockets, to the specified altitude above the earth (about 22,300 miles) and injecting it into orbit.

Fourth, we often try to understand a desirable result in terms of circumstances that are necessary to its occurrence. In such cases we may discover a necessary condition that is absent. For example, a certain amount of water is required if various crops—such as corn, hay, and cotton—are to flourish. In desert areas irrigation is practiced. Adequate water is not sufficient to ensure good crops; if the soil lacks certain nutriments, such as nitrogen, the crops will not be healthy. But without the required water, other steps, such as fertilization or crop rotation, will not yield bountiful harvests.

Fifth, explanations are sometimes sought in order to assign moral or legal responsibility for some happening—very often a harmful result. Situations of this sort may well provide the strongest case against deductivism in the realm of applied science. Operation Smoky is a good example. To ascertain whether the U.S. Army is responsible for the eight cases of leukemia among the soldiers who participated in that exercise, we want to determine whether exposure to intense radiation explains these cases of leukemia. In order to answer that question, we need a general statistical law connecting leukemia with exposure to radiation. This law is a required component of the explanans. We are *not* trying to explain some general statistical regularity; we are trying to explain *these particular cases* of leukemia. We know of no universal laws that would make it possible to explain these particular instances deductively, and we have no reason to suppose that any such universal regularity exists unbeknownst to us.

At this juncture the deductivist would seem to have three possible rejoinders. First, he or she might simply deny that we have explanations of such phenomena as these particular cases of Legionnaires' disease and leukemia. This tack, it seems to me, runs counter to well-established and reasonable usage. It is commonplace, and unobjectionable, to maintain that we can explain the occurrence of diseases even when we have no prospects of finding sufficient conditions for a victim to contract it.

Second, the deductivist might insist that phenomena of this sort always do have sufficient causes, and that they are amenable to D-N explanation. The explanations we can actually give are therefore to be viewed as partial D-N explanations; they are not completely without practical value as they stand, but they are not genuine explanations until they have been completed. Whoever subscribes to this stubborn metaphysical dogmatism deserves the title "deductive chauvinist pig."

Third, the deductivist might claim that our nondeductive 'explanations' are partial explanations even though, in some cases, it may be impossible in principle to complete them on account of the nonexistence of suitable universal laws. Such partial explanations, it might be maintained, have practical value even though they fall short of the ideals of explanation in the context of pure science.

There is a temptation to try to convict this third response as incoherent on the ground that "partial explanation" makes no sense where there is no possibility in principle of having complete explanations. Yet, it seems to me, that rejoinder would be philosophically unsound. If the relative sizes and distances of the sun, moon, and earth were just a little different, there might be no such thing as a total eclipse of the sun; nevertheless, there would be nothing strange in talking about partial eclipses and in assigning degrees of totality.

The appropriate strategy might rather be to accept this third move, pointing out that it hardly qualifies as deductivism. If the concept of partial explanation is to be serviceable, there must be standards in terms of which to judge which partial explanations are strong and which weak, which are useful and which useless. Requirements akin to Hempel's maximal specificity (1965a, pp. 399–400) or my maximal homogeneity (1984b, p. 37) would be needed to block certain kinds of partial explanations. In short, the deductivist would need to develop a theory of partial explanation that would be a direct counterpart of Hempel's I-S model, my S-R model, my statistical-causal concept, or any of several others. If the deductivist accepts the fact that something that is not a deductive argument and cannot possibly be made into a deductive argument can nevertheless be an acceptable partial explanation, it seems to me that he or she has given up the deductivist viewpoint and is simply substituting the expression "partial explanation" for the term "statistical explanation" as it is used by many philosophers who have rejected the deductivist viewpoint.

6. Conclusions

In this essay we visited explanatory deductivism in the context of Laplacian determinism and found it very much at home there. However, since we now have strong reasons to believe that our world does not conform to the deterministic model, we found it necessary to revisit explanatory deductivism in the modern context, where, quite possibly, some of the basic laws of nature are irreducibly statistical. Although explanatory deductivism does not reside as comfortably here, evicting it, we found, is no easy matter—as long as we confine our attention to pure science. The claim that every statistical explanation is either a D-S explanation of a statistical regularity or an incomplete D-N explanation of a particular fact proves difficult to dislodge.

When we turn our attention to applied science, however, the situation is radically different. Explanatory deductivism does not do justice to explanations in practical situations. An interesting parallel emerges. As I argue in "Rational Prediction" (Salmon,

1981b), the most decisive argument against inferential deductivism arises in connection with the use of scientific knowledge to make predictions that serve as a basis for practical decisions. Both types of deductivism are unsuited for the practical realm.

Even in the realm of pure science, it seems to me, both forms of deductivism are untenable. Inferential deductivism fails to allow for predictions—such as the claim that our expanding universe will eventually begin an era of contraction that will lead to a "big crunch"—which have no practical import but a great deal of intellectual fascination. Explanatory deductivism encounters several difficulties. One that has emerged in this essay concerns the relations between sufficient and necessary conditions. In the second section we looked at cases—Mackie's candy machines and Coffa–Salmon photon detectors—in which a conflict arises between the D-N demand for sufficient conditions to explain what happens and the intuitive demand for causal explanations, where the cause in question is a necessary but not sufficient condition.[8] Thus, there arises a serious tension between the deductivistic conception of scientific explanation and the causal conception even in the realm of pure science.

In several writings I attempt to compare and contrast three fundamental conceptions of scientific explanation, including the deductivistic and causal conceptions, in considerable detail.[9] In the context of Laplacian determinism they are virtually equivalent, and there is not much reason to prefer one to the other. In the modern context, in which at least the possibility of indeterminism must be taken seriously, the two conceptions diverge sharply. According to the causal conception, we explain facts (general or particular) by exhibiting the physical processes and interactions that bring them about. Such mechanisms need not be deterministic to have explanatory force; they may be irreducibly statistical. Causality, I argue, need not be deterministic; it may be intrinsically probabilistic. The benefit we obtain in this way is the recognition that we can provide scientific explanations of particular events that are not rigidly determined by general laws and antecedent conditions. As I argued in section 5, "Explanation in Applied Science," the availability of such explanations is required for the application of science in practical situations; it also seems to be faithful to the spirit of contemporary pure science. The notion that we can explain only those occurrences that are rigidly determined is a large and unneeded piece of metaphysical baggage.

The price we pay for the claim that phenomena that are not completely determined can be explained is the abrogation of Principle 1. As I have argued at some length (Salmon, 1984b, pp. 113–120; "Scientific Explanation: Three Basic Conceptions" [essay 20]), the price is not too high. Principle 1 is I believe, the explanatory deductive chauvinist's main bludgeon. Once it has been rendered innocuous, the chief appeal of explanatory deductivism is removed.

Notes

1. Even worse for the deductivist is the statistical-relevance (S-R) pattern of scientific explanation. Explanations conforming to that model not only fail to be deductive; in addition, they fail to qualify as arguments of any sort.

2. Bertrand Russell is the most distinguished proponent of this form of inferential deductive chauvinism; see Salmon (1974b).

3. According to Watkins (1984, p. 243), the probability that the proportion of undecayed atoms lies within four percent of .25 is greater than $1 - 10^{-(10^{15})}$ when $n > 10^{19}$.

4. Critics tend to agree, but they insist that such understanding is possible only through knowledge of the chromosomal processes; see, for example, Kitcher (1985, p. 634) and van Fraassen (1985, pp. 641–642).

5. Railton's D-N-P model (1978) has the great virtue of demanding reference to the mechanisms that bring about such indeterministic results.

6. See Watkins (1984, pp. 242–246) for a detailed analysis of the "beautiful cases."

7. Philip Kitcher (1985, p. 633) suggests that statistical explanations of particular events be considered incomplete, not on the ground that nature can or must furnish additional explanatory facts but on the ground that the explanatory value of such explanations falls short of that of D-N explanations. According to his terminology, explanations that are irreducibly statistical are incomplete, not because of our epistemic shortcomings but because nature does not contain additional factors in terms of which to render them "ideally complete." I find it more natural to speak of complete nondeductive or complete statistical explanations.

8. In Salmon (1971, pp. 58–62), I distinguish between relevant and irrelevant necessary conditions and between relevant and irrelevant sufficient conditions. Irrelevant conditions of both kinds lack explanatory import. In addition, I discuss the statistical analogues of necessary and sufficient conditions: necessary and sufficient conditions are simply limiting cases of these statistical relationships. It is argued that both the statistical analogues of sufficient conditions and those of necessary conditions have explanatory import, but only if they are statistically relevant to the explanandum. Because these statistical relationships play no role in the deductivist account, I have not discussed them in the text of this essay. Nevertheless, it seems to me, the fundamental answer to the question whether sufficient or necessary conditions have explanatory import should be based on the relevancy relations offered in that discussion.

9. In Salmon (1984b) I distinguish three major conceptions of scientific explanation, namely, epistemic, modal, and ontic. The view that all scientific explanations are arguments, either deductive or inductive, is identified as the inferential version of the epistemic conception; the doctrine that all explanations are deductive arguments represents the modal conception. The causal conception of scientific explanation is a version of the ontic conception. See "Comets, Pollen, and Dreams" (essay 3) and "Scientific Explanation: Three Basic Conceptions" (essay 20).

10

Explanatory Asymmetry

A Letter to Professor Adolf Grünbaum from His Friend and Colleague

June 1991

Dear Adi,

Following a venerable philosophical tradition, I am taking this occasion to address a profound problem in an open letter. Not long ago you asked me to tell you why—given the time-symmetry of most of the fundamental laws of nature—particular events cannot (accordingly to my lights) be explained by appeal to facts that obtain subsequent to their occurrences, but only by reference to those that obtained antecedently. This is, as I say, a profound question, and it has not been suitably addressed by many of those who want to exclude explanations in terms of subsequent facts. The following is my attempt to give that problem the serious attention it deserves. Unfortunately, the fact that a question is profound does not guarantee that the answer will be also; nevertheless, with my deepest respect, here is my best effort.

1. The Question

Can a particular event be explained by citing subsequent[1] conditions and events—along with appropriate laws—or do only antecedent conditions and events have explanatory import? For example, on 29 May 1919 there occurred an event of major importance to the history of science, namely, a total eclipse of the sun. The Isle of Principe in the Gulf of Guinea, West Africa, lay in its path of totality. Here observations were made to test Einstein's prediction of the bending of light passing near the limb of the sun. We know, of course, that a solar eclipse occurs at any given place if that place happens to be in the shadow cast by the moon. But what about the particular eclipse at Principe? Why did it occur at that particular place and time?

According to Sir Arthur Eddington's report ([1920] 1959, pp. 113–114), the Astronomer Royal called his attention to the forthcoming eclipse in March 1917. Using observational data concerning the relative positions and motions of the earth, sun, and moon, and applying the laws of motion he was able to deduce the alignment of those bodies at a particular time on 29 May 1919. If, subsequent to the eclipse, someone were to ask why it happened there and then, the data and the derivation of the Astronomer Royal could be offered in answer.

One reason for picking this example is that the explanation just mentioned conforms perfectly to the deductive-nomological (D-N) model of explanation, first articulated with precision and in detail by Carl G. Hempel and Paul Oppenheim ([1948] 1965) in their epoch-making essay. Another reason is that it admirably exemplifies the *explanation-prediction symmetry thesis* propounded in that essay (ibid., § 3). Precisely the same deductive argument that furnishes a prediction of the eclipse prior to its occurrence provides an explanation of its occurrence after the fact.

This particular example also serves to illustrate the opening question of this letter. Since the laws of motion are time-symmetric, the Astronomer Royal could have made observations of the sun-moon-earth system two years later—in 1921—and he could have deduced that the total eclipse had occurred on Principe on 29 May 1919. Can this latter deduction qualify as an explanation of the eclipse? A great many philosophers, myself included, would reject this suggestion. With Laplace we might say, "We ought to regard the present state of the universe as the effect of its antecedent state and as the cause of the state that is to follow" ([1820] 1951). Since it is generally agreed, we might continue, that causes can explain their effects, but effects cannot explain their causes, only the derivation of the occurrence of the eclipse from antecedent conditions can qualify as an explanation. The inference from subsequent conditions can qualify only as a retrodiction—not as an explanation as well. Nevertheless, two of the most profound philosophers of science of the twentieth century, you and Peter Hempel,[2] have disagreed with this doctrine. Given the superb philosophical credentials of the two of you, we should take a close look at your point of view.[3]

2. A Bit of History

The classic Hempel-Oppenheim ([1948] 1965) essay, a fountainhead from which much of the subsequent philosophical literature on scientific explanation has flowed, amazingly attracted virtually no attention for a full decade after its original publication. Then, following that lapse, a flurry of sharply critical papers appeared. One major focus of attention was the explanation-prediction symmetry thesis. Your paper "Temporally Asymmetric Principles, Parity between Explanation and Prediction, and Mechanism versus Teleology," in which you endeavored to clarify and defend the symmetry thesis, was published in 1962. You showed convincingly that many of the criticisms were based on misunderstandings, and you sought to correct the then current misinterpretations by such authors as N. R. Hanson (1959), Nicholas Rescher (1958), and Michael Scriven (1959).

According to Hempel and Oppenheim, a D-N explanation of a particular fact is a valid deductive argument. Its conclusion (the explanandum) states the fact to be explained; its premises (the explanans) present the explanatory facts. At least one law-statement must occur essentially among the premises, and the premises must be true. Hempel and Oppenheim never claimed that all legitimate scientific explanations fit the D-N model; on the contrary, they explicitly asserted that the sciences include acceptable inductive or statistical explanations as well, even though they did not attempt to provide a model for those of the latter sort ([1948] 1965, pp. 250–251). That project, undertaken in Hempel (1962a), yielded what later came to be called "the inductive-statistical (I-S) model of explanation."

Hempel's symmetry thesis has two parts. The first says that every scientific explanation could, under suitable circumstances, serve as a prediction. This part *seems* clearly to hold for D-N explanations. If the explanatory facts had been at our disposal before the occurrence of the fact to be explained (the explanandum), we would have been able to predict that fact, for we would have been in possession of true premises from which it follows deductively. It is worth noting that Peter intended the symmetry thesis to apply to I-S as well as D-N explanations.[4] In Hempel (1962b, pp. 10, 14), he maintained that either type of explanation of a particular event is an argument to the effect that the event to be explained was *to be expected* by virtue of the explanatory facts. This clearly implies the applicability of the first part of the symmetry thesis. In Hempel (1965a, pp. 367–368), he elevated this half of the symmetry thesis to a general condition of adequacy for any account of scientific explanation.

The second part of the symmetry thesis says that every legitimate scientific prediction could, under suitable circumstances, serve as an explanation. This obviously does not imply that every scientific prediction could serve as a D-N explanation. Many scientific predictions are probabilistic. According to the symmetry thesis, such predictions could, under suitable circumstances, serve as I-S explanations. As Israel Scheffler (1957) pointed out, however, a prediction is merely a statement about the future. As such, a prediction could not be an explanation, for an explanation, according to Peter, is an argument. The most that could be maintained is that legitimate scientific predictions are *the conclusions of* arguments that conform to the schemas of D-N or I-S explanation. This is the position I take you to have adopted (1962, pp. 157–158) in defending the symmetry thesis.

One major problem introduced in the Hempel-Oppenheim essay concerns the use of the term "antecedent." Just before listing the four general conditions of adequacy for deductive explanations, the authors remark that "the explanans falls into two subclasses; one of these contains certain sentences $C_1, C_2, \ldots, C_k$ which state specific *antecedent* conditions; the other is a set of sentences $L_1, L_2, \ldots, L_r$ which represent general laws" ([1948] 1965, p. 247; emphasis added). Two pages later, when they offer a schema for deductive explanation, they label the Cs "statements of *antecedent* conditions" (ibid., p. 249; emphasis added). One gets the impression that the Cs are supposed to refer to facts that obtain temporally prior to the explanandum-event E. When, however, we examine the conditions of adequacy, which are set forth between the two foregoing characterizations of the Cs, we see that no mention is made of temporal priority. One reason for *this* omission may be that the conditions of adequacy are intended to apply to explanations of laws as well as explanations of particular facts. In explanations of laws no premises are needed other than the Ls; consequently, the general conditions of adequacy do not make mention of the "antecedent conditions."

When we pass from their preliminary conditions of adequacy to their precise explication of deductive explanation, however, we learn that the analysis "will be restricted to the explanation of particular events, i.e., to the case where the explanandum, E, is a singular sentence" (ibid., p. 273). A footnote inserted at this point explains why an analysis of explanations of laws cannot be furnished. Given this restriction to explanations of particular events, it becomes necessary to include in the explanans a singular sentence C corresponding to the "antecedent conditions." In the formal explication, however, there is no requirement that C describe conditions that obtain prior to the occurrence of E.

At this point it is tempting to suppose that the omission of any condition on the temporal relationship between the Cs and E is a mere oversight—one that can easily be repaired. This supposition would, however, be unsound, for when Peter later addresses this issue directly, he denies that the Cs should be so restricted. After offering putative examples of explanations that cite "antecedent conditions" contemporaneous with or later than the explanandum, he comments:

> Any uneasiness at explaining an event by reference to factors that include later occurrences might spring from the idea that explanations of the more familiar sort . . . seem to exhibit the explanandum event as having been brought about by earlier occurrences; whereas no event can be said to have been brought about by factors some of which were not even realized at the time of its occurrence. Perhaps this idea also seems to cast doubt upon purported explanations by reference to simultaneous circumstances. But, while such considerations may well make our earlier examples of explanation, and all causal explanations, seem more natural or plausible, it is not clear what precise construal could be given to the notion of factors "bringing about" a given event, and what reason there would be for denying the status of explanation to all accounts invoking occurrences that temporally succeed the event to be explained. (Hempel, 1965a, 353–354)

He thus deliberately rejects the notion that antecedent conditions should be temporally antecedent to the events that are to be explained.

In your 1962 essay you introduced the technical term "H-explanation"—standing for "Hempelian explanation"—and defined it in a way that allows subsequent conditions to H-explain a given occurrence. Although this may seem strange from the standpoint of standard usage, we must realize that you were offering a characterization that is not only faithful to the *letter* of the 1948 paper but also faithful to the *spirit* of Hempel, as is shown by Peter's explicit statement of 1965.[5] When, as a consequence, you defined H-prediction—standing for "Hempelian prediction"—in a way that includes what we would normally consider retrodiction, you were only being faithful to the Hempel-Oppenheim assertion that "whatever will be said in this article concerning the logical characteristics of explanation or prediction will be applicable to either, even if only one of them should be mentioned" ([1948] 1965, p. 249). This is, of course, simply an explicit statement of the symmetry thesis.

What, then, is the difference between H-explanation and H-prediction? As you defined these terms, it coincides precisely with the following characterization by Hempel and Oppenheim: "The difference between the two [explanation and prediction] is of a

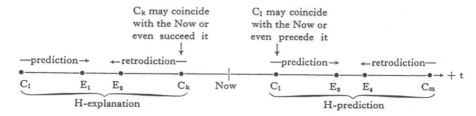

Figure 10.1. Reproduced from Adolf Grünbaum, "Temporally-Asymmetric Principles, Parity between Explanation and Prediction, and Mechanism versus Teleology," *Philosophy of Science*, vol. 29 (1962), p. 156, by permission of the University of Chicago Press.

pragmatic character. If E is given, i.e., if we know that the phenomenon described by E has occurred, and a suitable set of statements $C_1, C_2, \ldots, C_k, L_1, L_2, \ldots, L_r$ is provided afterwards, we speak of an explanation of the phenomenon in question. If the latter statements are given and E is derived prior to the occurrence of the phenomenon it describes, we speak of a prediction" (ibid., p. 249). Following the Hempel-Oppenheim text exactly, you offered figure 10.1.

On the one hand, as we see, whether a given event E is H-explained or H-predicted depends solely on whether E is before or after the 'now' of the person deriving E from a given set of conditions C_i (in conjunction with suitable laws). It does *not* depend on whether the Cs precede or follow E. On the other hand, as we also see, if the Cs precede E, we have an ordinary prediction (without the H prefix); the temporal relationship to the 'now' does not matter. Similarly, we have an (ordinary) retrodiction if the Cs follow E.[6] Thus, an H-prediction can be a retrodiction. If, for example, the Astronomer Royal had, in 1917, established the state that the sun-moon-earth system would assume in 1921, and had (still in 1917) derived the occurrence of the 1919 eclipse from the 1921 conditions, his inference would have been both a retrodiction and an H-prediction.

Although the foregoing terminological proposals may seem a bit odd from the standpoint of ordinary usage, they are perfectly legitimate as stipulative definitions, and they do reflect the theses propounded by Hempel and Oppenheim in 1948. In addition, they served you well in your efforts to clarify the Hempel–Oppenheim position and to expose misunderstandings of the symmetry thesis. For this we owe you a large debt of gratitude.

3. The Cheap Answer

In Salmon (1990b) I offered what I now recognize as an idiosyncratic and biased historical sketch of philosophical discussions of scientific explanation in the four decades beginning with the Hempel-Oppenheim paper.[7] Pointing out that many criticisms of the 'received view,' which included both the D-N and I-S models of explanation, had been articulated in terms of standard counterexamples, I employed the example of the eclipse to call attention to the failure of the Hempelian account to stipulate that the so-called antecedent conditions must be temporally prior to the explanandum-event. This counterexample then served as the basis for a negative answer to the question posed at the beginning of this letter. For a large majority of philosophers, I think, this type of argument is compelling. Our intuitions rebel at the supposition that bona fide explanations could violate that constraint. This strategy amounted to taking the easy way out of the problem.

4. Hempel's Argument

You and Peter have approached the intuitive appeal to counterexamples in different ways. He is willing to fight fire with fire—to pit counterexample against counterexample:

> It might . . . be argued that sometimes a particular event can be satisfactorily explained by reference to subsequent occurrences. Consider, for example, a beam of light that travels from a point *A* in one optical medium to a point *B* in another, which borders upon the first along a plane. Then, according to Fermat's principle of least time, the beam will follow a path that makes the traveling time from *A* to *B* a minimum as compared with alternative paths available. Which path this is will depend on the refractive indices of the two media; we will assume that these are given. Suppose now that the path from *A* to *B* determined by Fermat's principle passes through an intermediate point *C*. Then this fact may be said to be D-N explainable by means of Fermat's law in conjunction with the relevant data concerning the optical media and the information that the light traveled from *A* to *B*. But its "arrival at *B*," which thus serves as one of the explanatory factors, occurs only after the event to be explained, namely, the beam's passing through *C*. (Hempel, 1965b, p. 353)

The answer I would offer to Peter's example is based classically on the ability of the wave theory of light to explain the phenomena of geometrical optics (including Fermat's principle). A rather similar account can be given in quantum electrodynamics.

According to the wave theory, light emitted from *A* in the general direction of *B* passes through the interface between the two optical media in a large region surrounding *C. After the waves have traversed this interface,* destructive interference cancels out those portions of the waves not passing through *C,* yielding the appearance of a ray that travels from *A* through *C* to *B*. Peter is certainly correct in stating that a D-N explanation of the passage of the ray through *C* can be constructed (just as he has shown) using the arrival of light at *B* as an "antecedent condition," but this fact reveals a deficiency of the D-N model. The *causal explanation* offered by the wave theory is far more satisfactory because it exhibits the actual mechanisms involved in the production of the fact to be explained. And in *that* explanation we *do not* explain any event in terms of subsequent facts.

In Chapter 2 of his book *QED: The Strange Theory of Light and Matter* (1985), Richard Feynman uses precisely the same example Peter did to show how quantum electrodynamics deals with interactions between light and matter. Treating light as composed of particles (photons), he exploits the wave-particle duality to show how destructive interference yields the same result Fermat's principle does. As in the classical explanation in terms of the wave theory of light, Feynman emphasizes the fact that photons pass through the interface between the two media in all possible regions—not just at C—and he offers convincing experimental evidence to support that claim. In the introductory chapter he asserts, moreover, that QED explains all phenomena involving interactions between light and matter, and that its predictions are the most precise provided by any current physical theory.

One might be tempted to suppose that Peter chose an unfortunate example, and that a more apt example might be found by looking at other applications of variational principles such as Fermat's principle. That would be incorrect, I think. Max Born has argued in general that wherever such principles *seem* to offer an explanation of some physical fact in terms of subsequent conditions, there exists an explanation in terms of straightforward efficient causes. With regard to variational formulations in classical mechanics he wrote: "From now on the world is a mechanism, ruled by strict deterministic laws. Given the initial state, all further development can be predicted from the differential equations of mechanics. The minimum principles are not due to nature's parsimony but to human economy of thinking, as Mach said; the integral of action condenses a set of differential equations into one simple expression" ([1956] 1969, p. 124; see also pp. 55–79).

In the past I have argued that noncausal explanations such as Peter offered in this quoted passage are illegitimate. I no longer hold that view (Salmon, 1990b, § 5.1). It seems to me that explanations that involve unification, in the sense elaborated by Michael Friedman and Philip Kitcher, provide a type of understanding that complements, rather than excludes, causal explanation. The use of variational principles does unify an enormous variety of physical phenomena, but it does not provide a suitable basis for claiming that particular events can be explained in terms of subsequent conditions.[8]

5. Grünbaum's Argument

You have been much less sympathetic than Peter to the appeal to commonsense example and intuition:

> But, to my mind, the philosophical task before us is *not* the ascertainment of how the *words* "explain" and "predict" are used, even assuming that there is enough consistency and precision in their usage to make this lexicographic task feasible. And hence the verdict on the correctness of Hempel's symmetry thesis cannot be made to depend on whether it holds for what is taken to be the actual or ordinary usage of these terms. Instead, in this context I conceive the philosophical task to be both the elucidation and examination of the provision of scientific understanding of an *explanandum* by an *explanans* as encountered in actual scientific theory. (Grünbaum, 1962, p. 158)

In what way, you seem to ask, is the sort of use typically made of putative counterexamples different from a mere appeal to ordinary usage? In reference to another of the famous counterexamples—Sylvain Bromberger's flagpole—you retorted: "But is it not true after all that ordinary usage countenances the use of the term 'explanation' only in cases employing causal antecedents and laws of succession in the *explanans?* To this I say: this *terminological* fact is as unavailing here as it is philosophically unedifying" (ibid., p. 168). It is now time to attempt a philosophically deeper answer to our main question.

6. Where to Look for an Answer

It was never my intention to appeal to untutored usage to support criticisms of the 'received view,' but I do believe that sound philosophical explication must involve what Rudolf Carnap called "clarification of the explicandum"—a process he exemplified superbly in *Logical Foundations of Probability* (1950, chaps. 1, 2, and 4). This process often involves consideration of examples. Neglect of such preliminary clarification might result in an explication of something different from what we set out to explicate. In my opinion, most of the literature on scientific explanation in the 1950s and 1960s was seriously deficient in this respect.

The clearest clue you have given us by way of clarification, it seems to me, is found in a previously quoted passage, namely, "the elucidation and examination of the provision of scientific understanding of an *explanandum* by an *explanans* as encountered in actual scientific theory" (Grünbaum, 1962, p. 158). Since the term "scientific understanding" is far from clear and unambiguous, we still have some work to do. The best approach, I think, is to look at some examples from the history of science that seem to be universally agreed *by scientists and philosophers of science* to constitute bona fide explanations, at least with respect to the body of scientific knowledge available at the time. For example, the Newtonian synthesis has been hailed for its explanation of many different types of phenomena—planetary motions, comets and tides, to mention just a few. When Edmund Halley ([1687] 1947, p. xiv) wrote, in his "Ode to Newton":

> Now we know
> The sharply veering ways of comets, once
> A source of dread, nor longer do we quail
> Beneath appearances of bearded stars,

he seemed to be saying that we need not fear comets because now, thanks to Newton, we understand them.

One point to be noted immediately is that in many cases, as the foregoing examples illustrate, scientific understanding results when a general regularity is explained by derivation from even more general laws. This is particularly true of situations in which a number of apparently independent regularities are unified by subsumption under a unified theory. This feature of scientific understanding was persuasively elucidated by Michael Friedman (1974), who referred to the explananda as *phenomena.* As he emphasized, such phenomena are not particular facts but, rather, general regularities. Since these regularities are not localized in time, the symmetry thesis does not apply. The question whether "antecedent conditions" precede or succeed the explanandum-event does not arise because the explanandum is not an event. The explanans does not contain any Cs.

This point applies in a wide variety of cases. The Einstein–Smoluchowski theory of molecular bombardment explains the phenomenon of Brownian movement. Rutherford's planetary model of the atom enabled us to understand large angle scattering of alpha-particles, but did not help us to understand the stability of atoms. The kinetic theory of gases explains Avogadro's law. It would be easy, but pointless, to go on collecting examples of this sort. We must conclude, I believe, in agreement with Friedman, that theoretical unification, as exemplified by the Newtonian synthesis, does produce one type of scientific understanding, but this kind of understanding sheds no light whatsoever on the temporal relations between explanatory facts and the explanandum. The examples just cited have no bearing on the model of scientific explanation offered by Hempel and Oppenheim. As just noted, they offered no precise explication of explanations of laws— only explanations of particulars.

Does science provide understanding of particular facts or occurrences? One way to a positive answer might be to look at particular experiments that have been performed. Does science provide understanding of these? Consider the Michelson-Morley experiment. Its negative outcome baffled Michelson, but Lorentz proposed a contraction hypothesis to explain it. This explanation was later judged unsatisfactory and was replaced by the special theory of relativity. By postulating the constancy of the speed of light, this theory explains why, in general, attempts to ascertain the speed of any system relative to the luminiferous ether will have a null result. Or consider the Davisson-Germer experiment. Its result was first explained by the wave-particle duality of matter and later by quantum theory, from which one can deduce that material particles will, under certain circumstances, exhibit interference phenomena. In thinking about either one of these experiments, however, it is misleading to suppose (in spite of the singular grammatical form) that a single experiment is involved. Each encompasses a large series of experiments which established, among other things, that the results are reproducible. Hence, what is explained is not a particular fact but, rather, a general fact about experiments of particular types. Again, we must question whether bona fide explanations of particulars are involved. If not, these examples are not genuinely pertinent to our main question.

If these observations regarding scientific understanding are sound, then, apparently, theoretical science casts little light on the symmetry thesis. To find pertinent scientific material for our discussion we may have to abandon theoretical for applied science. In this domain, I believe, clear examples of scientific explanation result in genuine scientific understanding of particulars. The most obvious examples involve catastrophe, death, and destruction; in such contexts we are clearly concerned with particular cases.

Consider an example that has been widely discussed for approximately a decade, namely, the extinction of dinosaurs. Although it may have occupied a considerable stretch of space and time, it is a unique event. At a site near Gubbio, Italy, where the cretaceous-tertiary (K-T) boundary is exposed, Walter Alvarez examined a thin layer of clay and found it to be extraordinarily rich in iridium. With his father, Luis Alvarez, he proposed the hypothesis that this iridium anomaly could be explained in terms of a catastrophic collision of an asteroid or comet with the earth. This hypothesis has subsequently been rather strongly substantiated by additional evidence. They further hypothesized that the dire ecological consequences of this event resulted in the extinction of many living species, explaining the extinction, among others, of dinosaurs. This explanatory hypothesis has gained many adherents, though it is by no means as universally accepted as is the explanation of the iridium anomaly.

Although there are differences of opinion among experts regarding the acceptability of various proffered explanations of the iridium anomaly and of the extinction of dinosaurs, all seem clearly agreed that the search is for causal explanation in terms of temporally antecedent conditions. A convincing explanation will have to spell out the causal mechanisms whereby the explanandum is realized. I do not regard this claim as an appeal to the linguistic usage of scientists; it is a characterization of the practice of the experts in the field. It would obviously be easy to give many more examples of explanations of particular events—such as the *Challenger* shuttle disaster—in which the causal nature of the explanation is patent, but in examples of that sort it is possible to argue that a causal explanation is sought for the practical end of knowing how to prevent such things from occurring in the future, and not primarily for the sake of intellectual understanding. In such cases the issue of human control seems important, and indeed is important. The examples of the iridium anomaly and the dinosaur extinction show, however, that there are cases in which causal explanations are universally regarded as being of the appropriate type even where the possibility of human control is out of the question. Scientific understanding is the principal goal.

7. The Right Answer

In his posthumous book *The Direction of Time* (1956), Hans Reichenbach addressed various problems that you have dealt with at length under the heading of *temporal anisotropy*. Although you disagree with him on many fundamental points, I believe you are in accord on the following:

1. Our universe, at least in the present epoch, possesses an objective temporal anisotropy. Irreversible physical processes constitute a basis for this temporal anisotropy.
2. The fundamental laws of nature are temporally symmetric—leaving aside a few esoteric cases such as the law governing decay of the K^0-meson, which could hardly have a significant bearing on the pervasive temporal anisotropy of our world.
3. The temporal anisotropy of our world is de facto not de jure. The pervasive temporal anisotropy—what Reichenbach called "the direction of time"—is based on pervasive matters of fact, not on temporally asymmetric physical laws.
4. Although the laws are, within the limits just stated, time-symmetric, there are in nature de facto irreversible processes.

All of these seem to be sound.

Now, I take it that when you raise the question about the temporal anisotropy of explanation, you are prepared to distinguish two fundamentally different types of cases. In cases of the first type we are dealing with irreversible processes. Suppose, for example, that I associate with someone who has a cold, that as a result I contract a cold and lose my voice, and that in consequence I am unable to deliver a lecture to my class.[9] No one, I think, would be tempted to explain my cold by appealing to the subsequent condition of not being able to give my lecture. In nature, not being able to give a lecture does not lead causally to catching a cold. In contexts of this sort the temporal anisotropy of explanation seems to hold. The most high-powered medical research scientist would agree, I presume, with the ordinary person on the street. This is not an appeal to ordinary usage or untutored common sense.

In cases of the second type we have reversible processes; the eclipse example is a paradigm. Given the laws of mechanics, as well as what we know about general conditions in the universe, it appears to be entirely possible for there to be a solar system such as ours, but with the directions of motion of the earth about the sun and of the moon about the earth reversed. Leaving aside the extremely small frictional effects, the solar system constitutes a set of reversible processes.

When we consider the anisotropy of time, we try to take a global—or at least a large-scale—point of view. Given the facts about entropy and branch systems so clearly articulated by you and Reichenbach, we say that the temporal anisotropy is pervasive. In applies just as much to physical systems involving only reversible processes as to those that involve irreversible processes. We distinguish, consequently, earlier and later where lunar and planetary motions are concerned just as we do with regard to ice cubes melting in glasses of ginger ale. We do not say that temporal anisotropy prevails only in irreversible processes but not in reversible processes. To adopt a different policy with respect to time would play havoc with our most fundamental physical theories.

You and Reichenbach also agree on a causal theory of time, and, again, I think you are right. Time and causality go hand in hand. The anisotropy of time is deeply connected to the anisotropy of causality. Causes come before their effects, not after them. Now, if one agrees that causality is an indispensable component of scientific explanations of *particular events,* it is natural to suppose that the anisotropy of time and causality would be reflected in an anisotropy of scientific explanation.

Reichenbach (1956, § 18), in "Cause and Effect: Producing and Recording," offers a number of extremely suggestive hints concerning the relationships among time, causality, and explanation. In earlier sections he had discussed time direction in terms of thermodynamics and microstatistics; he then attempted to apply similar considerations at the macro-level:

> There exists an essential difference between microprocesses and macroprocesses. The former possess a natural shuffling mechanism given by the collisions of the molecules. The latter often do not possess any natural shuffling mechanisms . . . in other processes, the natural shuffling mechanism is so very slow that, at a given moment, the system remains practically unchanged. . . . This distinctive feature leads to peculiar consequences for macrostatistics: states of high order can here be preserved for a long time and can be observed conveniently. This is the reason why macrostatistics supplies what we call *records* and why, at the same time, it presents us with the key to the understanding of *causal explanation.* (ibid., pp. 149–150)

Reichenbach illustrates these considerations by means of the example of human footprints in the sand. Having defined order and disorder for macrosystems, he regards the sand with footprints as more highly ordered than smooth sand, and the footprints in the sand as records of a person having passed that way in the not too distant past.[10] He continues: "In addition to the clarification of the nature of records, the example of the footprints also helps us to analyze the meaning of *causal explanation.* Explanation in terms of causes is required when we meet with an isolated system displaying a state of order which in the history of the system is very improbable. We then assume that the system was not isolated at earlier times: explanation presents order in the present as a consequence of interaction in the past" (ibid., p. 151). He then offers an explication of cause and effect: *"The cause is the interaction at the lower end of the branch run through by an isolated system which displays order; and the state of order is the effect*" (ibid.). At this point in the text the editor (at my suggestion) inserted a footnote explaining that this explication is too narrow, but that an extension can be found a few pages later: "But in the sense of a transfer from relationships holding for irreversible processes in branch systems, the use of the word 'cause' is legitimate in application to macroprocesses governed entirely by the laws of mechanics" (ibid., p. 156).

As I said, Reichenbach offers hints about causality and explanation, but not a well worked out theory. Nevertheless, he argues that causal explanation always refers to the past (ibid., p. 152). And his subsequent discussion of the principle of the common cause (ibid., sec. § 19) reinforces the point. We explain improbable coincidences in terms of common causes, not in terms of common effects. The common cause temporally precedes the coincidence it is invoked to explain. Apparently, according to Reichenbach, anisotropy of time, anisotropy of causality, and anisotropy of explanation all go together, and are completely compatible with the time-symmetric character of the fundamental laws of nature. I think that this view is correct.

It is especially important to emphasize that the temporal anisotropy of explanation has nothing to do with the reversibility or irreversibility of the physical processes involved in the situation. The fact that there might be or might have been a solar system just about the same as ours, but with rotations reversed, is, to my mind, irrelevant to the fact that *in our solar system* the 1919 solar eclipse was a result of conditions in its past but not those in its future.

8. A Possible Bone of Contention

In this context let us recall Peter's remarks, quoted earlier, concerning temporally antecedent causes and "bringing about." In contrast to explanations invoking temporally subsequent conditions, "explanations of the more familiar sort . . . seem to exhibit the explanandum even as having been brought about by earlier occurrences; whereas no event can be said to have been brought about by factors some of which were not even realized at the time of its occurrence" (Hempel, 1965a, p. 353). He responds to this consideration by remarking that "it is not clear what precise construal could be given to the notion of factors 'bringing about' a given event" (ibid.). We could offer Reichenbach's account of *producing* as one way of furnishing a reasonably precise construal of that notion. His discussion of causal explanation could be offered as an answer to Peter's further query as to "what reason there would be for denying the status of explanation to all accounts invoking occurrences that temporally succeed the event to be explained" (ibid., pp. 353–354).

Although you and I have not discussed precisely this point, it occurs to me that you may feel intellectually ill at ease with the notions of *bringing about* and *producing.* Such terms may smack too much of the idea of *temporal becoming,* a concept of which you have been highly critical (e.g., Grünbaum, 1967, chap. 1). For example, when Peter wrote (but did not endorse) the claim that "no event can be said to have been brought about by factors some of which were not even realized at the time of its occurrence," he was expressing a thesis about explanation that depends on temporal becoming. Take away the notion of becoming, and we can say that the universe consists of events located at various spacetime points or regions. All are (timelessly) equally real. Some of them are earlier than a given occurrence that is to be explained and some of them are later. The question is whether any of those located at times later than the explanandum can have explanatory import with respect to it. On your view there could be no question about what is or is not "realized" at a given time; events do not come into being and gain reality or pass out of being and lose reality. To be real is to have a location at some place and time in the world.

A major point of disagreement between you and Reichenbach on time has centered on his thesis regarding the existence of an objective present—the 'now.' I find your arguments compelling; I have no desire to defend Reichenbach's view on this matter, and I would not appeal to the argument for explanatory anisotropy mentioned in the beginning of this section. It seems to me, nevertheless, that Reichenbach's most important insights on causality and explanation can stand without reliance on his claims about temporal becoming. For example, you and I can use such Reichenbachian concepts as branch systems and causal interactions without invoking temporal becoming in any way.[11] Moreover, Reichenbach's characterization of a conjunctive fork, which plays a key role in his theories of temporal anisotropy and causal explanations, involves only probability relations that do not involve temporal becoming. In reading Reichenbach's text one must, to be sure, exercise care in not allowing becoming to creep in, but I have no doubt that it can be done.

If the analysis offered in this letter is correct, then the *alleged* temporal anisotropy of scientific explanation is also the *actual* temporal anisotropy of scientific explanation.[12] In any case, this represents my serious and sincere effort to answer the question you posed—the one that constitutes the point of departure for this letter.

With deep affection and admiration,
Wesley C. Salmon

cc: Peter Hempel

Notes

1. Since the question to which this essay is devoted concerns explanation in terms of subsequent facts, the question concerning the explanatory status of simultaneous conditions and events will not be addressed.

2. Carl G. Hempel is known by colleagues and friends as Peter.

3. Because this is an *open* letter, I will review some aspects of the main question with which you are obviously completely conversant.

4. The footnotes in your (1962) seem to indicate that you were at least to some degree aware of the content of Hempel (1962a) prior to the publication of both of these papers.

5. I resist the temptation of attempting to explain your 1962 publication by reference to Peter's 1965 publication; I attribute it, rather, to your fundamental understanding, just prior to the time of writing his 1962 essay, of his position.

6. The term "H-retrodiction" is not defined; there is no need for it.

7. I was somewhat aware of this fact at the time, for I wrote, "We have arrived, finally, at the conclusion of our saga of four decades. It has been more the story of a personal odyssey

than an unbiased history. . . . My decisions about what to discuss and what to omit are, without a doubt, idiosyncratic, and I apologize to the authors of such works for my neglect" (Salmon 1990b, p. 180). Clearly you deserve such apologies, and I hereby offer them.

8. I am here using the term "phenomenon," as it is used in Friedman (1974), to designate a general fact rather than a particular occurrence.

9. This example is chosen to mirror the temporal relations in the eclipse example: Exposure to cold → contracting of cold → inability to lecture *correspond to* conditions in 1917 → conditions in 1919 (eclipse) → conditions in 1921.

10. Application of the explanation-prediction symmetry thesis for inductive-statistical explanations would lead to disastrous results if no temporal constraints were imposed, for then any event could be explained by a record of it. For example, relative widths of tree rings of logs found in an archaeological excavation could explain why a serious drought occurred in north-central Arizona at the end of the fourteenth century, for there is a strong (law-based) inductive inference from the dendrochronological data to the earlier climatic conditions. Likewise, without a temporal constraint, the iridium anomaly at the K-T boundary could explain why a massive body collided with Earth about 65 million years ago. Without appealing to matters of usage, we can say that such pseudo-explanations would not enhance our scientific understanding of the events in question.

11. In Salmon (1984b, p. 171) I characterize causal interactions roughly as intersections of processes (which can be represented by intersecting world-lines in spacetime diagrams) in which each process possesses some characteristic after the intersection that it did not possess prior to the intersection.

12. [This letter was originally published under the title "On the Alleged Temporal Anisotropy of Explanation."]

11

Van Fraassen on Explanation

With Philip Kitcher

T here should be no doubt about the fact that Bas van Fraassen has made substantial contributions to our current understanding of scientific explanation. But we believe that there is reason for doubt as to exactly what the contributions are. Chapter 5 of *The Scientific Image* (1980), "The Pragmatics of Explanation," offers the most detailed account of van Fraassen's view of explanation. We find both the title and the view ambiguous. The purpose of the present discussion is to underscore the difference between a theory of the pragmatics of explanation and a pragmatic theory of explanation. We believe that van Fraassen has offered the best theory of the pragmatics of explanation to date, but we shall argue that, if his proposal is seen as a pragmatic theory of explanation, then it faces serious difficulties.

1. Two Traditional Problems

Before we turn to van Fraassen's positive views, we want to consider his response to the tradition of theorizing about explanation. According to van Fraassen, there are two main problems "of the philosophical theory of explanation." These are "to account for legitimate rejections of explanation requests, and for the asymmetries of explanation" (p. 146). Van Fraassen's solution to the former problem seems to us to be ingenious and substantially correct. His treatment of the asymmetries of explanation we find deeply puzzling.

Within the mainstream of philosophical reflection about explanation, the problem of asymmetries arises because there are arguments that are closely related, that accord equally well with the conditions set down by models of explanation, and that differ dramatically in their explanatory worth. For present purposes assume either that some explanations (including the examples to be considered) are arguments or that some arguments (including those to be considered) provide a basis for acts of explanation.[1] Then the challenge is to differentiate between the argument that derives the length of a shadow from the height of a tower, the elevation of the sun, and the principles of optics and the argument that derives the height of the tower from the length of the shadow, the elevation of the sun, and the principles of optics. The former seems to be (a potential basis for) an explanation, whereas the latter does not.

One line of solution, hinted at by Carl G. (Peter) Hempel (1965a, pp. 252–253) in discussion of an analogous case, is to propose that there is no real difference between the two arguments and that the feeling of difference arises from anthropomorphic ideas from which we ought to liberate ourselves. This is not very convincing, and van Fraassen appears to adopt a more satisfactory method of dissolving the problem. *One* way to understand his fable "The Tower and the Shadow" is as an attempt to show that the claim of explanatory difference is shortsighted. Failing to appreciate that arguments are explanations (the basis of explanations) only relative to context, we assess the explanatory merits of the derivations by tacitly supposing contexts that occur in everyday life. With a little imagination, we can see that there are alternative contexts in which the argument we dismiss would count as explanatory.

In van Fraassen's story a character offers the following explanation of the height of a tower:

> That tower marks the spot where [the Chevalier] killed the maid with whom he had been in love to the point of madness. And the height of the tower? He vowed that shadow would cover the terrace where he first proclaimed his love, with every setting sun—that is why the tower had to be so high. (pp. 133–134)

Now, we grant that van Fraassen's story describes a context in which the utterance of these words constitutes an explanation for the position and height of the tower. But this will solve the *traditional* problem of the asymmetries of explanation only if one can claim that the argument underlying the quoted passage is the argument that the unimaginative have dismissed as nonexplanatory.

It seems obvious that this is not so. For the (D-N; deductive-nomological) argument that provides the basis for the act of explanation van Fraassen relates does not take the form of deducing the height of the tower from the length of the shadow (with the elevation of the sun and the principles of optics as the only other premises). Rather, we begin with some initial conditions about the psychological characteristics of the Chevalier: he wanted to build a tower with certain properties; he knew certain physical facts. Using general principles of rationality, we infer a statement to the effect that the Chevalier came to believe that if he built a tower of the appropriate height on the appropriate spot, it would meet his desiderata. Using yet another principle of rationality, we infer that the Chevalier built the tower to these specifications, and, using background principles about the stability of the height and position of such large physical objects, we conclude that the tower has the height and position it has.

It appears that an obvious way to interpret van Fraassen is mistaken: his story does not provide a context in which an argument wrongly dismissed as explanatory shows its explanatory worth. Moreover, since van Fraassen points out, quite explicitly, the dependence on desires (p. 132), we take him to appreciate that his story does not solve the traditional problem of the asymmetries of explanation. Instead, we construe him as claiming that the problem as we have posed it—a problem that talks about arguments and their merits as explanations (the bases of explanations)—is *mis*posed. Once the topic is approached in terms of van Fraassen's favored pragmatic machinery, we are to see that an *answer* that we might have considered inappropriate can have explanatory worth in the right context.

But this leaves us with puzzles. If we cannot formulate the traditional problem of the asymmetries of explanation in terms of arguments, then how is the problem to be formulated? Does an analogous problem arise within van Fraassen's own theory? Is it resolved by that theory? We shall return to these questions later.

2. Explanations as Answers

According to van Fraassen, an explanation is an answer to a question Q of the form "Why P_k?" where P_k states the fact to be explained—i.e., the explanandum (phenomenon). Any such question can be identified as an ordered triple $\langle P_k, X, R \rangle$, where P_k is called "the topic" of the question, $X = \{P_1, \ldots, P_k, \ldots\}$ is its contrast class, and R is its relevance relation. Such a question is posed in a context that includes a body of background knowledge K. Q also has a *presupposition,* namely,

(a) P_k is true;
(b)   each P_j in X is false if $j \neq k;$
(c)   there is at least one true proposition A that bears relation R to $\langle P_k, X \rangle$.

and (a) and (b) together constitute the *central presupposition* of Q. The why-question Q *arises* in the given context if K entails the central presupposition of Q and does not entail the falsity of (c). That is, it is altogether appropriate to raise Q even if we do not know whether there is a direct answer or not, provided the central presupposition is fulfilled.

If the question does not arise in the context, it should be rejected rather than answered directly. This can be done by offering a corrective answer, i.e., a denial of one or more parts of the presupposition. If the central presupposition is satisfied but (c) is in doubt, a corrective answer to the effect that (c) is false may be suitable.

If the question arises in the given context, it is normally appropriate to provide a *direct answer.* The canonical form of a direct answer to Q is

(∗) P_k in contrast to the rest of X because A.

The following conditions must be met:

(1) A is true.
(2) P_k is true.
(3)   No member of X other than P_k is true.
(4) A bears R to $\langle P_k, X \rangle$.

A is the *core* of the answer, for the answer can be abbreviated "Because A."

Since, typically, the person S_q who asks the question Q might be someone with a different body of knowledge from the respondent S_r, we might be tempted to say that two different contexts are involved. It seems more in keeping with van Fraassen's approach, however, to understand that S_q and S_r are operating in a common context with a common body of background knowledge K determined roughly by the state of science at the time. Thus, K may contain many propositions that neither the questioner nor the respondent knows. Moreover, S_q may have false beliefs that are in conflict with propositions in K. S_r may therefore offer corrective answers to flawed questions by pointing to items in K.

Whether A, the core of the answer to Q, is relevant depends solely on the relevance relation R. If A bears R to $\langle P_k, X \rangle$, then A is, by definition, the core of a relevant answer to Q. This way of stating the matter raises a difficulty. In his informal remarks van Fraassen repeatedly refers to R as a "relevance relation," but he incorporates no relevance require-ment on R in the formal characterization. Indeed, he points to the absence of any problematic constraint that would seek to capture "the inextricably modal or counterfac-tual element" (p. 143). Now, if R happens to be a relevance relation, then it is indeed correct to say that A is relevant to $\langle P_k, X \rangle$. But, as we shall now show, the lack of any constraints on "relevance" relations allows just about anything to count as the answer to just about any question.

3. Relevance Relations

Let P_k be any true proposition. Let X be any set of propositions such that P_k belongs to X and every member of X apart from P_k is false. Let A be any true proposition. Let R be $\{\langle A, \langle P_k, X \rangle \rangle\} \cup S$, where S is any set of ordered pairs $\langle Y, Z \rangle$ such that Y is a proposition and Z is $\langle V, W \rangle$ where V is a proposition and W a set of propositions, one of whose members is V.[2] Then there is a why-question $Q = \langle P_k, X, R \rangle$, and A is the core of a direct answer to Q. Moreover, it is easy to see that, with suitable restrictions on S (i.e., that S contain no $\langle Y, Z \rangle$ such that Y is true and Z is $\langle P_k, X \rangle$), A may be the core of the only direct answer to Q. Hence, for any true propositions P_k and A, there is a why-question with topic P_k such that A is the core of the only direct answer to that question. If explanations are answers to why-questions, then it follows that, for any pair of true propositions, there is a context in which the first is the (core of the) only explanation of the second.

We take it that this result is counterintuitive. Indeed, we would view it as a *reductio* of van Fraassen's account of explanation. How can it be avoided?

One way of blocking the trivialization we have outlined would be to impose restric-tions on relevance relations. We shall consider this possibility later. First, let us note that van Fraassen's theory of explanation comes in two parts: there is a thesis about what answers to why-questions are, and there is a thesis about how to evaluate answers to why-questions. Perhaps we can use the latter part of the story to defend against the trivialization that threatens the former.

According to van Fraassen, we evaluate answers to why-questions on three different grounds. We ask whether those answers are probable in light of our knowledge, we ask whether they *favor* the topic against the other members of the contrast class, and we ask whether they are made wholly or partially irrelevant by other answers that could be given. Using a notion that van Fraassen often employs in his informal remarks but does not define, let us say that an answer is *telling* if it scores well according to these criteria. More exactly, let us propose that an answer is more or less telling according to its performance on the three criteria. We shall be most interested in maximally telling answers. We shall call them *perfect* answers.

Notice that the theory of evaluating answers to why-questions allows us to compare different answers to the same questions. It does not enable us to assess the degree to which an answer to one question is more telling than an answer to another question. If the questions are of the contrived kind that we introduced at the beginning of this section, then there will be no more telling answer to them than the contrived answer. However, we may easily introduce a grading of questions by considering whether they admit of answers that favor their topic.

Let us say that questions are more or less *well founded* to the extent that they admit of telling answers. Thus, a question will be *maximally well founded* if it admits of a perfect answer. Suppose now that P_k is any true proposition, A any proposition, and X any set of two or more propositions such that P_k is its only true member. Let K be a set of propositions that includes both P_k and A, as well as the negations of all the other propositions in X. Then, we claim, there is a why-question whose topic is P_k, whose contrast class is X, such that A is an essential part of a perfect answer to that why-question.

To demonstrate this we need to examine in somewhat more detail van Fraassen's criteria for evaluating answers. On the first criterion, we award high marks to answers if they receive high probability in light of our background knowledge. A corollary of this is that, if the answer belongs to our background knowledge—as is often the case when we give scientific explanations—then it does as well as possible according to this criterion.

The second criterion (favoring) is less straightforward. Van Fraassen's idea is that the answer, to score well, should increase the distance between the probability of the topic and the probabilities of the other members of the contrast class. Typically, the answer *alone* will not redistribute probabilities in this way. Rather, the answer, taken together with certain auxiliary information, will redistribute the probabilities. However, we cannot suggest that the answer plus the total background knowledge K achieves this result; for, in cases where the topic and the negations of the other members of the contrast class belong to K, the suggestion would lead to immediate trivialization. Van Fraassen therefore suggests that the redistribution of probabilities be achieved by the answer in conjunction with "a certain part $K(Q)$ of K," where $K(Q)$ is supposed to be contextually determined.

We need not delve into the problems of deciding exactly what counts as singling out the topic within the contrast class, since we shall use a case in which van Fraassen explicitly recognizes that an answer is maximally successful. He writes: "If $K(Q)$ plus A implies B and implies the falsity of $C, \ldots, N$ then A receives in this context the highest marks for favoring the topic B" (p. 147). [There is a switch in notation here; the topic is B, the contrast class is $\{B, C, \ldots, N\}$].

Van Fraassen's third criterion concerns the availability of superior answers. The answer A loses marks if it has a rival that fares better, perhaps because the rival receives higher probability in light of background knowledge K, perhaps because the rival favors the topic more than A does, perhaps because the rival screens off A from the topic. Now,

A does not have to fear any rival if A belongs to K and if A plus $K(Q)$ implies the topic and the negations of the other members of the contrast class. For, under these circumstances, no rival can be more probable in light of K, no rival can do better at favoring the topic, and no rival can screen A off from the topic.

We conclude that any A belonging to K that, in conjunction with $K(Q)$, implies the topic P_k is a perfect answer to the question $\langle P_k, X, R \rangle$ (provided, of course, that it is an answer to this question).

Let us therefore define a "relevance relation" R as follows: we stipulate that R holds between B and $\langle P_k, X \rangle$ just in case P_k is a logical consequence of B. Let Z be the disjunction of all the propositions in X apart from P_k, and let B be the proposition

$$A \cdot (A \supset P_k) \cdot \sim Z$$

This proposition bears R to $\langle P_k, X \rangle$, and hence it counts as the core of a direct answer to the why-question $\langle P_k, X, R \rangle$. Moreover, by our earlier assumption, A, P_k, and $\sim Z$ belong to K. This means that all the conjuncts in B, and hence B itself, belong to K. Thus, B will be completely successful according to van Fraassen's first criterion for evaluating answers. Because $P_k \cdot \sim Z$ is a logical consequence of B, B maximally favors P_k—and we do not need to worry about how $K(Q)$ is selected since $P_k \cdot \sim Z$ is a consequence of B alone. Finally, because of this implication, there is no reason to fear that B will be screened off by some rival answer. Therefore, B is a perfect answer to $\langle P_k, X, R \rangle$.

We have devised one way of finding, for any pair of true propositions A, P_k, a why-question with P_k as topic to which there is a perfect answer with A as an essential part of its core. Moreover, once we see how the construction we have given is possible, it is easy to generate variations on the same theme. For example, if van Fraassen's account does not contain context-independent principles that preclude the possibility of assigning $(A \supset P_k) \cdot \sim Z$ to $K(Q)$, then it will be possible to claim that A is the core of a direct perfect answer to some question with P_k as topic.

We conclude that the machinery that van Fraassen introduces in his discussion of the evaluation of answers does not avail in protecting him against the kind of trivialization we presented at the beginning of this section. The moral is that, unless he imposes some conditions on relevance relations, his theory is committed to the result that almost anything can explain almost anything. Some kinds of relations R are silly, and why-questions that embody them are silly questions. If we pose silly questions, we should not be surprised to get silly answers.

4. Constraints on R?

Let us now consider a concrete example. Suppose S_q asks why John F. Kennedy died on November 22, 1963, where

$$P_k = \text{JFK died } 11/22/63, \quad X = \{\text{JFK died } 1/1/63, \text{ JFK died } 1/2/63,$$
$$\ldots, \text{JFK died } 12/31/63, \text{JFK survived } 1963\},$$

and R is a relation of astral influence. (One way to define R is to consider ordered pairs of descriptions of the positions of stars and planets at the time of a person's birth and propositions about that person's fate.) An answer with core A might consist of a *true* description of the positions of the stars and planets at the time of JFK's birth. Moreover, using astrological theory as background, one might be able to infer (at least with high probability) that JFK would die on 11/22/63.

We suggest that, in the context of twentieth-century science, the appropriate response to the question is rejection. According to our present lights, astral influence is not a relevance relation. We believe that the positions of the stars and planets on JFK's birthday have no effect on the probability of death on any particular day. Adding the knowledge of those positions does nothing to redistribute the probabilities of death among the members of the contrast class. The moral we draw here—as in the last section—is that van Fraassen's conditions (a)–(c) on answers to why-questions need to be supplemented by adding

(d) R is a relevance relation.

Moreover, we claim that (d) cannot be analyzed simply in terms of demanding that, if A bears R to $\langle P_k, X \rangle$, then A must redistribute the probabilities on X. For we can meet that demand by considering the proposition

$B = A$. (If A, then JFK died on 11/22/63) . JFK did not die on
1/1/63 . . . . JFK did not die on 11/21/63 . JFK did not die
on 11/23/63; . . . . . JFK did not survive 1963)

and defining the relation of astral influence R so that R contains $\langle B, \langle P_k, X \rangle \rangle$.

Once again, let us consider the question from the perspective of van Fraassen's account of evaluating answers. We note, first, that the true description of the positions of stars and planets at JFK's birth accords with our current scientific knowledge. So the answer gets high marks on this score. Second, we ask to what extent A favors P_k vis-à-vis the other members of X. On this criterion A fares poorly (although B, of course, does not). Perhaps an answer that negatively favored the topic might get still lower marks— though it is not clear to us that it should, since discovering a relevant factor seems better than offering an irrelevancy. Third, we must compare A with other answers to Q. This criterion has three parts. (1) Since A is true and since it belongs to our body of knowledge, no other answer can be more probable. (2) Since no astrological answer is relevant, all astrological answers equally fail to favor the topic. (3) Since every astrological answer is irrelevant, screening off is beside the point.

The result is that A is not telling. There is no telling answer to our original question. If we amend the question, we can produce a relative to which B is a maximally telling answer. In our view, both the questions ought to be rejected, and van Fraassen needs to supplement his theory of explanation with an account of relevance relations.

The astrological answer has a further twist, however. As van Fraassen explains, our general background knowledge K—suitably restricted to $K(Q)$ to avoid trivialization— furnishes a *prior distribution* of probabilities over the contrast class X. (Note that, in discussing favoring with respect to the contrived example of section 3, and answer B of this section, we were entitled to take $K(Q)$ to be any subset of K because we had no need of any additional premises in generating the most extreme distribution of probabilities over the contrast class.) Given A, we have a *posterior distribution* of probabilities over X.

(It should *not* be assumed that the prior distribution assigns equal probabilities to all members of *X;* surely, survival beyond 1963 was antecedently more likely than death on any given day, and surely some days are more dangerous than others in the life of a U.S. president.) *A* is the core of a relevant answer to *Q* only if addition of *A* to *K(Q)* would yield a posterior distribution different from the prior distribution. But what sorts of probabilities are these? If they are S_q's personal probabilities, then, given that S_q is a believer in astrology, we might well expect that knowledge of *A* would lead to a different distribution. So *A* would be relevant after all.

Van Fraassen might reply that astrological answers are debarred by his (frequently repeated)[3] remarks that explanations make use of accepted scientific theories. The astrological answers are precluded by the fact that they contain statements that are inconsistent with the background knowledge *K*. But this seems to mistake the purport of our examples. The statement *A belongs* to the background corpus; it is simply a report of the positions of the heavenly bodies at the time of JFK's birth, and we can assume that this report derives from the best current science. Of course, if the answer includes further bits of astrological theory designed to connect *A* with the statement that JFK died on 11/22/63, then van Fraassen will have grounds for ruling it out. But if the favoring of the topic is achieved solely through S_q's personal probabilities, then there is nothing in the answer to which van Fraassen can point as defective. Similarly, there is nothing in *B* that would be debarred on the basis of an appeal to background knowledge, for all the statements in *B* belong to the background corpus.

It should now be clear that these examples work by exploiting the laxity of the conditions on the relevance relation in order to reintroduce 'explanations' that van Fraassen hopes to debar by emphasizing the idea that good explanation must use good science. Unless there are constraints on genuine relevance relations, we can mimic the appeal to defiant beliefs in giving pseudo-explanations by employing deviant relevance relations. Hence, if van Fraassen is serious in his idea that genuine explanations must not make appeal to "old wives' tales," then he ought to be equally serious about showing that relevance is not completely determined by subjective factors. If we are talking about distributions and redistributions of personal probabilities, they must be subject to some kinds of standards or criteria. Coherence is one such criterion, but it cannot be sufficient. To be scientifically acceptable, the redistribution of probabilities must involve differences in objective probabilities (frequencies or propensities) in some fashion.

5. Traditional Problems Revisited

When van Fraassen explicitly discusses kinds of relevance relations, the kinds he picks out are fairly familiar from the literature on scientific explanation: we discover such relations as physical necessitation, being etiologically relevant, fulfilling a function, statistical relevance, and, in the fable "The Tower and the Shadow," a relation of intentional relevance. We have been arguing that there are some relations that ought not to be allowed in *any* context as genuine relevance relations. Thus, there appears to be a distinction to be drawn between the relations that can serve, in some context or another, as relevance relations (paradigmatically those relations that figure in van Fraassen's discussions) and those that cannot (such as the contrived relations of the last two sections).

How the distinction should be drawn depends on a very general issue about scientific explanation. Is there a set of genuine relevance relations that underlie the genuine why-questions for all sciences and for all times? Those who give an affirmative answer will see a full theory of explanation as offering a specification of the kinds of relevance relations that may underlie genuine why-questions. That specification would be strongly context-independent in that it would pick out the candidates for any given context of posing a why-question, and the candidates would always be the same.

But perhaps there is no such invariant set of genuine relevance relations. The set of genuine relevance relations may itself be a function of the branch of science and of the stage of its development. Consider the abandonment of teleological explanations in physics after the scientific revolution. This can be viewed as a modification of the set of relevance relations: in the context of Aristotelian physics the notion of teleological relevance was a genuine relevance relation; in the context of Newtonian physics it lost this status. Uniformitarians (those sympathetic to the view of the last paragraph) will deplore this relativism, contending that the notion of teleological relevance never was a genuine relevance relation and that its status was exposed during the scientific revolution. They will accuse relativists of confusing the variation in beliefs about relevance with the relativity of relevance itself.

We do not need to settle this dispute because, on both accounts, there is a nontrivial task of distinguishing genuine relevance relations from the contrived relations of the last two sections. Just as pluralists about literary works will insist that there are many interpretations of *Hamlet* while denying that any reader's fancy counts as an interpretation, so too relativists should concede that there are some relations that are not genuine relevance relations at any historical stage of any science. The most thorough version of a relativist account of explanation would consist in specifying those principles that determine, for each historical stage of each science, the selection of certain relations as genuine relevance relations. A more modest (and more sensible) approach would be to consider some particular science (or sciences) throughout some particular period and to identify the pertinent relevance relations. Thus, one might focus on contemporary physics and try to distinguish the associated genuine relevance relations from the residue of relations—the contrived, the discarded, and so forth.

Uniformitarians, ambitious relativists, and modest relativists all face the same kind of task. Although we do not know which version of the task he would wish to undertake, van Fraassen has remarked to us (in conversation) that he recognizes the importance of distinguishing genuine relevance relations and that he takes Aristotle's list of types of causes to be a promising start on drawing the distinction. We now want to suggest that completion of the task will require that van Fraassen solve most (if not all) of the traditional problems that have beset theories of explanation. For, depending on one's commitment on the large issue we have left unresolved, these problems take the form of showing why certain relations do not belong to the single set of genuine relevance relations that is associated with all sciences at all times, or of showing why certain relations do not belong to any of the sets of genuine relevance relations associated with different sciences at different times, or of showing, for some particular science(s) and period of interest, why certain relations do not belong to the associated set of genuine relevance relations. Henceforth, we intend that our presentations of problems should be systematically ambiguous among these forms.[4]

To simplify matters, we shall confine our attention to difficulties that arise in what Hempel would have viewed as deductive explanation. Consider the simple relation of derivation. This relation holds between A and $\langle P_k, X \rangle$ just in case there is a (first-order) derivation of P_k from A plus additional premises in $K(Q)$. We can define any number of relations by imposing constraints on the kinds of statements that should figure in the premises. Thus, to recall a famous Hempelian example, let P_k be the proposition that Horace is bald and R be the relation of Greenbury-school-board-derivation that holds between A and P_k just in case A is a conjunction of propositions one of whose conjuncts is the proposition that Horace belongs to the Greenbury school board, P_k is derivable from A, and there is no conjunct in A that could be deleted while still enabling P_k to be derivable from the result. Suppose that X includes the propositions that Horace is bald and that Horace is not bald. Let A be the proposition that Horace is a member of the Greenbury school board and that all members of the Greenbury school board are bald. $\langle P_k, X, R \rangle$ is a van Fraassen why-question to which A is a direct answer, and a perfect answer to boot.

We claim that the question we have just artificially constructed is not a genuine why-question and that A is no explanation of Horace's baldness. Moreover, we suggest that most (if not all) of the examples of nonexplanatory arguments that Hempel hoped to exclude—both those he succeeded in debarring and those that have caused persistent problems for the theory of D-N explanation—give rise to corresponding "relevance" relations that van Fraassen ought to exclude. As an illustration, let us return to his solution to the problem of the asymmetries of explanation in the light of what we have discovered about his treatment of why-questions.

The proposition that the tower was built on the spot where the Chevalier killed his beloved and that it was built of such a height that its shadow would fall across the terrace where he first vowed his love is relevant to the topic of the question "Why is the height of the tower h?" if we construe the relevance relation to be that of intentional relevance. That is just another way of putting the point that there is a perfectly good Hempelian argument that derives the height of the tower from premises about the Chevalier's attitudes and from psychological laws. But if we are moved by the traditional problem of the asymmetries of explanation, what we want to know is whether there is a context in which the statement "The length of the shadow is l" answers the question "Why is the height of the tower h?" in virtue of the fact that the assertion about shadow length, together with premises about the angle of elevation of the sun and the propagation of light [which may be relegated to the background $K(Q)$] favor the topic as against other propositions ascribing different heights. For that (or something very like it) is the translation into van Fraassen's idiom of the asymmetry problem that has bedeviled Hempel and his successors.

Now, unless we impose very delicate constraints on relevance relations, it is easy to contrive a maximally well founded question $\langle P_k, X, R \rangle$ such that the proposition ascribing shadow length will be the core of a perfect answer. The trick should be apparent by now: take P_k to be the proposition that ascribes the actual height to the tower, let X be a collection of propositions ascribing different heights, let R be the relation of *censored Hempelian derivation*—a relation that holds between A and $\langle P_k, X \rangle$ just in case there is a D-N argument that derives P_k from A plus additional premises in $K(Q)$. (Quite evidently, we could impose additional constraints so as to rule out the use of the psychological principles on which van Fraassen's account turns, and thus to ensure that the only available D-N arguments are those which invert the usual order of explanatory derivation). We take $K(Q)$ to be fixed in such a way as to include the proposition ascribing the elevation of the sun and the laws of propagation of light. This is surely quite reasonable, for some such $K(Q)$ will have to be allowed if we are to countenance the proposition that the height of the tower is h as the core of an answer to the question, Why is the length of the shadow l? So van Fraassen's theory allows explanations that correspond to those D-N explanations that intuitively "run the wrong way."

We suggest that this is a mistake. Just as the contrived questions of sections 3 and 4 should be eliminated by the imposition of constraints on relevance relations, so too the question of the last paragraph and its accompanying perfect answer ought to be banished. For otherwise van Fraassen's account of explanation will be deficient in exactly the way that Hempel's own treatment was. Every kind of asymmetry that arises for the D-N model can be generated within van Fraassen's framework. This means that, far from solving the problem of the asymmetries of explanation, van Fraassen presupposes a solution to that problem. Thus, if we are right, van Fraassen has offered a beautiful treatment of the pragmatics of explanation which should be viewed as a supplement, rather than a rival, to the traditional approaches to explanation.

6. Conclusion

As we have remarked (see note 3), there are many suggestions in van Fraassen's text that he does not intend to offer an 'anything goes' account of explanation. In the last section we have attempted to show that this intention ought to commit him to solving most (if not all) of the traditional problems of the theory of explanation. We want to conclude by considering an obvious question. If we interpret van Fraassen as supposing that there are constraints on why-questions and their answers, how does this affect the general argumentative strategy of *The Scientific Image?*

Van Fraassen's discussion of scientific explanation is part of an effort to show that theoretical virtues beyond the saving of the phenomena are pragmatic. That argument eliminates a certain strategy for defending theoretical realism. If the realist proposes that (1) there is an objective criterion of explanatory power that distinguishes among empirically equivalent theories, and (2) theories with greater explanatory power have a stronger title to belief, then the doctrine of *The Scientific Image* appears to oppose the proposal by denying (1).[5] If we are correct in our assessment of van Fraassen's position, then it seems that the realist can get at least as far as (1). For, if it is once granted that we can produce statements that favor the topic of a why-question but that do not stand in any objective relevance relation to that topic (or, more exactly, to the ordered pair of topic and contrast class), then it appears that a theory may save the phenomena without generating answers to why-questions founded on genuine relevance relations.

We have argued that, if he is to avoid the 'anything goes' theory of explanation, van Fraassen must offer a characterization of objective relevance relations that, in effect, overcomes the traditional problems of the theory of explanation. Now, within the traditional theories, there is ample room for prediction without explanations: we can have deductive arguments that fail to explain their conclusions, assemblages of statistical relevance relations that bestow high probability on a statement without explaining it. Once van Fraassen has introduced analogous distinctions within the theory of why-questions and their answers, through the provision of constraints on genuine relevance relations that separate mere favoring from the adducing of relevant information, there can be theories yielding statements that favor the set of topics in a given class (or even imply those topics) without generating answers to any genuine why-question with any of those topics. We would thus have the basis for claiming that such theories are objectively inferior to their rivals that do furnish explanatory answers.

The consequence would be that van Fraassen would have to revise his account of what it is to accept a scientific theory by adding the idea that acceptance involves believing that the theory has explanatory power as well as believing that it saves the phenomena (or, perhaps, believing that the theory offers the best tradeoff between saving the phenomena and having explanatory power). Indeed, he seems to take just this tack in the article cited in note 5. Since van Fraassen can still avail himself of a (different) distinction between acceptance and belief, this consequence should be seen as providing only the entering wedge for an argument for realism.

We conclude that, if van Fraassen avoids the Scylla of the 'anything goes' theory of explanation, then he is plunged into what he would view as the Charybdis of supposing that there is an objective virtue of theories distinct from their salvation of the phenomena. From our perspective, Scylla is (to say the least) uninviting, but Charybdis feels like the beginning of the way home.

Notes

This coauthored essay grew out of discussions with Philip Kitcher at the Minnesota Center for the Philosophy of Science during the fall of 1985. We would like to acknowledge the National Endowment for the Humanities for its support of an institute that made our collaboration possible. We are also grateful to Bas van Fraassen for some helpful clarifications of his ideas, made in response to an earlier draft. Parenthetical page references are to *The Scientific Image* (van Fraassen, 1980).

1. The idea that arguments might provide the basis for acts of explanation is suggested in Kitcher (1981) and is articulated in some detail in Railton (1981). Although one of us (WCS) rejects the thesis that explanations are arguments, or involve arguments in any essential way, this issue does not affect the present discussion of van Fraassen's views in any significant fashion.

2. [In fact, the ordered pair $\langle A, \langle P_k, X \rangle \rangle$ counts as a relevance relation consisting of a single ordered pair. We have added the set S to preclude avoidance of the difficulty by excluding such trivial relations.]

3. At the outset of his exposition of his theory of why-questions, van Fraassen remarks, "This evaluation [of answers] proceeds with reference to the part of science accepted as

'background theory' in that context" (p. 141). Earlier he had remarked that "To ask that . . . explanations be scientific is only to ask that they rely on scientific theories and experimentation, not on old wives' tales" (p. 129), and "To sum up: no factor is explanatorily relevant unless it is scientifically relevant; and among the scientifically relevant factors, context determines explanatorily relevant ones" (p. 126). In conclusion he says, "To call an explanation scientific is to say nothing about its form or the sort of information adduced, but only that the explanation draws on science to get this information (at least to some extent) and, more importantly, that the criteria of evaluation of how good an explanation it is, are being employed using a scientific theory" (pp. 155–156).

4. We are very grateful to an editor of the *Journal of Philosophy* who raised a question which prompted us explicitly to distinguish these three ways of pursuing the theory of explanation and, thus, substantially to improve on the formulations of an earlier draft.

5. It is clear from a subsequent paper (van Fraassen, 1983), in which he discusses Clark Glymour's views about explanation, that van Fraassen would also object to (2).

Part III

CAUSALITY

The essays in this part develop the details of the theory sketched in essay 1, "A New Look at Causality." The theory has two main features: first, it identifies causal connections with physical processes that transmit causal influence from one spacetime location to another; second, it incorporates probabilistic features of causality, keeping open the possibility that causality operates in indeterministic contexts.

Essay 12, "An 'At-At' Theory of Causal Influence," introduces *causal transmission*. To the best of my knowledge it is the first explicit treatment of this key concept to be found in the philosophical literature. Although at the time it was written I treated causal processes in terms of capacity to transmit marks, the basic concept of transmission works equally well in the conserved quantities theory introduced in essay 16, "Causality without Counterfactuals." In my view, the 'at-at' theory answers Hume's basic challenges to the concept of causality.

Essay 13, "Causal Propensities: Statistical Causality versus Aleatory Causality," argues that indeterministic causality cannot be adequately explicated by means of statistical relevance relations alone. Physical connections are also required. The same point applies to deterministic causality. This essay sets my view of causality apart from standard treatments in terms of abstract relations such as necessary condition, sufficient condition, and statistical relevance, which, in and of themselves, do not provide physical—or causal—*connections*.

Essay 14, "Probabilistic Causality," surveys the three classic theories offered by Hans Reichenbach (1956), I. J. Good (1961–62), and Patrick Suppes (1970). These were the only theories available in 1980, when this essay was first published. It points out severe problems with each and offers suggestions for a more satisfactory approach. Although I unqualifiedly reject *post hoc, ergo propter hoc,* the literature on probabilistic causality burgeoned in the 1980s and 1990s. The topic has become a major area of concern to many philosophers.

Essay 15, "Intuitions—Good and Not-So-Good," confronts conflicting intuitions about the character of probabilistic causality. Using several important additional examples, it addresses the responses of Good to the criticisms raised in the preceding essay. It exhibits complexities that arise when we try to accommodate sophisticated intuitions about probabilistic causality in an explicitly articulated theory.

Essay 16, "Causality without Counterfactuals," abandons the explication of causal processes in terms of capacity for mark transmission and substitutes transmission of conserved quantities. This new theory, based on the seminal ideas of Phil Dowe (1992c), overcomes a number of difficulties faced by my previous view. It eliminates a philosophically undesirable dependence on counterfactual conditions, it provides analyses of Y and λ types of causal interactions, and it suggests an avenue for avoiding problems about laws of nature.

Essay 17, "Indeterminacy, Indeterminism, and Quantum Mechanics," uses a historical approach to explore the extent to which quantum theory mandates indeterminism. It discusses explicitly the crucial distinction—often overlooked in the philosophical literature—between indeterminacy of physical quantities and causal indeterminism. It pursues at a more sophisticated level issues raised in essay 2, "Determinism and Indeterminism in Modern Science."

12

An "At-At" Theory of Causal Influence

To untutored common sense, and to many scientists uncorrupted by philosophical training, it is evident that causality plays a central role in scientific explanation. An appropriate answer to an explanation-seeking question beginning
with "why" will normally begin with "because," and the causal involvements of the
answer are usually not hard to find. In "Causal and Theoretical Explanation" (essay 7)
and "A Third Dogma of Empiricism" (essay 6) I have tried to exhibit some of the causal
aspects of scientific explanation and to offer rational grounds for insisting on the causal
component. This attempt to put the "cause" back into "because" does, however, go
against an influential philosophical tradition.

The concept of causality has been philosophically suspect ever since David Hume's
devastating critique (first published in 1739). Hume wrote:

> Here is a billiard ball lying on the table, and another moving toward it with rapidity.
> They strike; the ball which was formerly at rest now acquires a motion. This is as
> perfect an instance of the relation of cause and effect as any which we know either by
> sensation or reflection. Let us therefore examine it. It is evident that the two balls
> touched one another before the motion was communicated, and that there was no
> interval betwixt the shock and the motion. *Contiguity* in time and place is therefore a
> requisite circumstance to the operation of all causes. It is evident, likewise, that the
> motion which was the cause is prior to the motion which was the effect. *Priority* in time
> is, therefore, another requisite circumstance in every cause. But this is not all. Let us try
> any other balls of the same kind in a like situation, and we shall always find that the
> impulse of the one produces motion in the other. Here, therefore, is a *third* circum-
> stance, viz., that of *constant conjunction* betwixt the cause and the effect. Every object
> like the cause produces always some object like the effect. Beyond these three circum-
> stances of contiguity, priority, and constant conjunction I can discover nothing in this
> cause. (1740)

This discussion is, of course, more notable for factors Hume was unable to find than for those he enumerated. In particular, he could not discover any "necessary connections" relating causes to effects, or any "hidden power" by which the cause "brings about" the effect. This classic account of causation is rightly regarded as a landmark in philosophy.

The reputation of causality was further damaged by some oft-quoted remarks of Bertrand Russell at the beginning of his famous essay (written in 1913), "On the Notion of Cause," where he says:

> All philosophers, of every school, imagine that causation is one of the fundamental axioms or postulates of science, yet, oddly enough, in advanced sciences such as gravitational astronomy, the word "cause" never occurs. Dr. James Ward . . . makes this ground for complaint against physics: the business of those who wish to ascertain the ultimate truth about the world, he apparently thinks, should be the discovery of causes, yet physics never seeks them. To me it seems that . . . the reason why physics has ceased to look for causes is that, in fact, there are no such things. The law of causality, I believe, like much that passes muster among philosophers, is a relic of a bygone age, surviving, like the monarchy, only because it is erroneously supposed to do no harm. (1929, p. 180)

When Carl G. Hempel and Paul Oppenheim offered the first detailed elaboration of the deductive-nomological pattern of scientific explanation in their classic 1948 paper, they suggested that they were dealing with causal explanation ([1948] 1965, p. 250), but Hempel subsequently backed away from that interpretation (1965a, pp. 351–354). It seems clear that Hume's critique of causation has made philosophers—especially those with a scientific or empiricist bent—rather chary of the concept.

Modern physics has not, however, managed to avoid the concept of causality altogether. For example, it plays a fundamental role in Einstein's special theory of relativity ([1905] 1923). In that theory the speed of light constitutes an upper bound on the speed at which signals can be transmitted; light is what Hans Reichenbach ([1928] 1957) called a "first signal." A basic consequence of that theory is that no process capable of transmitting information can be propagated faster than light.[1] There are, nevertheless, certain pseudo-processes that can travel with arbitrarily high velocities, not limited by the speed of light. It thus becomes a matter of crucial importance to establish a criterion that will distinguish pseudo-processes from the kinds of genuine processes that are capable of transmitting signals or information. It seems natural to refer to the genuine processes as "causal processes," for it is by virtue of the ability of such processes to transmit causal influences that they can transmit signals or information. If genuine signals could be propagated with arbitrarily high speeds, absolute simultaneity would be reinstated. We see, then, that one of the most basic differences between classical mechanics and relativity theory hinges on this very distinction between pseudo-processes and genuine causal processes.

In "Causal and Theoretical Explanation" (essay 7) I placed heavy emphasis on the role of causal processes in scientific explanation, and I made much of the distinction between causal processes and pseudo-processes. Reichenbach's "mark method" was the criterion used to effect this distinction ([1928] 1957, §23). A simple example will illustrate the use of this method as well as the distinction between causal processes and pseudo-processes. Consider a rotating spotlight, mounted in the center of a circular room, which casts a spot of light on the wall. A light ray traveling *from* the spotlight *to* the wall is a causal process; the spot of light moving around the walls constitutes a pseudo-process. The former process occurs at the speed of light; the latter 'process' can go on at arbitrarily high velocities, depending on the size of the room and the rate of rotation of the light source. The speed of light places no restrictions on the velocity of the pseudo-process.[2]

The fact that the beam of light traveling from the light source to the wall is a causal process can be revealed by a simple experiment. If a red filter is interposed in the beam near its source, the color of the spot of the wall will be red. This 'mark' is transmitted along the beam. It is obvious how the transmission of such marks could be employed to send a message:

> Red if by land and blue if by sea.
> And I on the opposite shore will be
> Ready to ride and spread the alarm
> To every Middlesex village and farm.[3]

It is equally evident, I believe, that no information can be sent via the moving spot on the wall. If you are standing near the wall at one side of the room, and someone else is stationed at a diametrically opposite point, there is nothing you can do to the passing spot of light that will convey any information—e.g., "The British are coming!"—to the other person. Interposing a red filter may make the spot red in your vicinity, but the 'mark' will not be retained as the spot moves on.

There are two distinct causal ingredients in mark transmission. In the first place, the imposition of a mark involves a *causal interaction,* which is a localized affair. At some spatiotemporally restricted locale an interaction—such as interposition of a red filter—takes place. I am making no attempt to provide an analysis of causal interactions (but see "Why Ask, 'Why?'?" [essay 8]). In the second place, given a causal interaction that results in some modification of the 'process' in question—whether pseudo- or causal—there is the matter of propagation. Whether, in a given case, the result of the interaction will be transmitted is a question that can, in principle, readily be settled by experiment. We can produce interactions of one sort and another and see whether the resulting modifications in the process are preserved at other stages of the process.[4]

Hume, to the best of my knowledge, made no distinction between causal interactions and causal processes. Many of his examples, such as the collision between two billiard balls, are interactions. To whatever extent he did deal with causal processes, he probably regarded processes as continuous series of interactions. In the eighteenth century there was not much occasion to consider the distinction between causal processes and pseudo-processes. Such problems become acute, I believe, only when one begins to deal with finite speed limits on signal transmission, and the consequent undermining of absolute simultaneity.

Having made the distinction between causal processes and pseudo-processes, we must not forget that a pseudo-process may possess a high degree of regularity. The spot of light moving regularly around the wall provides a clear example. Still, we must insist on a fundamental difference: in the causal process the regularity is produced within the process itself, while in a pseudo-process the regularity is generated from sources external to the 'process'. In free space, light beams travel in regular and predictable paths without further assistance beyond the source that emits the light. This is a causal process. In contrast, the spot of light on the wall, once initiated, will not persist without external aid; if the spotlight at the center of the room is turned off, the spot will soon vanish, regardless of its prior history. Reichenbach's "mark method" is a criterion for distinguishing between processes of these two types.

In his later epistemological writings, Russell, though still scornful of naive philosophical conceptions of causation, attached considerable importance to causal processes.

> That there are such more or less self-determined causal processes is in no degree logically necessary, but is, I think, one of the fundamental postulates of science. It is in virtue of the truth of this postulate—if it is true—that we are able to acquire partial knowledge in spite of our enormous ignorance. (1948, p. 459)

He seems to have come close to a recognition of the significance of the capability of causal processes to transmit information.

> A "causal line," as I wish to define the term, is a temporal series of events so related that, given some of them, something can be inferred about the others whatever may be happening elsewhere. A causal line may always be regarded as the persistence of something—a person, a table, a photon, or what not. Throughout a given causal line, there may be constancy of quality, constancy of structure, or a gradual change of either, but not sudden change of any considerable magnitude. (ibid.)

Hume's three requirements for causal relations—contiguity, priority, and constant conjunction—are, at least roughly speaking, satisfied by causal lines as conceived by Russell. Unfortunately, pseudo-processes would also seem to qualify as causal lines under Russell's definition. To Hume's three criteria we must add a fourth— understandably overlooked by Hume: the ability to transmit a mark. This criterion will be used to distinguish those processes that transmit information or causal influence by virtue of their inner structure from those processes that exhibit a high degree of regularity but that do not transmit their structure internally.

When we characterize causal processes partly in terms of their ability to transmit a mark, we must face squarely the question whether we have violated the kinds of strictures Hume so emphatically expounded. He warned against the uncritical use of concepts such as power and necessary connection. Is not the *ability to transmit* a mark an example of just such a mysterious power?

Kenneth Sayre seems to be expressing misgivings on just this score when, after acknowledging the distinction between causal interactions and causal processes, he writes:

> [T]he causal process, continuous though it may be, is made up of individual events related to others in a causal nexus. . . . [I]t is by virtue of the relations among the members of causal series that we are enabled to make the inferences by which causal processes are characterized. [I]f we do not have an adequate conception of the relatedness between individual members in a causal series, there is a sense in which our conception of the causal process itself remains deficient. (1977, p. 206)

The "at-at" theory of causal transmission is an attempt to remedy this deficiency.

Does this remedy illicitly invoke the sort of concepts Hume proscribed? I think not. Ability to transmit a mark can be viewed as a particularly important species of constant conjunction—the sort of thing Hume recognized as observable and admissible. It is a matter of performing certain kinds of experiments. If we place a red filter in a light beam near its source, we can observe that the mark—redness—appears at all places to which the beam is subsequently propagated. This fact can be verified by experiments as often as we wish to perform them. If, contrariwise, we make the spot on the wall red by placing a filter in the beam at one point just before the light strikes the wall (or by any other means we might devise), we will see that the mark—redness—is not present at all other places in which the moving spot subsequently appears on the wall. This, too, can be verified by repeated experimentation. Such facts are straightforwardly observable.

The question can still be reformulated. What do we mean when we speak of transmission—how does the process *make* the mark appear elsewhere within it? There is, I believe, an astonishingly simple answer. The transmission of a mark from point A in a causal process to point B in the same process *is* the fact that it appears at each point between A and B without further interactions. If A is the point at which the red filter is inserted into the beam going from the spotlight to the wall, and B is the point at which the beam strikes the wall, then only the interaction at A is required. If we place a white card in the beam at any point between A and B, we will find the beam red at that point. We can, of course, arrange for a red mark to appear at points other than B in the pseudo-process, but only by means of additional interactions. Indeed, we could make the spot on the wall red throughout its journey by arranging for a red filter near the wall to travel along with the beam in such a way as to be always interposed in the beam just before it strikes the wall. (The same result could be accomplished, of course, by outfitting the spotlight itself with a red lens, but that would not count as *marking* the spot on the wall, for it would not constitute a *local* interaction with the spot that moves around the wall.) There is, however, a serious limitation on this possibility—one that brings us back to one of the fundamental differences between a causal process and a pseudo-process. A red filter is a physical object, and a moving physical object constitutes a causal process. If the spot on the wall moves at less than the speed of light, then the moving filter can keep up with the spot, but if the spot moves at a super-light velocity, it will be impossible in principle for the filter to keep up with it. The basic thesis about mark transmission can be stated as follows: *A mark that has been introduced into a process by means of a single intervention at point A is transmitted to point B if and only if it occurs at B and at all stages of the process between A and B without additional interventions.*

This account of mark transmission—which is the proposed foundation for the concept of propagation of causal influence—may seem too trivial to be taken seriously. I believe such a judgment would be mistaken. My reason lies in the close parallel that can be drawn between the foregoing solution to the problem of mark transmission and the solution of an ancient philosophical puzzle.

About 2,500 years ago Zeno of Elea enunciated some famous paradoxes of motion, including the well-known paradox of the arrow. This paradox was not adequately resolved until the early part of the twentieth century. To establish an intimate connection between this problem and our problem of causal transmission, two observations are in order. First, a physical object (such as the arrow) moving from one place to another constitutes a causal process, as can easily be demonstrated by application of the mark method—e.g., initials carved on the shaft of the arrow before it is shot are present on the shaft after it hits its target. Second, Zeno's paradoxes were designed to prove the absurdity not only of motion but also of process and change. Commenting on this relationship, Henri Bergson remarked that "every attempt to reconstitute change out of states implies the absurd proposition, that movement is made of immobilities" (quoted in Salmon, 1970a, p. 63). In response to this Bergsonian challenge, Russell replies (somewhat cryptically perhaps), "Weierstrass, by strictly banishing all infinitesimals, has at last shown that we live in an unchanging world, and that the arrow, at every moment of its flight, is truly at rest. The only point where Zeno probably erred was in inferring (if he did infer) that, because there is no change, therefore the world must be in the same state at one time as at another. This consequence by no means follows" (quoted in Salmon, 1970a, p. 23).

The solution of the arrow paradox to which Russell refers has been aptly called "the at-at theory of motion." Using the definition of a mathematical function supplied in the nineteenth century by Cauchy, it is pointed out that the mathematical description of motion is a function that pairs points of space with instants of time. To move from A to B is simply to occupy the intervening points at the intervening moments. It consists in being *at* particular points *at* the corresponding instants (hence the name of the theory). There is no additional question as to how the arrow *gets from* point A *to* point B; the answer has already been given: by being at the intervening points at the intervening moments. The answer is emphatically *not* that the arrow gets from A to B by zipping through the intermediate points at high speed.[5] Moreover, there is no additional question about how the arrow gets from one intervening point to another; the answer is the same, namely, by being at the points between them at the corresponding moments. And clearly there can be no question about how the arrow gets from one point to the next, for in a continuum there is no next point. I am convinced that Zeno's arrow paradox is a profound problem concerning the nature of change and motion, and that its resolution by means of the at-at theory of motion represents a distinctly nontrivial achievement. The fact that this solution can—if I am right—be extended in a direct fashion to provide a resolution of the problem of mark transmission is an additional laurel.

The at-at theory of mark transmission provides, I believe, an acceptable basis for the mark method, which can in turn serve as the means to distinguish causal processes from pseudo-processes. Causal processes play a fundamental role in physical theory, and as Russell correctly observed, their existence has profound epistemological significance. Causal processes are, of course, governed by natural laws; these laws constitute regularities whose presence can be empirically confirmed. Such regularities, presumably, represent the kinds of constant conjunctions to which Hume referred. The mark method may be said, roughly speaking, to provide a means for distinguishing causal regularities from other types of regularity in the world, including those that may be associated with pseudo-processes.

The world contains a great many types of causal processes: transmission of light waves, motions of material objects, transmission of sound waves, persistence of crystalline structure, etc. Processes of any of these types may occur without having any mark imposed. In such instances the processes still qualify as causal. *Ability* to transmit a mark is the criterion of a causal process; processes that are actually unmarked may be causal. Unmarked causal processes exhibit some sort of persistent structure; in such cases we say that the structure is transmitted within the causal process. Pseudo-processes may also exhibit persistent structure; in these cases we maintain that the structure is *transmitted not* by means of the process itself but by some other agency. The basis for saying that the regularity in the causal process is transmitted via the process lies in the ability of the causal process to transmit a modification in its structure resulting from an interaction—a mark. In offering the at-at theory of mark transmission as a basis for distinguishing causal processes from pseudo-processes, we have furnished an account of the transmission of information and the propagation of causal influence without appealing to any of the "secret powers" which Hume's account of causation so soundly proscribed.

Notes

I would like to express my gratitude to the National Science Foundation for support of research on scientific explanation and related topics.

1. I am ignoring speculations about "tachyons"—particles with velocities greater than light—because, first, if there were such things and *if they could be used to transmit signals,* the logical structure of special relativity would be seriously affected; and second, all efforts to discover tachyons have failed, so there is no reason to believe such things exist.

2. A dramatic example is given by the pulsar in the Crab nebula. Pulsars are believed to be rapidly rotating neutron stars which emit radiation in the radio frequency range, somewhat analogously to our rotating spotlight. This particular pulsar is located about 6500 light years away from us, and we receive about 30 pulses per second from it. Drawing a circle with the pulsar at the center and the distance to Earth as its radius, we get a circumference of approximately 41,000 light years. The "spot" of radiation, going around 30 times per second, traverses the circumference of this circle about 10^9 times per year. This amounts to a velocity of approximately 4×10^{13} times the speed of light (i.e., thousands of billions of times faster than light). (For an important application of this distinction in contemporary science, see "Quasars, Causality, and Geometry" [essay 25].)

3. With apologies to Henry Wadsworth Longfellow, as well as to my readers.

4. It is, of course, quite inessential that marks be the result of human intervention rather than other nonhuman or inanimate agency. Moreover, it would be a mistake to suppose that every interaction with a causal process produces a mark that will be transmitted. A process qualifies as causal if it is capable of transmitting some marks.

5. See Salmon (1970a, pp. 23–24) or (1975b, pp. 40–42) for a fuller discussion of the at-at theory of motion and its resolution of the paradox of the arrow. This discussion treats explicitly the bearing of the concept of instantaneous velocity on the problem. (In the first printing of Salmon [1970a] the line that should have been the fourth line of the last paragraph on p. 23 was inadvertently displaced to the top of the page: in later printings this error has been corrected.)

13

Causal Propensities

Statistical Causality versus Aleatory Causality

Like its immediate predecessor, this essay was written before my conversion to the conserved quantity theory. It explicates *causal processes* in terms of capacity for transmitting marks and *causal interactions* in terms of intersections of processes in which lasting modifications occur. This was a view I held prior to Phil Dowe's (1992c) exposition of the conserved quantity theory. Nevertheless, the present essay makes a point of crucial importance; it clearly distinguishes two fundamentally distinct approaches to causality of either the probabilistic or the deterministic variety. Whether causal processes are characterized in terms of mark transmission or transmission of conserved quantities, the notion that causal processes provide physical causal connections has deep philosophical import. Likewise, whether causal interactions are characterized in terms of mutual modifications of processes or in terms of exchanges of conserved quantities, the recognition of causal interactions as agents of causal modification also has basic philosophical significance. Whatever explications of causal processes and causal interactions are adopted, the utility of these concepts in analyzing cause-effect relations is unaffected. Moreover, the importance of the distinction between this physical approach and the approach that analyzes causality in terms of constant conjunction and/or statistical correlations cannot be overemphasized.

For many years I have been thinking about scientific explanation, especially statistical explanation. From the beginning I disagreed with Carl G. Hempel (1965b, §3) on this subject. He claimed that high probability is a requirement for acceptable statistical explanations; I argued that we need, instead, relations of statistical relevance (Salmon, 1971). At the same time, I was perfectly aware that statistical relations by themselves are not sufficient: in addition, we need to appeal to causal relations. For example, there is a strong correlation between the reading on a barometer and the occurrence of a storm, but the falling reading on the barometer does not produce the storm and does not explain it.

The explanation demands a cause. Since the explanation is statistical, the cause has to be of a probabilistic sort.

Around 1971 I hoped that it would be possible to define probabilistic causality in terms of statistical concepts such as Hans Reichenbach's *conjunctive fork* and his *screening-off* relation, but by the end of that decade I no longer saw any possibility of doing so. I gave my reasons in (essay 14) "Probabilistic Causality" and again in *Scientific Explanation and the Causal Structure of the World* (1984b, chap. 7). For example, Reichenbach used the conjunctive fork to define the relation of a common cause; however, there are events that constitute a conjunctive fork that does not contain a common cause.[1] It was necessary to define the concept of a *causal process* as a way to distinguish conjunctive forks in which there are bona fide common causes from those in which they do not exist. Causal processes are the key because they furnish the links between the causes and their effects.[2]

I found, moreover, that there are two types of causal forks. In addition to the conjunctive fork, we must define the *interactive fork,* in which two causal processes intersect each other. In this intersection both processes are modified, and the changes persist in these processes beyond the point of intersection. An interactive fork *constitutes* a causal interaction. It is not possible to define this type of causal fork in statistical terms.

It is extremely important to understand the profound difference between the screening-off relation and conjunctive forks on the one hand, and causal processes and causal interactions on the other. The former can be defined in statistical terms; the latter cannot be defined in this way. Causal processes and causal interactions are physical structures whose properties cannot be characterized in terms of relationships among probability values alone. Let us consider a few examples.

(1) There is a very small probability that an American man, selected at random, will contract paresis. There is a somewhat higher probability that an American man who has had sexual relations with a prostitute will contract paresis. The sexual relation with the prostitute is statistically relevant to paresis. In addition, there is a still higher probability that an American man who has contracted syphilis will develop paresis. Syphilis is also statistically relevant to paresis. However, for a man who has contracted syphilis, the relation with the prostitute is no longer statistically relevant. Syphilis screens off the visit to the prostitute from the paresis. Let

A = American men
B = having sexual relations with a prostitute
C = contracting syphilis
D = contracting paresis

Then

$$P(D|A) < P(D|A.B) < P(D|A.B.C) = P(D|A.C) \tag{1}$$

Formula (1) defines the screening-off relation; for a person of type A, C screens off B from D. We should take careful note of the completely statistical nature of this definition.

(2) Two young people take an outing in the country, and they collect some mushrooms. In the afternoon they cook the mushrooms and eat them. In the evening they suffer severe nausea. The common cause of their condition is the consumption of

poisonous mushrooms. For any person there is a small probability of becoming ill any evening; for any two people there is an extremely small probability that both will get sick on the same evening by chance. In this case, however, it is *not* by chance that they become ill the same evening, for there is a common cause. Here we have a conjunctive fork. Let

$$A = \text{illness of a woman}$$
$$B = \text{illness of a man}$$
$$C = \text{eating the mushrooms}$$

Then

$$P(A.B|C) = P(A|C) \times P(B|C) \tag{2}$$

$$P(A.B|\bar{C}) = P(A|\bar{C}) \times P(B|\bar{C}) \tag{3}$$

$$P(A|C) > P(A|\bar{C}) \tag{4}$$

$$P(B|C) > P(B|\bar{C}) \tag{5}$$

Let us observe, incidentally, that the following formula (6) can be derived from (2)–(5):[3]

$$P(A.B) > P(A) \times P(B) \tag{6}$$

This formula says that A and B are not statistically independent. When there is a common cause C, this cause explains the statistical dependence between A and B. Formulas (2)–(5) define the conjunctive fork. We should note again the completely statistical character of this definition.

(3) A causal process is defined as a process that has the ability to transmit a mark. For example, when a bullet is fired from a gun, the gun barrel makes marks on it. After the bullet leaves the barrel, the marks remain on the bullet. The traveling bullet is a causal process because it carries these marks. (The police use such marks to determine from what gun the bullet was fired.) It is very important that the mark persists without any additional interactions to impose the mark again.[4] Causal processes have the capacity to transmit information and causal influences. They constitute the causal links among events that occur at different locations in space and time.

There are many 'entities' that are similar to processes but that are not genuine causal processes. For example, a shadow moving across a wall is not a causal process. It is possible to modify the shadow at a certain point, but the change will not persist without additional interactions. I call phenomena of this type *pseudo-processes*. They cannot transmit information or causal influence.

The definition of causal processes is not statistical. It refers to the physical characteristics of the entities involved and to their positions in space and time. This type of definition stands in sharp contrast to the statistical definitions of the conjunctive fork and the screening-off relation.

(4) In the famous experiment of Arthur Compton, an X-ray photon collides with an electron. The photon and the electron are causal processes. When the photon and the electron collide, both are modified. The frequency, energy, and momentum of the photon are changed, as are the energy and momentum of the electron. (In this interaction energy and momentum are conserved.) Moreover, these changes persist beyond the collision.

This persistence is essential for a causal interaction; Compton scattering is an excellent example.

Sometimes two causal processes intersect with each other without entering into a causal interaction. For example, if two light rays intersect, they are superimposed at the point of intersection, but beyond this point the rays are not changed. Each one of them continues on as if nothing had happened; hence, their intersection does not constitute an interaction. Moreover, when two pseudo-processes intersect, or one causal process and one pseudo-process intersect, a causal interaction never occurs.

Obviously, the definition of a causal interaction, like the definition of a causal process, is not statistical. Like the latter, the former refers to the physical characteristics of the entities involved and to their positions in space and time. It seems to me that, of the two, the concept of a causal interaction is the more basic. The concept of a causal process ultimately depends on it, because to make a mark on a process requires a causal interaction. It seems to me, in addition, that these two concepts provide the foundation of our understanding of the concept of causality. In this respect I disagree with writers such as I. J. Good (1961–1962), Hans Reichenbach (1956), and Patrick Suppes (1970), who regard the statistical concepts as basic.

There is another extremely fundamental problem in this statistical approach to probabilistic causality. Everyone who follows this line of thought makes the relation of positive statistical relevance an essential characteristic of the concept of probabilistic cause. The basic idea is that the cause increases the probability of the effect. Although this idea has great intuitive plausibility, I think it is not acceptable. The reason is simple. There are many situations in which the cause has a negative relevance to the effect.

The most famous example was given by Deborah Rosen; it involves a game of golf (Suppes, 1970, p. 41). The player hits the ball very badly, but by chance it strikes the branch of a tree and falls into the hole. The probability of this result is extremely small— virtually infinitesimal. Nevertheless, there is a causal chain connecting the stroke of the golfer to the presence of the ball in the cup. Although this example was presented as the problem of an extremely unlikely event, it is actually an instance of the problem of negative relevance. It is much more probable that the ball would go directly into the hole than that it would do so after striking a branch of a tree.

Rosen's example is a product of fantasy, but there are many cases that are completely ordinary. Every time a result can come about in two or more different ways, the same problem arises if the probabilities of the various ways are not all equal. For example, suppose that a city needs a new bridge. There will soon be an election, in which there are two candidates for mayor. The probability that the first will win is 0.4; the probability that the second will win is 0.6. If the first candidate wins, the probability that the bridge will be constructed is 0.3; if the second candidate wins, the probability is 0.7. Before the election, therefore, the probability that the bridge will be built is 0.54. As it actually turns out, the first candidate becomes mayor and the bridge is built. The victory of the first candidate is negatively relevant to the construction of the bridge because the probability decreases from 0.54 to 0.3. Nevertheless, that victory is a link in the causal chain.

There are several ways to try to restore positive relevance in cases such as this; I have treated them in detail in Salmon (1984b, chap. 7). I do not believe that there is any method that works for every case, but I do not regard this as the most significant fact. The *most* fundamental fact is that, whether the relevance be positive or negative, there are causal processes that furnish the causal connections. It seems to me that the physical connections are more important than the statistical relations among the separate events. In addition, I have argued that causal processes constitute precisely the causal connections that David Hume was unable to find. I maintain that it is possible, using the concept of a causal process, to give a solution to Hume's problem of causality.

In 1957 Karl Popper proposed an interpretation of probabilities as physical propensities, and that interpretation has become quite popular among philosophers of science. However, as Paul Humphreys (1985) has shown, this concept does not satisfy the mathematical calculus of probability; therefore, it is not really an interpretation of probability. It seems to me, consequently, that we are not justified in replacing the older limiting frequency concept—which *does* satisfy the axioms of the probability calculus (with finite additivity)—with propensities. In spite of that fact, I believe that the propensity concept is extremely useful as a concept of probabilistic causality.

It is my considered opinion (along with most contemporary physicists, I think) that determinism is not true. In quantum theory, at least, there are events that are not determined; they are irreducibly statistical. Therefore, in the universe there are some events that are not completely causally determined; furthermore, they may not all be confined to the quantum domain. In this indeterministic context, if there is to be any sort of causality, it will have to be probabilistic.

Let us consider the causal interaction in Compton scattering. When a photon strikes an electron, the direction in which the electron will go is not determined. There is a probability distribution over all possible directions. In addition, in this collision the amount by which the frequency of the photon will change is not determined. There is a probability distribution over all possible amounts. Because of the conservation of energy and of momentum there is, of course, a perfect correlation between the direction of the electron and the change in frequency of the photon. But the pair of values is not determined.

The kind of causal interaction I am discussing is an intersection between two causal processes. Each process transmits various physical characteristics: energy, mass, frequency, momentum, etc. [This statement clearly anticipates Dowe's conserved quantities theory (1992c), but I did not see the possibility of such a theory when this essay was written.] In addition, these processes transmit probability distributions for interactions with the other processes that they encounter. It seems to me that these probability distributions constitute causal propensities. Thus, causal processes transmit propensities. In the context of quantum theory we associate amplitudes with systems; these amplitudes determine the probabilities of the various results when interactions occur. I have a strong temptation to equate the amplitudes with the propensities, but we must keep in mind that the probability is the square of the amplitude. In any case, a propensity is both a causal and a probabilistic tendency.

In the case of Compton scattering, the presence of the one process makes a difference when it meets the other. Without the photon the electron would not undergo scattering; without the electron the frequency of the photon would not change. However, there are some examples in which an effect can occur without an interaction with another system. For example, an atom in an excited state has a propensity to return to its ground state spontaneously, emitting a photon. The same atom in the same excited state has a different propensity to return to its ground state (by stimulated emission) if it encounters a photon of appropriate frequency. In this case a causal process—a photon—makes a probabilistic difference.

The work of Hume on causality has obviously profoundly influenced subsequent thought on this subject. His emphasis on constant conjunction and his rejection of hidden powers has drawn attention away from the mechanisms of causality. The statistical concepts, such as the conjunctive fork and the relation of screening off, follow in the tradition of Hume. Although they are not relations of constant conjunction, they are based on statistical regularities among diverse types of events. They do not refer to underlying mechanisms. I have no desire to say that these statistical relations are unimportant; on the contrary, they are indispensable, but they are not the most fundamental.

It seems to me that this neglect of underlying mechanisms constitutes, at least implicitly, a large part of the motivation of those philosophers who have abandoned the frequency interpretation of probability and adopted the propensity 'interpretation.' The former involves solely statistical regularities; the latter refers to chance set-ups or chance mechanisms. Nevertheless, those who have spoken in this manner of these mechanisms have not explained their nature. However, if we think of propensities as probabilistic causes, we can use the concepts of causal processes and causal mechanisms in order to explain the mechanisms of probabilistic causality.

Hume's most fundamental problem concerned the connections between causes and effects. Because I see causal processes as precisely these connections, I believe it is possible to resolve that problem by means of a satisfactory analysis of causal processes. I tried to give such an analysis in "An 'At-At' Theory of Causal Influence" (essay 12), and I explained it at greater length in Salmon (1984a, chap. 5).

The first point I should like to make is heuristic. Hume and most other philosophers who have discussed causality have thought of a cause and its effect as distinct events. If they are separated in space, or time, or both, one tries to find intermediate events that provide a connection between them. By this means one tries to construct a causal chain. Nevertheless, if there is a problem of the relation between the original cause and its effect, precisely the same problem arises among the intermediate events. What is the power of one link in the causal chain to produce the next one?

It seems to me that this mode of thinking carries with it unnecessary difficulties. Although I have spoken—as almost everyone does—of causal chains and their links, I would advise thinking of a thread or a cord instead of a chain. They are continuous; they are not composed of links. Thus, there is no question about the power to produce the next event because no such next event exists. In a causal process, the causal influence is transmitted continuously. We need to understand clearly the nature of this continuous causal transmission. The concept of transmission of a mark is the crucial point.

I have defined the causal process in terms of the capacity to transmit a mark. We can consider any change in a process as a mark: recall the marks on the bullet. A causal interaction imposes the mark. This action is local; it happens in the process. If the mark persists after the interaction, without any other interactions, the process transmits the mark. If the mark is made at point P, and if this mark is present in the same process at a subsequent point Q and at every point between P and Q, the process has transmitted the mark from *P* to *Q*. The mark is transmitted in virtue of being at every point between *P* and *Q*. As I explain in the preceding essay, this thesis is completely analogous to the famous "at-at" theory of motion of Bertrand Russell. The aim of Russell's theory was to resolve Zeno's ancient paradox of the arrow. The purpose of my "at-at" theory of causal transmission is to resolve Hume's problem of causal connections. In my opinion Russell's theory is extremely profound; it made my task easy.

It is possible, I believe, to resolve Hume's problem of causal connections without abandoning empiricism. If we have any type of process, we can devise an experiment to discover if this type of process can transmit a mark. Experience has shown, for example, that electromagnetic waves and material objects (whether moving or at rest) are important types of causal processes. A shadow is not a causal process. Even if a process that has the ability to transmit a mark is not doing so, it still qualifies as a causal process. These processes can also transmit signals, information, energy, and causal influence. Causal processes provide the causal connections between that which happens at one space-time location and that which occurs at others.

There are two fundamental types of causal action—production and propagation (see "Causality: Production and Propagation" [essay 18]). Causal interactions are the agents of production: they produce changes in processes that intersect one another. Causal processes are the agents of propagation; they transmit causal influence throughout the universe. We must keep in mind that the concepts of which I am speaking are probabilistic. Therefore, when two causal processes intersect, there is a distribution of probabilities over the different possible results, including the possibility of no interaction at all. For example, when two light rays intersect, there is an extremely small probability that two photons will collide with each other as particles—thus interacting causally—but almost always no interaction occurs. In addition, when a causal process proceeds without interacting with any other process, it is possible for a spontaneous change to occur. For example, a person may suffer a heart attack without any immediate external cause, or a free neutron may decay spontaneously, yielding an electron, a proton, and an antineutrino.

As the examples we have been considering show, the causal concepts I have been discussing apply at many levels. Compton scattering is a microcosmic phenomenon of the quantum domain. The examples of paresis and the poisonous mushrooms involve processes of human physiology and the actions of microscopic organisms. The example of the golf player deals with ordinary middle-sized physical objects, similar to the famous billiard balls of Hume. The construction of the bridge in the city involves complex social and political processes. Whether one treats a process as simple or complex depends on pragmatic aspects of the situation.

In the macrocosm, it seems to me, there are two types of causal mechanisms—processes and interactions. In the microcosm, where quantum theory is applicable, another mechanism appears to operate. Often called *the collapse of the wave function,* this mechanism is not clearly understood even at present. The fundamental problem was exhibited by Einstein, Podolsky, and Rosen (1935); it is aggravated by Bell's theorem and the Aspect experiment (Mermin, 1985). To many people this raises deeply perplexing questions about action at a distance. We will not completely understand the causal structure of the world until we understand this mechanism.

In this discussion we have seen two ways of approaching causality in an indeterministic context. I suggest that we use the expression "probabilistic causality" as the generic term for any type of indeterministic causality. The first approach to this concept places fundamental emphasis on constant conjunctions and statistical regularities—on concepts such as screening off, the conjunctive fork, and positive statistical relevance. Some authors who pursue this way also allude to relations in space and time, but they do not mention causal mechanisms. Let us call the concept that results from this approach *statistical causality*. I do not believe that this approach can provide an adequate analysis of causality. The second way places primary emphasis on the mechanisms of causality—such as processes and interactions—but does not disdain statistical regularities. In the detailed analysis of any particular process or interaction, it is necessary to appeal to laws of nature, such as conservation of energy or conservation of momentum.[5] The laws invoked may be universal or statistical. The principal emphasis, however, is always on the mechanisms, and these mechanisms obviously need not be deterministic. Let us call the concept that results from this approach *aleatory causality*.[6] We stand a better chance of reaching an adequate understanding of causality, I have tried to argue, by adopting this latter approach—that is, by focusing primarily on the mechanisms rather than primarily on the values of probabilities.

Notes

1. In "Probabilistic Causality" (essay 14) an example, originally owing to Ellis Crasnow, is given.

2. Additional reasons were given by Richard Otte (1981).

3. See Salmon (1984b, p. 160 n. 2) for the proof.

4. Hans Reichenbach used the idea of mark transmission to distinguish causal processes from pseudo-processes (which he called "unreal sequences"), but he did not analyse or use the concept of a causal process as I do in this essay.

5. Notice, however, that in essay 16, "Causality without Counterfactuals," I suggest that it may not be necessary to appeal to *laws* of conservation; *facts* of conservation may suffice.

6. I use the term "aleatory" in the spirit of Paul Humphreys (1981, 1983). The notion of aleatory causality offered in this essay fits harmoniously, I believe, with his account of aleatory explanation. For a fuller account, see his book *The Chances of Explanation* (1989).

14

Probabilistic Causality

Although many philosophers would be likely to brand the phrase "probabilistic causality" a blatant solecism, embodying serious conceptual confusion, it seems to me that probabilistic causal concepts are used in innumerable contexts
of everyday life and science. We hear that various substances are known to cause cancer
in laboratory animals—see the label on your favorite diet soft-drink can—even though
there is no presumption that every laboratory animal exposed to the substance developed
any malignancy. We say that a skid on a patch of ice was the cause of an automobile
accident, though many cars passed over the slick spot, some of them skidding on it,
without mishap. We have strong evidence that exposure to even low levels of radiation
can cause leukemia, though only a small percentage of those who are so exposed actually
develop leukemia. I sometimes complain of gastric distress as a result of eating very
spicy food, but such discomfort is by no means a universal sequel to well-seasoned
Mexican cuisine. It may be maintained, of course, that in all such cases a fully detailed
account would furnish invariable cause-effect relations, but this claim would amount to
no more than a declaration of faith. As Patrick Suppes has ably argued, it is as pointless
as it is unjustified (1970, pp. 7–8).

There are, in the philosophical literature, three attempts to provide theories of probabilistic causality; Hans Reichenbach (1956), I. J. Good (1961–1962), and Patrick Suppes
(1970) have offered reasonably systematic treatments.[1] In the vast philosophical literature on causality they are largely ignored. Moreover, Suppes makes no mention of
Reichenbach (1956), and Good gives it only the slightest note (II, p. 45),[2] though both
offer brief critical remarks on some of his earlier work. Suppes makes the following
passing reference to Good's theory: "After working out most of the details of the
definitions given here in lectures at Stanford, I discovered that a closely related analysis
of causality had been given in an interesting series of articles by I. J. Good (1961, 1962),
and the reader is urged to look at Good's articles for a development similar to the one

given here, although worked out in rather different fashion formally and from a different viewpoint" (1970, p. 11). Even among those who have done constructive work on probabilistic causality, there is no sustained discussion of the three important extant theories.

The aim of the present essay is to take a close critical look at the proposals of Good, Reichenbach, and Suppes. Each of the three is, for reasons I shall attempt to spell out in detail, seriously flawed. We shall find, I believe, that the difficulties arise from certain rather plausible assumptions about probabilistic causality, and that the objections lead to some rather surprising general results. In the concluding section I briefly sketch what seem to me the appropriate ways of circumventing the problems associated with these three theories of probabilistic causality.

1. Good's Causal Calculus

Among the three theories of probabilistic causality, Good's appears to be the least familiar to philosophers. One reason for this neglect—over and above the fact that most philosophers ignore the very concept of probabilistic causality—may be the rather forbidding mathematical style of Good's presentation. Fortunately, the aspects of his theory that give rise to the fundamental objections can be extracted from the heavy formalism and presented in a fashion that makes them intuitively easy to grasp. I offer two basic objections. The first objection concerns the manner in which Good attempts to assign a degree of strength to a causal chain on the basis of the strengths of the individual links in the chain. He seems to be unaware of any problem in this connection. The second objection, which Good's theory shares with those of Reichenbach and Suppes, concerns cases in which an effect is brought about in an improbable fashion. Both Good and Suppes are aware of this rather familiar difficulty, and they try to deal with it in ways that are different but complementary. I argue that their answers are inadequate.

The basic materials with which we shall work are familiar enough. We suppose that there are aggregates of events, denoted by E, F, G, . . . (with or without subscripts), among which certain physical probability relations hold.[3] The particular events are located in space-time. Like Suppes, but unlike Reichenbach, Good stipulates that cause temporally precede their effects—that is, temporal priority is used to define causal priority, not vice versa. Good's aim is to examine certain types of networks of events that join an initial event F to a final event E, usually by way of various intermediate events G_i, and to define a measure $\chi(E{:}F)$ of "the degree to which F caused E" or "the contribution to the causation of E provided by F" (I, p. 307). The specification of this measure, and various related measures, involves 24 axioms and 18 theorems, some of which are relatively abstruse.

One particularly important special case of a causal net is a *causal chain*. In a causal chain, all of the constituent events $F = F_0, F_1, F_2, . . . , F_n = E$ are linearly ordered. It is assumed that the adjacent events F_i and F_{i+1} are spatiotemporally contiguous (or approximately so), that they do not overlap too much, and that F_{i+1} does not depend on the occurrence of any event in the chain prior to F_i (II, p. 45). In order to arrive at the measure χ for a wider class of nets, Good defines a measure $S(E{:}F)$ of the strength of the causal chain joining F to E. A particularly simple type of causal chain is one consisting only of the two events F and E. A measure of the strength of a chain of this sort can be used, according to Good, to define a measure of the strength of longer chains. It is this aspect of Good's approach—the attempt to compound the strengths of the individual "links" of the chain in order to ascertain the strength of the entire chain—that is the locus of the first problem.

In order to initiate the enterprise, Good introduces a measure $Q(E:F)$ which is to stand for "the tendency of F to cause E" (I, p. 307). This informal rendering in words of the import of Q is, I think, seriously misleading, for Q turns out to be no more nor less than a measure of statistical relevance. Events of type A are statistically relevant to events of type B, it will be recalled, if the occurrence of an event of type A makes a difference to the probability that an event of type B will occur. We say that the relation is one of positive relevance if the occurrence of a member of A increases the probability that a member of B will occur; we speak of negative relevance if the occurrence of A decreases the probability of B. As we all recognize, a mere correlation does not necessarily constitute a causal relation—not even a tendency to cause. The falling barometric reading has no tendency at all to cause a storm, though the barometric reading is highly relevant statistically to the onset of stormy weather. [This criticism is incorrect; see essay 15.]

There are, of course, many different measures of statistical relevance. Although the first five axioms, A1–A5, do not fix the precise form of Q, they do show what sort of measure it is. According to A1, $Q(E:F)$ is a function of $P(E)$, $P(E|F)$, and $P(E|\bar{F})$ alone; according to A5, $Q(E:F)$ has the same sign as $P(E|F) - P(E)$; and according to A3 and A4, $Q(E:F)$ increases continuously with $P(E|F)$ if $P(E)$ and $P(E|\bar{F})$ are held constant, and it decreases continuously as $P(E|\bar{F})$ increases if $P(E)$ and $P(E|F)$ are held constant. Q may be a real number, it may assume the value $+\infty$ or $-\infty$, or under special circumstances it may be indeterminate. However, Q need not be the simplest sort of statistical relevance measure—such as $P(E|F) - P(E)$, $P(E|F)/P(E)$, $P(E|F) - P(E|\bar{F})$, or $P(E|F)/P(E|\bar{F})$—for A1–5 allow that it may be a function of all three of the above-mentioned probabilities. When Good does choose a particular form for $Q(E:F)$, however, he adopts one which is a function of $P(E|F)$ and $P(E|\bar{F})$, but which is independent of $P(E)$ (I, pp. 316–317). Suppes, in his definition of *prima facie cause* (1970, p. 12), and Reichenbach, in his definitions of *causal betweenness* (1956, p. 190) and *conjunctive fork* (ibid., p. 159), use relevance measures which are functions of $P(E|F)$ and $P(E)$.[4] What is important in this context is that every theory of probabilistic causality employs a statistical relevance measure as a basic concept; for present purposes the precise mathematical form that is chosen is of secondary significance. Good's relevance measure $Q(E:F)$ is used to furnish the strength $S(E:F)$ of the chain that consists only of the events F and E (A10; I, p. 311).

The next problem is to characterize the strength of the causal chain from F to E when there are other intermediate events in the chain between F and E. We know that the strength of each link in the chain, $F_i \rightarrow F_{i+1}$, is simply $Q(F_{i+1}:F_i)$, and that Q is a statistical relevance measure. The trick is to figure out how—if it is possible to do so at all—to compound the strengths of the individual links $s_0, s_1, \ldots, s_{n-1}$ so as to get the strength of the chain itself. Good proceeds (A11; I, p. 311) to make the simplest reasonable assumption, namely, that the strength $S(E{:}F)$ of the chain is some function $\phi(s_0, s_1, \ldots, s_{n-1})$ of the strengths of the individual links. Although brief justificatory remarks accompany some of the axioms, this one has none; perhaps Good regarded it as too obvious to need comment. In spite of its initial plausibility, this assumption seems to me untenable, as I shall now try to show by means of a simple counterexample.

Consider the following simple two-stage game. To play, one first tosses a fair tetrahedron with sides marked 1, 2, 3, and 4. If the tetrahedron comes to rest on any side other than 4—i.e., if side 4 shows after the toss—one draws from a deck that contains 16 cards, 12 of which are red and 4 of which are black; if side 4 does not show, one draws from a deck containing 4 red and 12 black. Suppose that, on a given play, one tosses the tetrahedron (event F) and it comes to rest with side 4 showing, so that one draws from the first above-mentioned special deck (event G), with the result that one gets a red card (event E). This is a simple three-event chain,[5] and all of the constituents are events that actually obtain, as Good demands (II, p. 45). We inquire about the degree to which F caused E.

In the first place, we must construe the situation in such a way that F is positively relevant to G and G is positively relevant to E; otherwise, as Good shows in theorem T2 (I, p. 311), there is no causal chain. Let us therefore assume that $P(G|\bar{F}) = 0$; that is, the only way to get a chance to draw from the special deck is to enter the game and toss the tetrahedron. Thus, $P(G|F) > P(G|\bar{F})$ and $Q(G{:}F) \neq 0$. We can now use the theorem on total probability

$$P(E|F) = P(G|F){\cdot}P(E|F.G) + P(\bar{G}|F){\cdot}P(E|F.\bar{G}) \tag{1}$$

to calculate the probability of drawing a red card (event E) given that the player has tossed the tetrahedron (event F). Since causal chains, as defined by Good, possess the Markov property,

$$P(E|G) = P(E|F.G) \text{ and } P(E|\bar{G}) = P(E|F.\bar{G}) \tag{2}$$

the theorem on total probability can be rewritten in a simplified form:

$$P(E|F) = P(G|F){\cdot}P(E|G) + P(\bar{G}|F){\cdot}P(E|\bar{G}) \tag{3}$$

Using this equation and the stipulated probability values, we find that $P(E|F) = {}^{10}\!/_{16}$.

For purposes of comparison, let us consider another game just like the foregoing except that different decks of cards are used. To play, one tosses a fair tetrahedron (event F'), and if side 4 shows, one draws from a deck containing 14 red cards and 2 black cards (event G'). If the tetrahedron comes to rest on side 4, one draws from a deck containing 10 red cards and 6 black cards (event $\bar{G}'$). In this game the probability of drawing a red card (event E') equals ${}^{13}\!/_{16}$. It is easily seen that, in this game, as in the other, the toss of the tetrahedron is positively relevant to the draw from the favored deck, and the draw from that deck is positively relevant to getting a red card. In Good's notation, $Q(G'{:}F') > 0$ and $Q(E'{:}G') > 0$. Assume that a player of the second game had tossed the tetrahedron with the result that side 4 shows, and that this player has drawn a red card from the favored deck. We have two causal chains: $F \to G \to E$ (first game) and $F' \to G' \to E'$ (second game). Let us compare them.

We must now take account of the particular form Q assumes in Good's causal calculus; it is given in T15 (I, p. 317) as

$$Q(E{:}G) = \log \frac{P(\bar{E}|\bar{G})}{P(\bar{E}|G)} = \log \frac{1 - P(E|\bar{G})}{1 - P(E|G)} \qquad (4)$$

In the first game, $P(\bar{E}|\bar{G}) = \frac{3}{4}$ and $P(\bar{E}|G) = \frac{1}{4}$; hence, $Q(E{:}G) = \log 3$. In the second game, $P(\bar{E}'|\bar{G}') = \frac{3}{8}$ and $P(\bar{E}'|G') = \frac{1}{8}$; thus $Q(E'{:}G') = \log 3$. Clearly $Q(G{:}F) = Q(G{:}F')$, since $P(G|F) = P(G'|F')$ and $P(G|\bar{F}) = P(G'|\bar{F}')$. Therefore, the corresponding links in the two chains have equal strength. We have already noted, however, that $P(E|F) \neq P(E'|F')$—that is, the probability that a player who tosses the tetrahedron in the first game will draw a red card is not equal to the probability that a player who tosses the tetrahedron in the second game will draw a red card. It is easily seen, moreover, that the statistical relevance of F to E is not the same as the statistical relevance of F' to E'. We begin by noting that the only way in which a red card can be drawn in either game is by a player who has commenced the game by tossing the tetrahedron; consequently, $P(E|\bar{F}) = P(E'|\bar{F}') = 0$. Using the previously established values, $P(E|F) = \frac{10}{16}$ and $P(E'|F') = \frac{13}{16}$, we find that $P(E|F) - P(E|\bar{F}) = \frac{10}{16}$, while $P(E'|F') - P(E'|\bar{F}') = \frac{13}{16}$. Given both the difference in probability and the difference in statistical relevance between the first and last members of the two chains, it seems strange to say that the causal strengths of the two chains are equal. If, however, ϕ is made a function of the Q-values of the individual links, this is the consequence we are forced to accept.[6]

In order to bring out the import of this argument, I should like to apply it to an example that is a bit less artificial and more concrete than the tetrahedron-cum-card game. Suppose that two individuals, Joe Doakes and Jane Bloggs, suffer from sexual disabilities. Joe is impotent and Jane is frigid. Each of them decides to seek psychotherapy. There are two alternative types of therapy available, directive or nondirective. When Joe seeks out a psychotherapist (event F), there is a probability of $\frac{3}{4}$ that he will select a directive therapist and undergo that type of treatment (event G), and a probability of $\frac{1}{4}$ that he will select a nondirective therapist and undergo that type of treatment (event $\bar{G}$). If he is treated by a directive therapist, there is a probability of $\frac{3}{4}$ that he will be cured (event E), and if he is treated by a nondirective therapist, there is a probability of $\frac{1}{4}$ that he will be cured. Given these values, there is a probability of $\frac{10}{16}$ that he will be cured, given that he undertakes psychotherapy.

When Jane seeks out a psychotherapist (event F'), there is a probability of $\frac{3}{4}$ that she will select a directive therapist (event G'), and a probability of $\frac{1}{4}$ that she will select a nondirective therapist (event $\bar{G}'$). If she is treated by a directive therapist, there is a probability of $\frac{7}{8}$ that she will be cured (event E'), and if she is treated by a nondirective therapist, the probability of a cure is $\frac{5}{8}$. Given these values, there is a probability of $\frac{13}{16}$ that she will be cured, given that she undertakes psychotherapy.

Joe and Jane each undertake psychotherapy, each is treated by a directive psychotherapist, and each is cured. Thus, we have two causal chains, $F \rightarrow G \rightarrow E$ and $F' \rightarrow G' \rightarrow E'$. We may assume, moreover, that both chains have the Markov property, formulated in equation (2), for it is reasonable to suppose that people who undergo psychotherapy on account of these problems do so voluntarily, so $G = F.G$ and $G' = F'.G'$. The question is, on what basis, if any, would we be warranted in claiming that the two chains have the same strength—i.e., that the degree to which the seeking out of psycho-therapeutic treatment caused the cure is the same for both.

The appropriate response, it seems to me, is that not enough information has been given to answer the question about the relative strengths of the two chains. We need to know, at the very least, what the probability of a cure would be if psychotherapy were not undertaken. We could, of course, make the patently unrealistic assumption that the probability of a cure in the absence of psychotherapy is zero in both cases. This assumption gives this psychotherapy example precisely the same probabilistic structure as the tetrahedron-cum-cards example. Given this assumption, it seems intuitively unreasonable to say that the psychotherapy contributed equally strongly to the cures in the two cases, for the degree of relevance of the cause to the effect differs in the two cases. If we make other assumptions—e.g., that there is quite a high probability (say ¾) of remission without psychotherapy in the case of frigidity, but a low probability (say $\frac{1}{100}$) of remission without psychotherapy in the case of impotence—then it seems intuitively clear that the causal contribution to the cure in the case of Joe is much greater than it is in the case of Jane. For Jane's problem, the probability of cure if she undergoes therapy ($\frac{13}{16}$) is only slightly higher than the probability of spontaneous remission ($\frac{12}{16}$), but for Joe's problem, the probability of a cure if he undergoes psychotherapy ($\frac{10}{16}$) is much greater than the probability of spontaneous remission ($\frac{1}{100}$). Other assumptions about spontaneous remission could alter the situation dramatically. In the light of these considerations, I am extremely skeptical about the possibility of arriving at a suitable measure of the strength S of a causal chain in terms of any function of the Q-values of its individual links. I agree with Good that the strength of a causal chain cannot be measured in terms of the statistical relevance of the initial member F to the final member E alone, but I do not believe that this factor can be ignored. To attempt to determine the strength of the chain without comparing the probability of E when F is present with the probability of E when F is absent seems quite futile; we can evidently alter the strength of a causal chain by altering nothing about the chain except the probability $P(E|\bar{F})$.

In order to assign Q-values to the links of a chain $F \to G \to E$, we may use the following probability values (see equation [4]):

$$P(G|F), \; P(G|\bar{F}), \; P(E|G), \; P(E|\bar{G}) \tag{5}$$

Since causal chains are assumed to have the Markov property, we can use these values with equation (3) to compute the values of $P(E|F)$ and $P(E|\bar{F})$:

$$P(E|F) = P(G|F){\cdot}P(E|G) + P(\bar{G}|F){\cdot}P(E|\bar{G}) \tag{6}$$

$$P(E|\bar{F}) = P(G|\bar{F}){\cdot}P(E|G) + P(\bar{G}|\bar{F}){\cdot}P(E|\bar{G}) \tag{7}$$

As our examples have shown, the relation between $P(E|F)$ and $P(E|\bar{F})$ is not a function of the Q-values of the links of the chain. We must therefore conclude that the transition from the four basic probability values (5) to the two statistical relevance measures sacrifices information essential to the determination of the strength of the causal chain.

It is important to recognize that this consequence is *not* a result of the particular measure of statistical relevance adopted by Good; it is, instead, a result of rather general features of statistical relevance relations. If a relevance measure is defined as $P(B|A)/P(B|\bar{A})$, or as any function of that ratio, it is easy to construct counterexamples along precisely the same lines as those given earlier (simply by adjusting the makeup of the two decks of cards in the game) to show that the relevance of A to B and the relevance of B to C do not determine the relevance of A to C. The same may be said for any relevance measure defined in terms of the difference $P(B|A) - P(B|\bar{A})$.

I obviously have not considered every possible form of statistical relevance relation, but it seems intuitively evident that the foregoing sorts of arguments can be applied to statistical relevance relations quite generally. Thus, at this point I would conjecture the general proposition that the strengths of causal chains cannot be measured by the statistical relevance relations of their adjacent members alone. In this sense, it seems, *the strength of a causal chain cannot be a function of the strengths of its individual links.* It appears, again as a general conjecture, that one needs the individual probability values that are used to define the relevance measures in order to deal adequately with the strengths of the causal chains. Too much information is thrown away when these probability values are combined to form statistical relevance measures.

There is a second difficulty, which Good's theory shares with those of Suppes and Reichenbach. Suppose, going back to the first tetrahedron-cum-card game, that one tosses the tetrahedron and it lands on side 4. One must draw from the deck that has a paucity of red cards: nevertheless, one draws a red card. According to Good's calculus, the three events F, $\bar{G}$, E do not form a causal chain (II, p. 45). On *any* reasonable relevance measure, $\bar{G}$ is not positively relevant to E, for $P(E|\bar{G}) < P(E|G)$. This result seems to me to be untenable; if $F \to G \to E$ qualifies as a causal chain when this sequence of events occurs, so must $F \to \bar{G} \to E$ when it occurs. Good has an answer to this problem; we shall examine it in connection with the theories of Reichenbach and Suppes.[7]

2. Reichenbach's Macrostatistical Theory

Unlike Good and Suppes, who attempt to provide analyses of probabilistic causality for their own sake, Reichenbach develops his analysis as a part of his program of implementing a causal theory of time. Thus, in contrast to the other two authors, he does not build into his definitions the stipulation that causes are temporally prior to effects. Instead, he attempts to construct a theory of causal relations that will yield a causal asymmetry which can then be used to define a relation of temporal priority. Two of the key causal concepts introduced in this construction are the relation of *causal betweenness* and the structure known as a *conjunctive fork*. The main use of the betweenness relation is to establish a linear time order; the conjunctive fork is employed to impose a direction or asymmetry upon the linear time order. In the present discussion I do not attempt to evaluate the temporal ramifications of Reichenbach's theory; instead, I confine my attention to the adequacy of the causal concepts as such.

Reichenbach's formal definition of causal betweenness, translated from his notation into a standard notation, reads as follows (1956, p. 190):

An event B is *causally between* the events A and C if the relations hold:

$$1 > P(C|B) > P(C|A) > P(C) > 0 \tag{8}$$

$$1 > P(A|B) > P(A|C) > P(A) > 0 \qquad (9)$$

$$P(C|A.B) = P(C|B) \qquad (10)$$

Together with the principle of *local comparability of time order,* the relation of causal betweenness can, according to Reichenbach, be used to construct causal nets and chains similar to those mentioned by Good in his causal calculus. Unlike Good, however, Reichenbach does not attempt a quantitative characterization of the strengths of such chains and nets. It is worth noting that formulas (8) and (9) embody several statistical relevance relations: A is relevant to the occurrence of C, but B is more highly relevant to C; conversely, C is relevant to the occurrence of A, but B is more highly relevant to A. Moreover, according to (10), B screens A off from C and C off from A—that is, B renders A and C statistically irrelevant to each other. A chain of events $A \rightarrow B \rightarrow C$ thus has the Markov property (equation [2]) which Good demanded of his causal chains.

The inadequacy of Reichenbach's definition of causal betweenness was pointed out by Clark Glymour, in conversation, a number of years ago, when he was a graduate student at Indiana University. The cases we discussed at that time were similar in principle to an excellent example, owing to Deborah Rosen, reported by Suppes (1970, p. 41):

> [S]uppose a golfer makes a shot that hits a limb of a tree close to the green and is thereby deflected directly into the hole, for a spectacular birdie. . . . If we know something about Mr. Jones' golf we can estimate the probability of his making a birdie on this particular hole. The probability will be low, but the seemingly disturbing thing is that if we estimate the conditional probability of his making a birdie, given that the ball hit the branch, . . . we would ordinarily estimate the probability as being still lower. Yet when we see the event happen, we recognize immediately that hitting the branch in exactly the way it did was essential to the ball's going into the cup.

If we let A be the event of Jones teeing off, B the event of the ball striking the tree limb, and C the event of the ball dropping into the cup at one under par for the hole, we have a violation of Reichenbach's condition (8), for $P(C|B) < P(C|A)$. The event B is, nevertheless, causally between events A and C.[8] Various retorts can be made to this purported counterexample. One could maintain (see von Bretzel, 1977, p. 182) that sufficiently detailed information about the physical interaction between the ball and the branch might enable us to raise the conditional probability of the ball going into the hole, given these precisely specified physical circumstances, above the conditional probability of Jones's making a birdie given only that he tees off. As von Bretzel himself notes, this kind of response seems ad hoc and artificial, and there is no good reason to suppose that it would take care of all such counterexamples even if it were adequate for this particular one. Indeed, it seems to me that many examples can be found that are immune to dismissal on these grounds.

Rosen's colorful example involves a near-miraculous occurrence, but we do not need to resort to such unusual happenings in order to find counterexamples to Reichenbach's definition of causal betweenness. The crucial feature of Rosen's example is that Jones makes his birdie "the hard way." Since much in life happens "the hard way," we should be able to find an abundance of everyday counterexamples; in fact, we have already considered one. When the game of tetrahedron tossing and card drawing was used in the previous section to raise the second objection to Good's causal calculus, we looked at the case in which the player drew the red card and won the prize "the hard way." In that case the tetrahedron came to rest on side 4, forcing the player to draw from the deck with a smaller proportion of red cards. As the original game was set up, one's initial probability of drawing a red card is $^{10}\!/_{16}$, but if one is required, as a result of the toss, to draw from the less favorable deck, the probability of drawing a red card is only $\frac{1}{4}$. Nevertheless, when the player whose toss fails to show side 4 succeeds in drawing a red card from the unfavorable deck, the draw from the unfavorable deck is causally between the toss and the winning of the prize. Drawing a red card from a deck that contains 4 red and 12 black cards can hardly be considered a near-miracle.

Once we see the basic feature of such examples, we can find others in profusion. The expression "the hard way" is used in the game of craps, and this game provides another obvious example.[9] One wins if one throws 7 or 11 on the first toss; one loses if one throws 2, 3, or 12 on the first toss. If the first toss results in any other number, that is one's "point," and one wins if in subsequent tosses one makes one's point before throwing a 7. The probability of the shooter's winning in one or another of these ways is just slightly less than one half. A player who throws 4 on the initial toss clearly reduces the chances of winning (this conditional probability is $\frac{1}{3}$), but nevertheless can win by making the point. Throwing 4 is, however, causally between the initial toss and the winning of the bet on that play.

A pool player has an easy direct shot to sink the 9-ball, but chooses, for the sake of the subsequent position, the much more difficult play of shooting at the 2-ball and using it to put the 9-ball in the pocket. The initial probability of sinking the 9-ball is much greater than the probability of getting the 9-ball in the pocket if the cue-ball strikes the 2-ball, but the collision with the 2-ball is causally between the initiation of the play and the dropping of the 9-ball into the pocket. Similar examples can obviously be found in an enormous variety of circumstances in which a given result can occur in more than one way, and in which the probabilities of the result differ widely, given the various alternative ways of reaching it. The attempt to save Reichenbach's definition of causal betweenness by ad hoc devices appears to be a hopeless undertaking. We shall see, however, that Good suggests a method for handling such examples, and that Rosen offers a somewhat different defense on behalf of Suppes.

Reichenbach's definition of *conjunctive fork* does not fare much better. The basic motivation for introducing this concept is to characterize the situation in which an otherwise improbable coincidence is explained by appeal to a common cause. There are many familiar examples—e.g., the explanation of the simultaneous illness of many residents of a particular dormitory in terms of tainted food in a meal they all shared. Reichenbach defines the conjunctive fork in terms of the following formulas (1956, p. 159), which I have renumbered and translated into standard notation:

$$P(A.B|C) = P(A|C) \times P(B|C) \tag{11}$$

$$P(A.B|\bar{C}) = P(A|\bar{C}) \times P(B|\bar{C}) \tag{12}$$

$$P(A|C) > P(A|\bar{C}) \tag{13}$$

$$P(B|C) > P(B|\bar{C}) \tag{14}$$

In order to apply these formulas to the foregoing example, we may let A stand for the illness of Smith on the night in question, B the illness of Jones on the same night, and C the presence of spoiled food in the dinner served at their dormitory that evening.

The following example, owing to Ellis Crasnow, shows the inadequacy of Reichenbach's formulation. Brown usually arrives at the office about 9:00 A.M., fixes a cup of coffee, and settles down to read the morning paper for half an hour before beginning any serious business. Upon occasion, however, Brown arrives at 8:00, and the factotum has already brewed a fresh pot of coffee, which is served immediately. On precisely the same occasions, some other person meets Brown at the office and they begin work quite promptly. This coincidence—the coffee's being ready and the other person being at the office—demands explanation in terms of a common cause. As it happens, Brown usually takes the 8:30 bus to work in the morning, but on those mornings when the coffee is prepared for Brown's arrival and the other person shows up, Brown takes the 7:30 bus. It can plausibly be argued that the three events, A (the coffee's being ready), B (the other person showing up), and C (Brown taking the 7:30 bus), satisfy Reichenbach's requirements for a conjunctive fork. Clearly, however, Brown's bus ride is not a cause either of the coffee's being made or the other person's arrival. The coincidence does, indeed, require a common cause, but that event is a telephone appointment made the preceding day.

The crucial feature of Crasnow's counterexample is easy to see. Brown arises early and catches the 7:30 bus if and only if an early appointment was previously arranged. The conjunctive fork is constructed out of the two associated effects and another effect which is strictly correlated with the bona fide common cause. When we see how this example has been devised, it is easy to find many others of the same general sort. Suppose it is realized before anyone actually becomes ill that spoiled food has been served in the dormitory. The head resident may place a call to the university health service requesting that a stomach pump be dispatched to the scene; however, neither the call to the health service nor the arrival of the stomach pump constitutes a genuine common cause, though either could be used to form a conjunctive fork.[10]

Inasmuch as two of Reichenbach's key concepts—causal betweenness and conjunctive fork—are unacceptably explicated, we must regard his attempt to provide an account of probabilistic causality as unsuccessful.

3. Suppes's Probabilistic Theory

In spite of his passing remark about Good's causal calculus, Suppes's theory bears a much more striking resemblance to Reichenbach's theory than to Good's. As mentioned earlier, Suppes and Good agree in stipulating that causes must, by definition, precede their effects in time, and in this they oppose Reichenbach's approach. But here the similarities between Good and Suppes end. Like Reichenbach, and unlike Good, Suppes does not attempt to introduce any quantitative measures of causal strength. Like Reichenbach, and unlike Good, Suppes frames his definitions in terms of measures of probability, without introducing any explicit measure of statistical relevance. It is evident, of course, that considerations of statistical relevance play absolutely fundamental roles in all three theories, but as I commented regarding Good's approach, the use of statistical relevance measures instead of probability measures involves a crucial sacrifice of information. In addition, Suppes introduces a number of causal concepts, and in the course of defining them, he deploys the relations of positive statistical relevance and screening off in ways that bear strong resemblance to Reichenbach. A look at several of his most important definitions will exhibit this fact.

In definition (1) (p. 12) an event B is said to be a *prima facie cause* of an event A if B occurs before A and B is positively relevant, statistically, to A.[11] Suppes offers two definitions of spurious causes (pp. 23, 25), the second of which is the stronger and is probably preferable.[12] According to this definition (3), an event B is a *spurious cause* of an event A if it is a prima facie cause of A and it is screened off from A by a partition of events C_i that occur earlier than B. We are told (p. 24), though not in a numbered definition, that a *genuine cause* is a prima facie cause that is not spurious. These concepts can easily be applied to *the* most familiar example. The falling barometer is a prima facie cause of a subsequent storm, but it is also a spurious cause, for it is screened off from the storm by atmospheric conditions that precede both the storm and the drop in barometric reading.

There is a close similarity between Suppes's definition of spurious cause and Reichenbach's definition of conjunctive fork. It is to be noted first, as Reichenbach demonstrates (1956, pp. 158–160), that

$$P(A.B) > P(A) \times P(B) \tag{15}$$

follows from relations (11)–(14). Therefore, A and B are positively relevant to each other. If A and B are not simultaneous, then one is a prima facie cause of the other. Second, Reichenbach's relations (11) and (12) are equivalent to screening-off relations. According to the multiplication axiom,

$$P(A.B|C) = P(A|C) \times P(B|A.C) \tag{16}$$

therefore, it follows from (11) that

$$P(A|C) \times P(B|C) = P(A|C) \times P(B|A.C) \tag{17}$$

Assuming $P(A|C) > 0$, we divide through by that quantity, with the result

$$P(B|C) = P(B|A.C) \tag{18}$$

which says that C screens off A from B. In precisely parallel fashion, it can be shown that (12) says that $\bar{C}$ screens off A from B. But, $\{C, \bar{C}\}$ constitutes a partition, so B is a spurious cause of A or vice versa.[13] Suppes does not define the concept of conjunctive fork. Since he assumes temporal priority relations already given, he does not need conjunctive forks to establish temporal direction, and since he is not concerned with scientific explanation, he does not need them to provide explanations in terms of common causes. Nevertheless, there is a considerable degree of overlap between Reichenbach's conjunctive forks and Suppes's spurious causes.

Although Reichenbach defines conjunctive forks entirely in terms of relations (11)– (14), without imposing any temporal constraints, his informal accompanying remarks (1956, pp. 158–159) strongly suggest that the events A and B occur simultaneously, or nearly so. One might be tempted to suppose that Reichenbach wished to regard A and B as simultaneous to a sufficiently precise degree that a direct causal connection between them would be relativistically precluded. Such a restriction would, however, make no real sense in the kinds of examples he offers. Since the velocity of light is approximately 1 foot per nanosecond (1 nsec = 10^{-9} sec), the onsets of vomiting in the case of two roommates in the tainted food example would presumably have to occur within perhaps a dozen nanoseconds of each other.

Reichenbach's basic intent can be more reasonably characterized in the following manner. Suppose events of the types A and B occur on some sort of clearly specified association more frequently than they would if they were statistically independent of each other. Then, if we can rule out a direct causal connection from A to B or from B to A, we look for a common cause C which, along with A and B, constitutes a conjunctive fork. Thus, if Smith and Jones turn in identical term papers for the same class—even if the submissions are far from simultaneous—and if careful investigation assures us that Smith did not copy directly from Jones and also that Jones did not copy directly from Smith, then we look for the common cause C (e.g., the paper in the sorority or fraternity file from which both of them plagiarized their papers). It is the absence of a direct causal connection between A and B, not simultaneous occurrence, that is crucial in this context. Thus, in Reichenbach's conjunctive forks A may precede B or vice versa, and hence, one may be a prima facie cause of the other.

Suppes does not introduce the relation of causal betweenness, but he does define the related notions of direct and indirect causes. According to definition (5) (1970, p. 28), an event B is a *direct cause* of an event A if it is a prima facie cause of B and there is no partition C_i temporally between A and B that screens B off from A. A prima facie cause that is not direct is *indirect*. Use of terms such as "direct" and "indirect" strongly suggests betweenness relations. Suppes's definition of indirect cause clearly embodies a condition closely analogous to formula (10) of Reichenbach's definition of causal betweenness, but Suppes does not invoke the troublesome relations (8) and (9) that brought Reichenbach's explication to grief. It appears, however, that Suppes's theory faces similar difficulties.

Let us take another look at Rosen's example of the spectacular birdie. Again let A stand for Jones teeing off, B for the ball striking the tree limb, and C for the ball going into the cup. If this example is to be at all relevant to the discussion, we must suppose that A is a prima facie cause of C, which requires that $P(C|A) > P(C)$. We must, therefore, select some general reference class or probability space with respect to which $P(A)$ can be evaluated. The natural choice, I should think, would be to take the class of all cases of teeing off at that particular hole as the universe.[14] We may then suppose that Jones is a better than average golfer; when Jones tees off, there is a higher probability of a birdie than there is for golfers in general who play that particular course. We may further assume that A is a genuine cause of C, since there is no plausible partition of earlier events that would screen A off from C. Certainly B cannot render A a spurious cause of C, for B does not even happen at the right time (prior to A).

There is a more delicate question of whether A is a direct or indirect cause of C. We may reasonably assume that B screens A off from C, for presumably it makes no difference which player's shot from the rough strikes the tree limb. It is less clear, however, that B belongs to a partition, each member of which screens A from C. In other cases birdies will occur as a result of a splendid shot out of a sand trap, or sinking a long putt, or a fine chip shot from the fairway. In these cases, it seems to me, it would not be irrelevant that Jones, rather than some much less accomplished player, was the person who teed off (A). It might be possible to construct a partition B_i that would accomplish the required screening off by specifying the manner in which the ball approaches the cup, rather than referring merely to where the ball came from on the final shot. But this ploy seems artificial. Just as we rejected the attempt to save Reichenbach's definition of causal betweenness by specifying the physical parameters of the ball and the branch at the moment of collision, so also, I think, must we resist the temptation to resort to similar physical parameters to find a partition that achieves screening off. We are, after all, discussing a golf game, not Newtonian particle physics, as Suppes is eager to insist. The most plausible construal of this example, from the standpoint of Suppes's theory, is to take A to be a direct cause of C, and to deny that the sequence A, B, C has the Markov property. In contrast to Good and Reichenbach, Suppes does not require causal sequences to be Markovian.

The crucial problem about B, it seems to me, is that it appears not to qualify even as a prima facie cause of C. It seems reasonable to suppose that even the ordinary duffer has a better chance of making a birdie $P(C)$ than Jones has of getting the ball in the hole by bouncing it off of the tree limb $P(C|B)$. In Suppes's definitions, however, being a prima facie cause is a necessary condition of being any kind of cause (other than a negative cause). Surely, as Suppes himself remarks, we must recognize B as a link in the causal chain. The same point applies to the other examples introduced earlier to show the inadequacy of Reichenbach's definition of causal betweenness. Since the crapshooter has a better chance of winning at the outset $P(C)$ than of winning by getting 4 on the first toss $P(C|B)$, shooting 4 is not even a prima facie cause of winning. Even though Suppes desists from defining causal betweenness, the kinds of examples that lead to difficulty for Reichenbach on that score result in closely related troubles in Suppes's theory.

The fundamental problem at issue here is what Rosen (1978, p. 606) calls "Suppes' thesis that a cause will always raise the probability of the effect." Although both Suppes (1970, p. 41) and Rosen (1978, p. 607) sometimes refer to it as the problem of unlikely or improbable consequences, this latter manner of speaking can be confusing, for it is *not* the small degree of probability of the effect, given the cause, that matters; it is the *negative statistical relevance* of the cause to the occurrence of the effect that gives rise to the basic problem. While there is general agreement that positive statistical relevance is not a sufficient condition of direct causal relevance—we all recognize that the falling barometric reading does not cause a storm—the question is whether it is a necessary condition. Our immediate intuitive response is, I believe, that positive statistical relevance is, indeed, a necessary ingredient in causation, and all three of the theories I am discussing make stipulations to that effect. Reichenbach (1956, p. 201) assumes "that causal relevance is a special form of positive [statistical] relevance." Suppes makes positive statistical relevance a defining condition of prima facie causes (1970, p. 12), and every genuine cause is a prima facie cause (ibid., p. 24). Good incorporates the condition of positive statistical relevance into his definition of causal chains (1961–62, II, p. 45).

In a critical note on Suppes's theory, Germund Hesslow (1976) challenges this fundamental principle:

The basic idea in Suppes's theory is of course that a cause raises the probability of its effect, and it is difficult to see how the theory could be modified without upholding this thesis. It is possible however that examples could be found of causes that lower the probability of their effects. Such a situation could come about if a cause could lower the probability of other more efficient causes. It has been claimed, e.g., that contraceptive pills (C) can cause thrombosis (T), and that consequently there are cases where C_t caused T_t'. [The subscripts t and t' are Suppes's temporal indices.] But pregnancy can also cause thrombosis, and C lowers the probability of pregnancy. I do not know the values of $P(T)$ and $P(T|C)$ but it seems possible that $P(T|C) < P(T)$, and in a population which lacked other contraceptives this would appear a likely situation. Be that as it may, the point remains: *it is entirely possible that a cause should lower the probability of its effect.* (1976, p. 291; Hesslow's emphasis)

Rosen defends Suppes against this challenge by arguing:

[B]ased on the available information represented by the above probability estimates, we would be hesitant, where a person suffers a thrombosis, to blame the person's taking of contraceptive pills. But it does not follow from these epistemic observations that a particular person's use of contraceptive pills lowers the probability that she may suffer a thrombosis, for, unknown to us, her neurophysiological constitution (N) may be such that the taking of the pills definitely contributes to a thrombosis. Formally,

$$P(T|C.N) > P(T)$$

represents our more complete and accurate causal picture. We wrongly believe that taking the pills always lowers a person's probability of thrombosis because we base our belief on an inadequate and superficial knowledge of the causal structures in this medical domain where unanticipated and unappreciated neurophysiological features are not given sufficient attention or adequate weighting. (1978, p. 606)

Rosen comments on her own example of the spectacular birdie in a similar spirit: "Suppes' first observation in untangling the problems of improbable consequences is that it is important not to let the curious event be rendered causally spurious by settling for a superficial or narrow view" (ibid., p. 608). As I have indicated, I do not believe that this is a correct assessment of the problem. If the causal event in question—e.g., the ball striking the branch—is negatively relevant to the final outcome, it is not even a prima facie cause. A fortiori, it cannot achieve the status of a spurious cause, let alone a genuine cause. Rosen continues:

[I]t is the angle and the force of the approach shot together with the deflection that forms our revised causal picture. Thus we begin to see that the results are unlikely only from a narrow standpoint. A broader picture is the more instructive one. (ibid.)

As a result of her examination of Hesslow's example, as well as her own, she concludes that it is a virtue of Suppes's probabilistic theory to be able to accommodate "unanticipated consequences" (ibid.).

Rosen's manner of dealing with the problem of causes that appear to bear negative statistical relevance relations to their effects (which is similar to that mentioned by von Bretzel) might be called *the method of more detailed specification of events.* If some event C, which is clearly recognized as a cause of E, is nevertheless negatively relevant to the occurrence of E, it is claimed that a more detailed specification of C (or the circumstances in which C occurs) will render it positively relevant to E. I remain skeptical that this approach—though admittedly successful in a vast number of instances—is adequate in general to deal with all challenges to the principle of positive statistical relevance.

Good was clearly aware of the problem of negative statistical relevance, and he provided an explicit way of dealing with it. His approach, which differs from Rosen's, might be called *the method of interpolated causal links*. In an appendix (1961–1962, I, p. 318) designed to show that his χ cannot be identified with his Q, he offers an example along with a brief indication of his manner of dealing with it:

> Sherlock Holmes is at the foot of a cliff. At the top of the cliff, directly overhead, are Dr Watson, Professor Moriarty, and a loose boulder. Watson, knowing Moriarty's intentions, realizes that the best chance of saving Holmes's life is to push the boulder over the edge of the cliff, doing his best to give it enough horizontal momentum to miss Holmes. If he does not push the boulder, Moriarty will do so in such a way that it will be nearly certain to kill Holmes. Watson then makes the decision (event F) to push the boulder, but his skill fails him and the boulder falls on Holmes and kills him (event E).
>
> This example shows that $Q(E|F)$ and $\chi(E{:}F)$ cannot be identified, since F had a tendency to prevent E and yet caused it. We say that F was a cause of E because there was a chain of events connecting F to E, each of which was strongly caused by the preceding one.

This example seems closely related to the remark, later appended to theorem T2 (see note 7), to the effect that a cut chain can be uncut by filling in more of the details. Good could obviously take exception to any of the examples I have discussed on the ground that the spatiotemporal gaps between the successive events in these chains are too great. He could, with complete propriety, insist that these gaps be filled with intermediate events, each of which is spatiotemporally small, and each of which is contiguous with its immediate neighbors (see I, pp. 307–308; II, p. 45). I am not convinced, however, that every "cut chain" that needs to be welded back together can be repaired by this device[15]; on the contrary, it seems to me that size is not an essential feature of the kinds of examples that raise problems for Suppes's and Reichenbach's theories. We can find examples, I believe, that have the same basic features, but that do not appear to be amenable to Good's treatment.

Consider the following fictitious case, which has the same statistical structure as the first tetrahedron-cum-card example. We have an atom in an excited state which we shall refer to as the 4th energy level. It may decay to the ground state (0th level) in several different ways, some of which involve intermediate occupation of the 1st energy level. Let $P(m \rightarrow n)$ stand for the probability that an atom in the mth level will drop directly to the nth level. Suppose we have the following probability values[16]:

$$P(4 \rightarrow 3) = \tfrac{3}{4} \qquad P(3 \rightarrow 1) = \tfrac{3}{4}$$
$$P(4 \rightarrow 2) = \tfrac{1}{4} \qquad P(2 \rightarrow 1) = \tfrac{1}{4} \tag{19}$$

It follows that the probability that the atom will occupy the 1st energy level in the process of decaying to the ground state is $\tfrac{10}{16}$; if, however, it occupies the 2nd level on its way down, then the probability of its occupying the 1st level is $\tfrac{1}{4}$. Therefore, occupying the 2nd level is negatively relevant to occupation of the 1st level. Nevertheless, if the atom goes from the 4th to the 2nd to the 1st level, that sequence constitutes a causal chain, in spite of the negative statistical relevance of the intermediate stage. Moreover, in view of the fact that we cannot, so to speak, "track" the atom in its transitions from one energy level to another, it appears that there is no way, even in principle, of filling in intermediate "links" so as to "uncut the chain." Furthermore, it seems unlikely that the Rosen method of more detailed specification of events will help with this example, for when we have specified the type of atom and its energy levels, there are no further facts that are relevant to the events in question. Although this example is admittedly fictitious, one finds cases of this general sort in examining the term schemes of actual atoms.[17]

There is another type of example that seems to me to cause trouble for both Reichenbach and Suppes. In a previous discussion of the principle of the common cause, "Why Ask, 'Why?'?" (essay 8), I suggested the need to take account of *interactive forks* as well as conjunctive forks. Consider the following example. Pool balls lie on the table in such a way that the player can put the 8-ball into one corner pocket at the far end of the table if and almost only if the cue-ball goes into the other far corner pocket. Being a relative novice, the player does not realize that fact; moreover, this player's skill is such that there is only a 50–50 chance of sinking the 8-ball. Let us make the further plausible assumption that, if the two balls drop into the respective pockets, the 8-ball will fall before the cue-ball does. Let event A be the player attempting that shot, B the dropping of the 8-ball into the corner pocket, and C the dropping of the cue-ball into the other corner pocket. Among all of the various shots the player may attempt, a small proportion will result in the cue-ball's landing in that pocket. Thus, $P(C|B) > P(C)$; consequently, the 8-ball's falling into one corner pocket is a prima facie cause of the cue-ball's falling into the other pocket. This is as it should be, but we must also be able to classify B as a spurious cause of C. It is not quite clear how this is to be accomplished. The event A, which must surely qualify as a direct cause of both B and C, does not screen B off from C, for $P(C|A) = \frac{1}{2}$ while $P(C|A.B) = 1$.

It may be objected, of course, that we are not entitled to infer, from the fact that A fails to screen off B from C, that there is no event prior to B that does the screening. In fact, there is such an event—namely, the compound event that consists of the state of motion of the 8-ball and the state of motion of the cue-ball shortly after they collide. The need to resort to such artificial compound events does suggest a weakness in the theory, however, for the causal relations among A, B, and C seem to embody the salient features of the situation. An adequate theory of probabilistic causality should, it seems to me, be able to handle the situation in terms of the relations among these events without having to appeal to such ad hoc constructions.

4. A Modest Suggestion

In the preceding sections I have invoked counterexamples of three basic types to exhibit the most serious inadequacies of the theories of Good, Reichenbach, and Suppes. The first type consists simply of situations in which a given result may come about in more than one way. Since there are usually "more ways than one to skin a cat," such counterexamples cannot be considered weird or outlandish; on the contrary, they typify a large class of actual situations. Theories of probabilistic (or any other sort of) causality that cannot handle cases of this kind are severely inadequate. The second type consists of what may generally be classified as causal interactions. In "Why Ask 'Why?'?" (essay 8) I offered Compton scattering as an example of an interactive fork.[18] Other examples include the interaction between a ray of white light and a red filter, the deformations of fenders of automobiles that bump into one another, and the mutual arousal of lovers who exchange a passionate kiss. Examples of this sort are also far from esoteric; the notion of causal interaction is central to the whole concept of causality. The third type is illustrated by Crasnow's example, which shows how easily a bona fide common cause can give rise to a triple of events that satisfy all of Reichenbach's conditions (11)–(14) for conjunctive forks but that embody spurious causal relations. Again, the situations in which this can occur are far from exceptional.

It seems to me that the fundamental source of difficulty in all three of the theories I have discussed is that they attempt to carry out the construction of causal relations on the basis of probabilistic relations among discrete events without taking account of the physical connections among them. This difficulty, I believe, infects many nonprobabilistic theories as well. When discrete events bear genuine cause-effect relations to one another—except, perhaps, in some instances in quantum mechanics—there are spatiotemporally continuous causal processes joining them.[19] It is my view that these processes transmit causal influence (which may be probabilistic) from one region of space-time to another. Thus, a golf ball flying through the air is a causal process connecting the collision with the tree branch to the dropping of the ball into the cup. Various types of wave phenomena, including light and sound, and the continuous space-time persistence of material objects, whether moving or stationary, are examples of causal processes that provide causal connections between events. In essay 12 I attempt to develop an 'at-at' theory of causal influence which explicates the manner in which causal processes transmit such influence. In essays 7 and 8 I say something about the role played by causal processes in causal explanations.

There is a strong tendency on the part of philosophers to regard causal connections as being composed of chains of intermediate events, as Good brings out explicitly in his theory, rather than spatiotemporally continuous entities that enjoy fundamental physical status, and that do *not* need to be constructed out of anything else. Such a viewpoint can lead to severe frustration, for we are always driven to ask about the connections among *these* events, and interpolating additional events does not seem to mitigate the problem. In his discussion of Locke's concept of power, Hume ([1748] 1955, §7, pt. 1) seems to have perceived this difficulty quite clearly. I am strongly inclined to reverse the position, and to suggest that we accord fundamental status to processes. If only one sort of thing can have this status, I suggest that we treat events as derivative. As John Venn remarked more than a century ago, "Substitute for the time-honoured 'chain of causation,' so often introduced into discussion upon this subject, the phrase a 'rope of causation,' and see what a very different aspect the question will wear" (1866, p. 320).

It is beyond the scope of this essay to attempt a rigorous construction of a probabilistic theory of causality, but the general strategy should be readily apparent. To begin, we can easily see how to deal with the three basic sorts of counterexamples I have discussed. First, regarding Rosen's example, let us say that the striking of the limb by the golf ball is causally between the teeing-off and the dropping into the hole because there is a spatiotemporally continuous causal process—the history of the golf ball—that connects the teeing-off with the striking of the limb, and connects the striking of the limb with the dropping into the hole. Second, we can handle the pool ball example by noting that the dropping of the 8-ball into the pocket is not a genuine cause of the cue-ball falling into the other pocket because there is no causal process leading directly from the one event to the other. Third, we can deal with the Crasnow example by pointing out that the telephone appointment made by Brown constitutes a common cause for the coffee's being ready and for the arrival of the business associate because there is a causal process that leads from the appointment to the making of the coffee and another causal process that leads from the appointment to the arrival of the other person. However, there are no causal processes leading from Brown's boarding of the early bus to the making of the coffee or to the arrival of the other person.

In this essay I have raised three fundamental objections to the theories of probabilistic causality advanced by Good, Reichenbach, and Suppes. Taking these objections in the reverse order to that in which they were introduced, I should like to indicate how I believe they ought to be handled.

(1) In the interactive fork—e.g., Compton scattering or collisions of billiard balls— the common cause C fails to screen off one effect A from the other effect B. In Suppes's theory, if A bears the appropriate temporal relation to B, A will qualify as a prima facie cause of B, but because of the failure of screening off, it cannot be relegated to the status of spurious cause. When we pay appropriate attention to the causal processes involved, the difficulty vanishes. As I have already remarked, the presence of causal processes leading from C to A and from C to B, along with the absence of any direct causal process going from A to B, suffices to show that A is not a genuine cause of B.

Reichenbach's theory of common causes needs to be supplemented in two ways. First, we must recognize that there are two types of forks, conjunctive and interactive. Conjunctive forks are important, but they do not characterize all types of situations in which common causes are operative. Second, the characterization of both types of forks needs to incorporate reference to connecting causal processes in addition to the statistical relevance relations among triads of events. In the case of conjunctive forks, the appeal to causal processes enables us to overcome the problem raised by Crasnow's counter-example.

(2) The most difficult problem, it seems to me, involves the dictum that cause-effect relations must always involve relations of positive statistical relevance. I believe that the examples already discussed show that this dictum cannot be accepted in any simple and unqualified way; at the same time, it seems intuitively compelling to argue that a cause that contributes probabilistically to bringing about a certain effect must at least raise the probability of that effect vis-à-vis some other state of affairs. For example, in the tetrahedron-cum-card game, once the tetrahedron has been tossed and has landed on side 4, the initial probability of drawing a red card in the game is irrelevant to the causal process (or sequence[20]) that leads to the draw of a red card from the deck that is poorer in red cards. What matters is that a causal process has been initiated that may eventuate in the drawing of a red card; it makes no difference that an alternative process might have been initiated that would have held a higher probability of yielding a red card.

Once the tetrahedron has come to rest, one of two alternative processes is selected. There is an important sense in which it possesses an *internal* positive relevance with respect to the draw of a red card. When this example was introduced, I made the convenient but unrealistic simplifying assumption that a draw would be made from the second deck if and only if the tetrahedron toss failed to show side 4. However, someone who has just made a toss on which the tetrahedron landed on side 4 might simply get up and walk away in disgust, without drawing any card at all. In this case, of course, one is certain not to draw a red card. When we look at the game in this way, we see that, given the result of the tetrahedron toss, the probability of getting a red card by drawing from the second deck is greater than it is by not drawing at all; thus, drawing from the second deck *is* positively relevant to getting a red card.

In order to deal adequately with this example, we must restructure it with some care. In the first place, we must choose some universe U that is not identical with F, for otherwise $P(E|F) = P(E)$ and $P(G|F) = P(G)$, in which case F is not positively relevant either to its immediate successor G or to the final effect E. Thus, the player may choose to play the game (event F)—perhaps a fee must be paid to enter the game—or may choose not to play (event $\bar{F}$). If one chooses not to play, then one has no chance to draw from either deck, and no chance to draw a red card. We must now distinguish between the outcome of the toss of the tetrahedron and the draw from the associated deck of cards. Let G = the tetrahedron lands showing side 4, $\bar{G}$ = the tetrahedron does not show side 4; H_1 = the player draws from the first deck, H_2 = the player draws from the second deck, H_3 = the player does not draw from either deck. As before, E = the player draws a red card, $\bar{E}$ = the player does not draw a red card.[21]

Now, suppose that the following chain of events occurs: $F \to \bar{G} \to H_2 \to E$. We can say that each event in the chain is positively relevant to its immediate successor in the sense that the following three relations hold:

$$P_U(\bar{G}|F) > P_U(\bar{G}) \tag{20}$$

$$P_F(H_2|\bar{G}) > P_F(H_2) \tag{21}$$

$$P_{\bar{G}}(E|H_2) > P_{\bar{G}}(E) \tag{22}$$

We begin with the probabilities taken with respect to the original universe U (the subscript U is vacuous in [20] but it is written to emphasize the point); when F has occurred, our universe is narrowed to the intersection $F \cap U$ (hence the subscript F in [21]); when $\bar{G}$ has occurred, our universe is narrowed to the intersection $\bar{G} \cap F \cap U$ (hence the subscript $\bar{G}$ in [22]). Even though this chain has the Markov property, guaranteeing that

$$P_U(E|H_2) = P_{\bar{G}}(E|H_2) \tag{23}$$

the successive narrowing of the universe after each event in the chain has occurred does reverse some of the relevance relations; in particular, although

$$P_U(E|H_2) < P_U(E) \tag{24}$$

nevertheless, as (22) asserts

$$P_{\bar{G}}(E|H_2) > P_{\bar{G}}(E)$$

because the prior probability of E is quite different in the two universes. The equality (23) obviously does not prevent the reversal of the inequalities between (22) and (24).[22]

[This analysis is called "the method of successive reconditionalization" in Salmon (1984b).]

A similar approach can be applied to Rosen's example of the spectacular birdie and to Good's example of Dr. Watson and the boulder. Once the player has swung on the approach shot, and the ball is traveling toward the tree and not toward the hole, the probability of the ball's going into the hole if it strikes the tree limb is greater—given the general direction it is going—than if it does not make contact with the tree at all. Likewise, once Watson has shoved the rock so that Moriarty cannot get his hands on it, the probability of Holmes's death if Moriarty pushes the boulder over the cliff is irrelevant to the causal chain.

While the foregoing approach, which combines features of Rosen's method of more detailed specification of events with Good's method of interpolation of causal links, seems to do a fairly satisfactory job of handling macroscopic chains—which, no matter how finely specified, can always be analyzed in greater detail—it is doubtful that it can deal adequately with the example of the decaying atom introduced earlier. If the atom is in the 4th energy level, it will be recalled, there is a probability of $^{10}/_{16}$ that it will occupy the 1st energy level in the process of decaying to the ground state. If it drops from the 4th level to the 2nd level in the course of its decay, then there is a probability of $\frac{1}{4}$ that it will occupy the 1st level. Thus, occupation of the 2nd level is negatively relevant to its entering the 1st level. However, in a given instance an atom *does* go from the 4th level to the 2nd, from the 2nd level to the 1st, and then to the ground state. In spite of the fact that being in the 2nd level is negatively relevant to being in the 1st, the causal chain *does* consist in occupation of the 4th, 2nd, 1st, and ground levels. Occupation of the 2nd level is the *immediate* causal antecedent of occupation of the 1st level. The atom, in sharp contrast to the golf ball, cannot be said—subsequent to its departure from the 2nd level and prior to its arrival at the 1st level—to be "headed toward" the 1st level rather than the ground state. Interpolation of intermediate events just does not work in situations of this sort.[23]

An appropriate causal description of the atom, it seems to me, can be given in the following terms. Our particular atom in its given initial state (e.g., the 4th energy level) persists in that condition for some time, and as long as it does so, it retains a certain probability distribution for making spontaneous transitions to various other energy levels, i.e., $P(4 \rightarrow 3) = \frac{3}{4}$; $P(4 \rightarrow 2) = \frac{1}{4}$. Of course, if incident radiation impinges on the atom, it may make a transition to a higher energy by absorbing a photon, or its probability of making a transition to a lower energy level may be altered owing to the phenomenon of stimulated emission. But in the absence of outside influences of that sort, it simply *transmits* the probability distribution through a span of time. Sooner or later it makes a transition to a lower energy level (say the 2nd), and this event, which is marked by emission of a photon of characteristic frequency for that transition, transforms the previous process into another, namely, that atom in the state characterized by the 2nd energy level. This process carries with it a new probability distribution, i.e., $P(2 \rightarrow 1) = \frac{1}{4}$; $P(2 \rightarrow 0) = \frac{3}{4}$. If it then drops to the 1st level, emitting another photon of suitable frequency, it is transformed into a different process which has a probability of 1 for a transition to the ground state. Eventually it drops into the ground state.

It is to be noted that the relation of positive statistical relevance does not enter into the foregoing characterization of the example of the decaying atom. Instead, the case is analyzed as a series of causal processes, succeeding one another in time, each of which transmits a definite probability distribution—a distribution that turns out to give the probabilities of certain types of interactions. Transmission of a determinate probability distribution is, I believe, the essential function of causal processes with respect to the theory of probabilistic causality.[24] Each transition event can be considered as an intersection of causal processes, and it is the set of probabilities of various outcomes of such interactions that constitute the transmitted distribution. While it is true that the photon, whose emission marks the transition from one process to another, does not exist as a separate entity prior to its emission, it does constitute a causal process from that time on. The intersection is like a fork where a road divides into two distinct branches—indeed, it qualifies, I believe, as a bona fide interactive fork. [This type of causal interaction is treated explicitly in "Causality without Counterfactuals" (essay 16).] The atom with its emitted photons (and absorbed photons as well) exemplifies the fundamental interplay of causal processes and causal interactions upon which, it seems to me, a viable theory of probabilistic causality must be built. The thesis that causation entails positive statistical relevance is not part of any such theory.

(3) As we have seen, Good has undertaken a more ambitious project than Reichenbach and Suppes, for he has attempted to construct a quantitative measure of the degree to which one event causes another. I am inclined to think that this effort is somewhat premature—that we need to have a much clearer grasp of the qualitative notion of probabilistic causality before we can hope to develop a satisfactory quantitative theory. I have indicated why I believe Good's particular construction is unsatisfactory, but that does not mean that a satisfactory measure cannot be developed. As I indicated earlier, I think it will need to employ the individual probability values $P(E|F)$ and $P(E|\bar{F})$ instead of Good's relevance measure $Q(E:F)$, or any other measure of statistical relevance. The details remain to be worked out.

The essential ingredients in a satisfactory qualitative theory of probabilistic causality are, it seems to me: (1) a fundamental distinction between causal processes and causal interactions, (2) an account of the propagation of causal influence via causal processes, (3) an account of causal interactions in terms of interactive forks, (4) an account of causal directionality in terms of conjunctive forks, and (5) an account of causal betweenness in terms of causal processes and causal directionality. The 'at-at' theory of causal influence gives, at best, a symmetric relation of causal connection. Conjunctive forks are needed to impose the required asymmetry upon connecting processes.

If an adequate theory of probabilistic causality is to be developed, it will borrow heavily from the theories of Reichenbach and Suppes; these theories require supplementation rather than outright rejection. Once we are in possession of a satisfactory qualitative theory, we may be in a position to undertake Good's program of quantification of probabilistic causal relations. These goals are, I believe, eminently worthy of pursuit. [Readers who are interested in the pursuit should consult Humphreys (1989), Eells (1991), and Hitchcock (1993). Hitchcock's treatment is especially sophisticated and elegant.]

Notes

This material is based on work supported by the National Science Foundation under grant no. SOC-7809146. The author wishes to express his gratitude for this support, and to thank I. J.

Good, Paul Humphreys, Merrilee H. Salmon, Patrick Suppes, and Philip von Bretzel for valuable comments on an earlier version of this essay.

1. Both Good and Reichenbach published earlier discussions of probabilistic causality, but both authors regard them as superseded by the works cited here.

2. The roman numerals in references to Good refer to the respective parts of his two-part article.

3. Throughout this essay I use italic capital letters to designate classes of individuals or events. I construe physical probabilities as relative frequencies, but those who prefer other concepts of physical probability can easily make any adjustments they deem appropriate. In certain contexts, where no confusion is likely to arise, I speak of the occurrence of an event A instead of using the more cumbrous expression "occurrence of an event which is a member of the class A."

4. $P(E|F)$, $P(E)$, and $P(E|\bar{F})$ are independent probabilities; no one of the three can be deduced from the other two. In contexts such as the present, we shall assume that neither $P(F)$ nor $P(\bar{F})$ vanishes, for if they did, there would be problems about whether the conditional probabilities $P(E|F)$ and $P(E|\bar{F})$ are well defined. Given this assumption about $P(F)$ and $P(\bar{F})$, it is easily shown that

$$P(E|F) > P(E) > P(E|\bar{F}) \text{ if } P(E|F) > P(E|\bar{F}),$$

$$P(E|F) = P(E) = P(E|\bar{F}) \text{ if } P(E|F) = P(E|\bar{F}),$$

$$P(E|F) < P(E) < P(E|\bar{F}) \text{ if } P(E|F) < P(E|\bar{F});$$

thus, the sign of $P(E|F) - P(E|\bar{F})$ is the same as the sign of $P(E|F) - P(E)$. For purposes of developing a qualitative theory of probabilistic causality, it does not matter much whether one takes a relevance measure defined in terms of $P(E|F)$ and $P(E|\bar{F})$ or one defined in terms of $P(E|F)$ and $P(E)$, for positive relevance, irrelevance, and negative relevance are all that matter. For purposes of developing a quantitative theory of probabilistic causality, this choice of relevance measures is extremely important. In the cases of positive and negative relevance, the foregoing relations tell us that $P(E)$ lies strictly between $P(E|F)$ and $P(E|\bar{F})$, but from the values of these latter two probabilities, we cannot tell where $P(E)$ lies within that interval—i.e., whether it is close to $P(E|F)$, close to $P(E|\bar{F})$, or nearly midway between. If, in addition to $P(E|F)$ and $P(E|\bar{F})$, we are also given the value of $P(F)$, then we can answer that question, for

$$P(\bar{F}) = 1 - P(F) = \frac{P(E|F) - P(E)}{P(E|F) - P(E|\bar{F})}$$

Suppose that F is a bona fide probabilistic cause of E—e.g., F might be some particular disease and E might be death. An epidemiologist who asks how serious a disease F is might well be concerned with the value of $P(E) - P(E|\bar{F})$, that is, the amount by which the actual death rate in the actual population exceeds the death rate in a similar population which is free from the disease F. Since

$$P(E) - P(E|\bar{F}) = P(F)[P(E|F) - P(E|\bar{F})]$$

that quantity is a function of two factors, namely, $P(F)$, which tells us how widespread the disease is in the population at large, and $P(E|F) - P(E|\bar{F})$, which tells us how greatly an individual's chance of death is enhanced by contracting the disease. The same overall effect might be the result of two different situations. In the first place, F might represent a disease such as cancer of the pancreas, which does not occur with especially high frequency, but which is almost always fatal when it does occur. In the second place, F might represent a disease such as influenza, which occurs much more widely, but which is not fatal in nearly such a high percentage of cases. Notice that this latter consideration is measured not by $P(E|F)$ but by $P(E|F) - P(E|\bar{F})$; we are concerned not with the probability that someone who has influenza will die, but rather with the difference made by influenza to the probability of death.

If, along with Good, we are interested in measuring the degree to which F caused E in an actual causal chain of events, then it seems clear that we are concerned with $P(E|F) - P(E|\bar{F})$, or some other function of these two probabilities, for we want to be able to say in individual cases to what degree F contributed causally to bringing about the result E. Since we are not concerned primarily with the overall effect of F in the population at large, measures which are functions of $P(E)$ and $P(E|F)$ or of $P(E)$ and $P(E|\bar{F})$ are not suitable.

5. For the moment we need not regard the tetrahedron's coming to rest with side 4 showing and the draw from the predominantly red deck as two distinct events inasmuch as one happens (by the rules of the game) if and only if the other does.

6. If we use Good's own relevance measure Q, it is easily shown (see equation [4]) that $Q(E:F) = \log {}^{16}\!/\!_5 \neq Q(E':F') = \log {}^{16}\!/\!_3$. In the text I avoided the use of Good's Q, for I did not want to give the impression that I am arguing the trivial point that $Q(E:F)$ is not a function of $Q(E:G)$ and $Q(G:F)$. The main point is, rather, that in measuring the strengths of causal chains, we cannot afford to neglect the statistical relevance of the first event to the last.

7. The requirement that each event in the chain be positively relevant to its immediate successor appears in two places—in the formal definition of causal chains (II, p. 45) and in theorem T2 (I, p. 311), which says, "φ vanishes if the chain is cut, i.e., if any of the links is of zero strength." In "Errata and Corrigenda" Good adds a gloss on T2: "It is worth noting that a 'cut' chain can often be uncut by filling in more detail." In section 4 I consider whether this stratagem enables Good to escape the basic difficulty that Reichenbach and Suppes face.

8. In most cases, of course, the shot from the tee is not the one which strikes the branch, for there are few, if any, par 2 holes. However, the fact that there are other strokes does not alter the import of the example with respect to Reichenbach's definition of causal between-ness.

9. The basic features of this game are given clearly and succinctly by Copi (1972, pp. 481–482). One whose point is 4, for example, is said to make it "the hard way" if one does so by getting a double 2, which is less probable than a 3 and a 1.

10. The day after I wrote this paragraph, an announcement was broadcast on local radio stations informing parents that students who ate lunch at several elementary schools might have been infected with salmonella, which probabilistically causes severe gastric illness. Clearly the consumption of unwholesome food, not the radio announcement, was the common cause of the unusually high incidence of sickness within this particular group of children.

11. In defining many of his causal concepts, Suppes uses conditional probabilities of the form $P(B|A)$. Since, according to the standard definition of conditional probability $P(B|A) = P(A \cdot B)/P(A)$, this probability would not be well defined if $P(A) = 0$. Suppes explicitly includes in his definitions stipulations that the appropriate probabilities are non-zero. In my discussion I assume, without further explicit statement, that all conditional probabilities introduced into the discussion are well defined.

12. Suppes refers to these as "spurious in sense one" and "spurious in sense two." Since I adopt sense two uniformly in this discussion, I do not explicitly say "in sense two" in the text.

13. In an easily overlooked remark (1956, p. 159), Reichenbach says, "If there is more than one possible kind of common cause, C may represent the disjunction of these causes." Hence, Reichenbach recognizes the need for partitions finer than $\{C,\bar{C}\}$, which makes for an even closer parallel between his notion of a conjunctive fork and Suppes's notion of a spurious cause.

14. We cannot let A = the universe, for then $P(C|A) = P(C)$ and A could not be even a prima facie cause.

15. Paul Humphreys has provided a theorem that has an important bearing on the question of the mending of cut chains. In any two-state Markov chain, the statistical relevance of the first to the last member is zero if and only if at least one link in the chain exhibits zero relevance, and the statistical relevance of the first to the last member is negative only if an odd number of links exhibit negative relevance. The first member of a two-state Markov chain is positively relevant to the last if and only if no link has zero relevance and an even number (including none) of the links exhibit negative relevance. In other words, the signs of the relevance measures of the links multiply exactly like the signs of real numbers. Thus, it is impossible for a two-state Markov chain whose first member is negatively relevant to its last, or whose first member is irrelevant to its last, to be constructed out of links all of which exhibit positive relevance—just as it is impossible for the product of positive real numbers to be zero or negative. It may, however, be possible to achieve this goal if, in the process of interpolating additional events, the two-state character is destroyed by including new alternatives at one or more stages.

16. We assume that the transition from the 3rd to the 2nd level is prohibited by the selection rules.

17. See, for example, the cover design on the well-known elementary text (Wichmann, 1967), which is taken from the term scheme for neutral thallium. This term scheme is given in fig. 34A, p. 199.

18. This example is important because, in contrast to billiard ball collisions, there is no plausible ground for maintaining that a deterministic analysis of the interaction is possible in principle; it appears to constitute an instance of irreducibly probabilistic causation. It is possible to argue, as Bas van Fraassen kindly pointed out to me, that Compton scattering does exhibit some of the anomalous characteristics of the Einstein-Podolsky-Rosen situation, but these features of it do not, I believe, affect the simple use of the example in this context.

19. I do not believe that quantum indeterminacy poses any particular problems for a probabilistic theory of causality, or for the notion of continuous causal processes. This quantum indeterminacy is, in fact, the most compelling reason for insisting on the need for probabilistic causation. The really devastating problems arise in connection with what Reichenbach called "causal anomalies"—such as the Einstein-Podolsky-Rosen problem— which seem to involve some form of action-at-a-distance. I make no pretense of having an adequate analysis of such cases.

20. Actually, in this example as in most others, we have a *sequence of events* joined to one another by a *sequence of causal processes*. The events, so to speak, mark the ends of the segments of processes; they are the points at which one process joins up with another. Events can, in most if not all cases, be regarded as intersections of processes.

21. Notice that, in restructuring this example, I have removed it from the class of two-state Markov chains, for there are three alternatives H_i. The theorem of Paul Humphreys, mentioned in note 15, is therefore not applicable.

22. In this context we must take our relevance relations as signifying a relationship between such probabilities as $P(E|F)$ and $P(E)$, rather than $P(E|\bar{F})$ and $P(E|F)$, for $P_F(E|\bar{F})$ is undefined.

23. If we take F to stand for occupation of the 4th level, G occupation of the 3rd level, $\bar{G}$ occupation of the 2nd level, and E occupation of the 1st level, we cannot say whether or not

$$P(\bar{G}|F) > P(\bar{G})$$

for it depends upon the universe selected. If we take as our universe the set of neutral thallium atoms in highly excited states, it may be less probable that an atom in the 4th level will occupy the 2nd level than it is for thallium atoms in general in higher energy levels. But this question seems to be beside the point; we are concerned with what happens to an atom, given that it occupies the 4th level. In that case we can say unequivocally that

$$P_F(E|\bar{G}) < P_F(E)$$

therefore, the positive relevance requirement fails even if we conditionalize at every stage.

There are at least two plausible rebuttals to this argument against the positive relevance requirement. It might seem altogether reasonable to deny that the atom provides a causal chain, for the transitions "just happen by chance"; nothing *causes* them.

I. J. Good, in private correspondence, has argued that my treatment of the example is oversimplified. He points out, quite correctly, that I have used a simplified version of his notation throughout, ignoring the more detailed notation he introduced in his articles (esp. I, p. 309). He suggests that we need to compare not only atoms in the 3rd level with atoms in the 2nd level, but also the presence of atoms with the situation in which no atom is present at all. I have a great deal of sympathy with this approach, but I am inclined to feel that it is better to emphasize causal transmission of a definite probability distribution than to insist on positive relevance.

24. In this simple example the probabilities remain constant as the process goes on, but this does not seem to be a general feature of causal processes. In other cases the probabilities change in a lawful way as the process progresses. A golf ball in flight loses energy and momentum, and this changes its probability of breaking a pane of glass interposed in its path.

15

Intuitions—Good and Not-So-Good

In the preceding essay, "Probabilistic Causality," I offered a critical survey of what I regarded as the three significant theories of probabilistic causality available at that time, namely, those of I. J. Good, Hans Reichenbach, and Patrick Suppes. Both Good and Suppes responded to that article (Good, 1980, 1985; Suppes, 1984, chap. 3), and I took the subject up again in (Salmon, 1984b, chap. 7). The purpose of this essay is to continue that discussion. As I see it, we have arrived at a point at which basic intuitions about probabilistic causality clash.[1] It may not be possible to resolve the conflicts, but I hope at least to clarify the issues.

Good (1961–1962) set forth an ambitious and complicated quantitative theory of probabilistic causality—the kind of theory that, if sound, would certainly be wonderful to have. He defined two measures—$Q(E:F)$, the degree to which F tends to cause E; and $\chi(E:F)$, the degree to which F actually causes E. My first criticism was that Q is nothing but a measure of statistical relevance, and that it fails to distinguish between what Suppes (1970) called genuine causes and spurious causes. That criticism was based on a misunderstanding, as (Good, 1980) pointed out; for this I want simply to retract the criticism and apologize to him. Moreover, even if I had been right in my interpretation, it would have been a simple matter to repair the defect.

Another issue raised in "Probabilistic Causality" is not as easily resolved. In Good's causal calculus, the strength S of a causal chain is a function of the Q-values of its links. I offered two counterexamples designed to show that this functional relation cannot hold. Good (1980) responded that he simply did not see in what way my examples support my case. I had thought that the force of the counterexamples was intuitively obvious. But then, as we all know, one person's counterexample is another person's *modus ponens*. So let's look at the counterexamples.

The first putative counterexample involves a comparison between two simple games. In the first game the player begins by tossing a tetrahedron with sides marked 1, 2, 3, 4

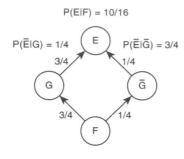

Tetrahedron - Card Game
First Game

Psychotherapy - Joe Doakes

Figure 15.1a

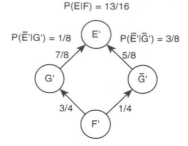

Tetrahedron - Card Game
Second Game

Psychotherapy - Jane Bloggs

Figure 15.1b

(event F). If the tetrahedron comes to rest with side 4 showing, the player draws from a deck containing 16 cards, 12 red and 4 black (event G). If the tetrahedron comes to rest with side 4 on the bottom, the player draws from another deck containing 16 cards, 4 red and 12 black (event $\bar{G}$). Drawing a red card constitutes a win (event E). In this game the probability of winning on any play is $^{10}/_{16}$ (see fig. 15.1a).

The second game is just like the first except for the makeup of the two decks. In this game the first deck contains 14 red and 2 black cards; the other deck contains 10 red cards and 6 black. The events in this game are designated by primed letters. In this game the probability of winning on any play is $^{13}/_{16}$ (see fig. 15.1b).

Suppose, now, that one play of each game has occurred, and in each case the 4 showed on the toss of the tetrahedron, with the result that the player drew from the favorable deck. Moreover, each player drew a red card. Thus, we have two chains of events: $F \rightarrow G \rightarrow E$ and $F' \rightarrow G' \rightarrow E'$.

In that essay I argued as follows. According to Good's definition,

$$Q(Y{:}X) = \log[P(\bar{Y}|\bar{X})/P(\bar{Y}|X)]$$
$$= \log\{[1 - P(Y|\bar{X})]/[1 - P(Y|X)]\}$$

On the reasonable supposition that the only way a player gets to draw from either deck in either game is by entering and tossing the tetrahedron, we have

$$P(G|F) = P(G'|F') = {}^{3}\!/_{4}; \qquad P(G|\bar{F}) = P(G'|\bar{F}') = 0$$

Since

$$[1 - P(G|\bar{F})]/[1 - P(G|F) = [1 - P(G'|\bar{F}')]/[1 - P(G'|F')]$$

it follows that

$$Q(G{:}F) = Q(G'{:}F') \qquad (= \log 4)$$

From the given probability values we calculate that

$$Q(E:G) = \log\{[1 - P(E|\bar{G})]/[1 - P(E|G)]\}$$
$$= \log[(\tfrac{3}{4})/(\tfrac{1}{4})] = \log 3$$

and

$$Q(E':G') = \log\{[1 - P(E'|\bar{G}')]/[1 - P(E'|G')]\}$$
$$= \log[(\tfrac{3}{8})/(\tfrac{1}{8})] = \log 3$$

Therefore,

$$Q(E:G) = Q(E':G')$$

Consequently, on Good's measure S of strength—according to which the strength of a causal chain is a function of the Q-values of its links—the causal strength of the chain $F \to G \to E$ equals that of $F' \to G' \to E'$. Comparing the two games, it seemed intuitively evident to me that the strength of the primed chain is greater than that of the unprimed chain. After all, E can come about in either of two ways, and E' can come about in either of two ways, and for each way the probability of E' is greater than the probability of E. In addition, given our stipulation that the only way one can win in either game is by tossing the tetrahedron and drawing from the appropriate deck, the statistical relevance of F' to E' is greater than that of F to E. Good informed me in a letter (15 June 1979) that he was unmoved by this example.

In an attempt to make my objection more compelling, I offered a second putative counterexample involving two cases. Although there are fundamental differences between the first and second examples, the causal chains have the same associated probabilities in both examples (see fig. 15.1).

Suppose that two individuals, Joe Doakes and Jane Bloggs, suffer from sexual dysfunctions. Joe is impotent and Jane is frigid. Each of them decides to seek psychotherapy. There are two alternative types of therapy available, directive and nondirective. When Joe seeks out a psychotherapist (event F), there is a probability of $\tfrac{3}{4}$ that he will select a directive therapist and undergo that type of treatment (event G), and a probability of $\tfrac{1}{4}$ that he will select a nondirective therapist and undergo that type of treatment (event $\bar{G}$). If he is treated by a directive therapist, there is a probability of $\tfrac{3}{4}$ that he will be cured (event E), and if he is treated by a nondirective therapist, there is a probability of $\tfrac{1}{4}$ that he will be cured. Given these values, there is a probability of $^{10}\!/_{16}$ that he will be cured, given that he undertakes psychotherapy.

When Jane seeks out a psychotherapist (event F'), there is a probability of $\tfrac{3}{4}$ that she will select a directive therapist (event G'), and a probability of $\tfrac{1}{4}$ that she will select a nondirective therapist (event $\bar{G}'$). If she is treated by a directive therapist, there is a probability of $\tfrac{7}{8}$ that she will be cured (event E'), and if she is treated by a nondirective therapist, the probability of a cure is $\tfrac{5}{8}$. Given these values, there is a probability of $^{13}\!/_{16}$ that she will be cured, given that she undertakes psychotherapy.

Joe and Jane each undertake psychotherapy, each is treated by a directive therapist, and each is cured. [They meet, fall in love, and live happily ever after.] Thus, we have two causal chains, $F \to G \to E$ and $F' \to G' \to E'$ . . . . The question is, on what basis, if any, would we be warranted in claiming that the two chains have the same strength—i.e., that the degree to which the seeking out of psychotherapeutic treatment caused the cure is the same for both? ("Probabilistic Causality")

In this example, the initial cause F is the decision to undertake psychotherapy. It does not involve the selection of a therapist and the choice of a particular type of treatment. Ann Landers frequently advises people to "get counseling," without specifying anything about the type. Indeed, the advice, often, is to talk to a member of the clergy, a family doctor, a psychologist, a psychiatrist, or a social worker. It is highly unspecific. The idea behind my example is to ask how efficacious in overcoming either of the dysfunctions is the taking of such general advice as "get psychotherapy." In this example I intend to confine the types of treatment considered to those offered by psychiatrists or clinical psychologists. Arbitrarily, perhaps, results of anything else fall into the category of spontaneous remission.

According to Good's causal calculus the strengths of the two chains must be equal, for the same reasons as in the first example. He apparently finds this conclusion unobjectionable. I suggested that we cannot answer the question about the causal efficacy of psychotherapy in these two cases until we know the probabilities of spontaneous remission for the two conditions. Let us therefore arbitrarily stipulate some values: $1/100$ for impotence and $3/4$ for frigidity. Given these values, I claimed that the strength of the causal chain in Joe's case is greater than it is in Jane's, since the probability of a cure with the aid of psychotherapy for Joe ($10/16$) is much greater than the rate of spontaneous remission, while the probability of a cure with the aid of psychotherapy for Jane ($13/16$) is only slightly greater than the spontaneous remission rate. We should not be misled by the words "cure" and "remission"; each refers simply to the fact that the problem went away. There is no reliable way to say in advance whether the cure will last. We cannot legitimately construe "cure" to mean something permanent and "remission" to refer to a temporary condition that will be followed by a recurrence of the problem.

Good was no more moved by the second example than by the first. He remarks, "I do not see in what way his examples support his case except that what is said three times sounds true" (1980, p. 302). He suggests that my belief in the difference in strength of the two chains results from confusing Q (the tendency to cause) with χ (the strength of the causal connection). The Q-values are different: $Q(E:F) = 0.971$ while $Q(E':F') = 0.288.^2$ We both agree that the tendency of psychotherapy to eliminate a sexual dysfunction is different for Bloggs and Doakes, but we disagree about the degree to which psychotherapy actually caused the cure in the two cases. I want to try to understand this disagreement.

In an effort better to understand Good's theory—in particular, the distinction between Q and χ—let us now consider some of the examples he has offered. Good (1961–1962) gives an example to show that Q and χ cannot be identified:

Sherlock Holmes is at the foot of a cliff. At the top of the cliff, directly overhead, are Dr Watson, Professor Moriarty, and a loose boulder. Watson, knowing Moriarty's intentions, realises that the best chance of saving Holmes's life is to push the boulder over the edge of the cliff, doing his best to give it enough horizontal momentum to miss Holmes. If he does not push the boulder, Moriarty will do so in such a way that it will be nearly certain to kill Holmes. Watson then makes the decision (event F) to push the boulder, but his skill fails him and the boulder falls on Holmes and kills him (event E).

This example shows that $Q(E:F)$ and $\chi(E:F)$ cannot be identified, since F had a tendency to prevent E and yet caused it. We say that F was a cause of E because there was a chain of events connecting F to E, each of which was strongly caused by the preceding one. (Good, 1961–1962, pt. 1, p. 318)

This example, which I have reproduced in its entirety, is the only concrete illustration furnished by Good (1961–62) of the difference between Q and χ. While it has some heuristic value, it suffers from the fact that degrees of causal efficacy are not involved. Watson's decision fully failed to prevent E; indeed, it fully caused E, given the conditions provided in the example. So it does not help much in our trying to understand the Bloggs–Doakes example. The fact that Moriarty would have killed Holmes if Watson had done nothing has no bearing on the fact that Watson killed Holmes. One important difference between this example and the Bloggs–Doakes example is the fact that the cause of Holmes's death is readily identifiable as an act of Watson rather than an act of Moriarty. The Bloggs–Doakes example defies such an analysis.

In Good (1980) another example is offered to illustrate the distinction between Q and χ, namely, the distinction between murder and attempted murder. It is easy to see how an attempt at murder has a tendency to cause the death of the victim, and to see how the strength of that tendency might be quantified in degrees, depending on the skill and motivation of the perpetrator, the conditions under which the attempt is made, and steps taken to prevent the murder. If, however, the attempt is successful and the victim is killed, it is not easy to see how to assign a degree to the contribution of the murderer to the death of the victim other than to say simply that the murderer did it. Even if the murderer is a professional assassin, and even if the person who took out the contract on the life of the victim hires several other professional assassins as well to make sure that the victim dies, it is hard to see how—if they all work independently—the successful assassin should fail to get full credit and the others no credit whatever.

One complication that might arise, of course, is for two assassins to shoot the victim at precisely the same moment, each shot being sufficient to cause immediate death by itself. This makes the example the same in principle as another of Good's examples, namely, the firing squad. We shall look at that example in a moment. Such examples are cases of overdetermination, about which much has been written. As we shall see, Good has an interesting method for dealing with them.

The situation would be somewhat more complicated if two assassins were to work together. Suppose, for example, that the intended victim is in a building with only two exits. One assassin enters through the front door; the other waits outside the back door. As the killers arrive on the scene, it is impossible to predict which one will actually kill this victim. If he remains in his office, he will confront the first assassin and be killed; if he makes a run for it through the back door the other assassin will kill him. In this example, both common sense and the law hold both assassins responsible. I am not at all sure how we should quantitatively apportion the blame. If, in this case, both were indispensable, then perhaps it is reasonable to divide the responsibility equally between them. If, however, there is some small chance that the assassin who enters the front door would be successful without the aid of the other assassin, while the one waiting outside the back door has no chance of success working alone, then it would probably be reasonable to assign to the first assassin a higher causal contribution than to the second.

Where basic intuitions seem to conflict, it is useful—indeed, indispensable—to consider a variety of examples. Prior to Good's PSA paper (Good, 1985), he had furnished a paucity of examples to illustrate his basic concepts; in fact, the only ones with which I was familiar were the little Sherlock Holmes story and the unelaborated mention of the distinction between murder and attempted murder. Fortunately, his PSA paper as well as his principal contribution to Skyrms and Harper (1988) contain a number of useful ones.

One of his examples, which follows naturally from the foregoing elaboration of the distinction between attempted murder and murder, involves a firing squad with two marksmen, both crack shots. When the captain gives the order to fire (event F), both shoot (events G_1 and G_2). Each shot would, by itself, be sufficient to cause death (event E). In this case there are two causal chains from F to E, and each has maximal strength (positive infinity). Again, we get no real feeling from such examples as to the way to assign nonextreme degrees of causation.

That drawback is overcome in an ingenious variation on the firing squad example. Suppose the squad has three members, all crack shots, each of whom uses a standard six-shot revolver. When the captain gives the order to fire, each marksman spins the cylinder of the weapon (as in Russian roulette), aims at the condemned, and pulls the trigger. If a bullet is in the firing position when the trigger is pulled, a fatal shot is fired. The condemned person dies if and only if at least one shot is fired (it is possible, of course, that more than one shot is fired). In this case we have more than one causal chain, and each contains links of nonextreme strength. It provides a good example of a causal net.[3]

Now that we have before us several examples (owing to Good or myself) that have appeared in the literature, let us attempt to examine them in a more systematic manner. At the outset a simple but basic distinction must be made. There are cases in which two or more causes acting together bring about a result. For example, reaching the goal in a fund drive is achieved because gifts are received from many sources, and each gift constitutes a definite portion of the amount raised. Here it is natural to quantify the degree to which each cause contributed to the effect. There is, of course, nothing probabilistic about this example.

In a different sort of case we have two or more causes, any one of which might, by itself, bring about a result. The potential causes are mutually exclusive; if one operates, the others do not. For instance, on any given morning I may walk to my office, drive, or take a bus. When I adopt one mode of transport, I reject the others for that day. This example is also nonprobabilistic.

The Russian roulette firing squad exemplifies the first situation probabilistically. There are three intermediate causes that may be said to cooperate in bringing about a result; all three marksmen are present and follow orders for any given "execution" (see fig. 15.2a). As I construe this example, the condemned person may survive the carrying out of the sentence, because the order to fire is given only once for any person sentenced to face that sort of firing squad. Thus, neither singly nor in concert do the marksmen inevitably produce death. As we shall see shortly, this example lends itself nicely to quantitative analysis.

Some of our examples fit the pattern of several different alternative causes, only one of which is present in any given case (see fig. 15.2b). For instance, both versions of the tetra-card game exemplify it, for on each play of both games the player draws from one deck only. Good's example of the pinball machine also fits this case, for there are many possible routes a ball can travel before it drops out of play at the bottom. On any given play, of course, it follows only one route. The two versions of the psychotherapy example may also seem to fit, for Bloggs and Doakes undergo only one form of therapy

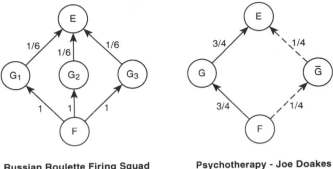

Russian Roulette Firing Squad   **Psychotherapy - Joe Doakes**

Figure 15.2a   **Figure 15.2b**

each. It is probably more realistic, however, to consider the factors—whatever they may be—that sometimes lead to spontaneous remission to be additional probabilistic causes that are also present when a person undergoes psychotherapy. It would be implausible to suppose that the decision to undertake psychotherapy eliminates them.

Let us see how situations of both types are handled. Good has remarked on several occasions that an analogy between causal nets and electrical circuits has provided fruitful intuitions. Indeed, he introduces the notions of resistance R and strength S of the elements of causal nets (where strength is analogous to electrical conductance). In circuit theory resistance and conductance are reciprocal; in Good's causal calculus their counterparts are not quite reciprocal, but as one increases, the other decreases in a precisely defined way.[4]

Let us look at the analogy. In circuit theory the resistance of a circuit composed of several elements connected in series is equal to the sum of their resistances. The conductance of a circuit composed of several elements connected in parallel is equal to the sum of their conductances. These relations are intuitively obvious—as well as being experimentally confirmed (see fig. 15.3). If a circuit has three parallel paths of equal conductance C, the conductance of the circuit is $3C$. Exploiting the well-worn analogy between electricity and hydraulics, we may say that three times as much "juice" can flow through all three channels as can flow through one alone. Indeed, the electrical current consists of a flow of electrons. Thinking now of a causal net, we see that the strength of the causal connection in a net containing three chains of strength S in parallel should be equal to $3S$, the sum of the strengths of three chains. Good stipulates, however, that the chains must be mutually independent (1961–1962, pt. 1, pp. 313–314).

Let us apply these ideas to some examples. The simple firing squad with two members who fire fatal shots if and only if the captain gives the command to fire is a net consisting of two chains in parallel. Since each probability is unity, all of the links have positive infinity as their degree of strength. Since the strength of the net is equal to the sum of the strengths of the chains, the strength of the net is also positive infinity. Overdetermination thus poses no problem for the causal calculus.

Let us now consider a nondegenerate example. In the case of the Russian roulette firing squad, event F, the command to fire, gives rise to G_1, G_2, and G_3 with a probability

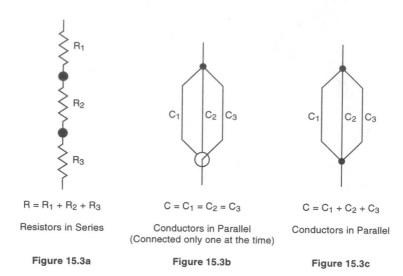

R = R₁ + R₂ + R₃	C = C₁ = C₂ = C₃	C = C₁ + C₂ + C₃
Resistors in Series	Conductors in Parallel (Connected only one at the time)	Conductors in Parallel
Figure 15.3a	**Figure 15.3b**	**Figure 15.3c**

of one. Consequently, the links between F and each of the three have maximal strength of positive infinity. Each event G_i has a probability of $\frac{5}{6}$ to produce a fatal shot. At this point we must exercise a bit of care. If we take E to stand for the death of the condemned person, and if we compute $P(E|G_i)$, we must recall that, if one marksman pulls the trigger, all three do. Thus, the probability that the condemned dies if one marksman pulls the trigger is the probability that at least one of the three fires a fatal shot, i.e.:

$$P(E|G_i) = 1 - (\tfrac{5}{6})^3 = 0.421$$

This is clearly the wrong probability to use to evaluate the causal contribution of a given marksman to the death of the condemned, for the fact that one marksman pulls the trigger makes no causal contribution to the fact that the others do also; it is the command of the commissar that causes each to pull his or her trigger. It would be an obvious mistake to use that value to calculate the strength of one of the chains, and then to multiply by three to get the strength of the entire net. Instead, we might write E_i to signify that a fatal shot was fired by G_i. E is then equivalent to the nonexecutive disjunction of the E_i. We calculate the Q-value of each link as follows:

$$Q(E_i|G_i) = \log\{[1 - P(E_i|\bar{G}_i)]/[1 - P(E_i|G_i)]\} = 0.182$$

In Good's causal calculus the strength of a single link is equal to the Q-value of that link, so $S(E:G_i) = 0.182$. Given the maximal value of the first link, the strength of any one of the three chains $F \to G_i \to E_i$ is also 0.182. The strength of the entire net is the sum of the strengths of the three chains connected in parallel, namely, 0.547. That value represents $\chi(E:F)$, the actual causal contribution of the commissar's order "Fire!" to the death of the condemned person. In this case, it happens, $Q(E:F)$, the tendency of the captain's order to cause death, has the same value, as can be shown by computing the value of log $\{[1 - P(E|\bar{F})]/[1 - P(E|F)]\}$.

Let us also look at the case of Jane Bloggs. Using the probability values as given, we can easily compute $Q(G':F') = \log 4 = 1.39$ and $Q(E':G') = \log 3 = 1.10$.[5] For the separate links, $S(G':F') = Q(G':F')$ and $S(E':G') = Q(E':G')$. Since the two links are connected in series, we can get the resistance of the chain by computing the sum of the resistances of the links. Strength and resistance are related by the formulas

$$R(Y:X) = -\log[1 - e^{-S(Y:X)}]$$
$$S(Y:X) = -\log[1 - e^{-R(Y:X)}]$$

According to my calculations,

$$R(G':F') = 0.405; \qquad R(E':G') = 0.288$$

which yields

$$S(E':F') = 0.693$$

as the strength of the chain $F' \rightarrow G' \rightarrow E'$.

Let us now consider Jane's cousin Amy who suffered from the same sexual dysfunction as Jane, undertook psychotherapy, and got over her frigidity. In contrast to Jane, however, Amy sought help through nondirective therapy. In her case we have the chain $F' \rightarrow \bar{G}' \rightarrow E'$; we find that $Q(\bar{G}':F') = 0.288$, but $Q(E':\bar{G}')$ has a negative value (-1.10). Any link that has a negative Q-value has, by definition, a strength of zero, and any chain that has such a link also has zero strength. In Good's calculus no chain is stronger than its weakest link. If we consider only the case of Jane Bloggs, this result may not seem troubling. But what are we to say of Amy? Is there, in her case, no causal chain? The intuitive answer is that there is indeed a causal chain. We shall have to develop a sounder method for evaluating cases of this sort.[6]

The question just raised is the problem of negative relevance, which was vividly posed in Suppes (1970) in terms of Deborah Rosen's famous case of the spectacular birdie.[7] It is further discussed in the preceding essay and in Salmon (1984b) and Suppes (1984). A similar example is given by Good (1985). In both examples a skillful golfer makes a shot, with the intention of getting the ball in the hole (Rosen) or on the green (Good). In Rosen's example, the shot is actually quite poor, but the ball hits a tree branch and is deflected into the hole. In Good's example, the tee shot might have been good or poor (he does not say which), but (in either case) the ball strikes a bird and falls to the ground, where it is picked up by a chimpanzee who carries it to the green and drops it there. That is called "making it the hard way."[8]

Good uses this example to illustrate the distinction between Q and χ; he remarks, "Here $Q(E:F)$ is appreciable but $\chi(E:F)$ is negligible." Why is the Q-value appreciable? Perhaps because this particular shot is excellent; but for the bird it was almost certain to land on the green. Why is the χ-value negligible? Because the causal chain from stroke to bird to chimp to green has some very weak links. When the ball hits the bird, the probability that it will get to the green is tiny, for it is most unlikely that there will be a chimp or other messenger there to get it from the spot at which it landed to the green. Moreover, when the chimp picks up the ball, the probabilities may overwhelmingly favor its carrying the ball back to the language lab rather than depositing it on the green. In Good's causal calculus a theorem states that no causal chain is stronger than its weakest link, and this chain has some very weak links.

This example seems to illustrate quite well (qualitatively) the distinction between Q and χ, but there is one major question: Is it a causal chain? To qualify as a causal chain, a series of events must sustain suitable relations of positive relevance to one another; in particular, each event in the series must be positively relevant to its immediate successor. The first event in Good's example is the tee stroke, and the second is the striking of the bird. It is reasonable to say that the ball is more likely to hit the bird if it is driven from the tee than if it is not. The third event is the landing of the ball (somewhere on the fairway or in the rough). I find it difficult to believe that the probability of the ball's landing in about the same place, given that it hits the bird, is not pretty small. That event must be quite sensitive to the position and attitude of the bird, and, for a bird in flight, those parameters change very quickly and quite irregularly. But the question is not whether that probability value is small; it is a question of relevance. So we must ask what if the ball had not struck the bird? Well, if the immediately preceding event is the tee stroke, then one might suspect that it is more probable that the shot from the tee would put it there than that it would fall in that general location, given that it has struck the bird. Of course, we lack statistics on results of collisions of golf balls with flying birds (and experiments would obviously be difficult to conduct), so I could be wrong about this matter.

Given that the ball drops to the ground where it does, it will take a miracle to get it to the green before it comes to rest, but we are not disappointed: enter the chimp. The next event is either the entry of the chimp or the chimp's picking up the ball. Either way we seem to be in trouble, for the striking of the bird by the ball is certainly irrelevant to the presence of the chimp in that general vicinity, and it is also irrelevant to the chimp's picking up the ball. If the chimp has picked up the ball, is that positively relevant to the later presence of the ball on the green? The answer must be affirmative, for given the place the ball has reached after colliding with the bird, there is no other way the ball will get to the green. So even if the probability of the chimp's carrying the ball to the green is not large—it is much more likely that the chimp will carry the ball back to the language lab—it is still positively relevant.

In the preceding essay, in discussing several examples of the foregoing kind that seem to involve chains with negatively relevant links, I formulated two (not mutually exclusive) strategies that have been adopted by various authors. I called them *the method of interpolated causal links* and *the method of more detailed specification of events*. I tried to show that they are inadequate to restore positive relevance to all links of causal chains. I have suggested (Salmon, 1984b, pp. 196–202) that an approach called *the method of successive reconditionalization* would be more successful. Although Good's approach, which conditionalizes not only on the given event but also on the state of the universe just prior to that event, does not embody precisely this method of circumventing the problem of negative relevance, it can easily be modified so as to do so.

In his earliest presentation of the causal calculus, Good emphasized that the notation I have been using thus far is condensed, and that all probabilities are to be taken as conditional upon the state of the universe U and all of the laws of nature H. Hence, $P(G|F)$ should be construed as $P(G|F.H.U)$, where U is the state of the universe just before F occurs; similarly, $P(E|G)$ stands for $P(E|G.H.U')$, where U' is the state of the universe just before G occurs. It was because of my failure fully to appreciate the import of this interpretation of the probability expression that I failed to realize in "Probabilistic Causality" that Good does have the distinction between genuine and spurious causes built in. A small (but important) modification would make it possible for his causal calculus to incorporate the method of successive reconditionalization as well. What is required is to stipulate that the probabilities are to be conditioned on the state of the universe at the time of the event in question instead of just prior to its occurrence.

For purposes of illustration let us apply this approach to the Bloggs case. Our previous calculations of $Q(G':F')$ and $Q(\bar{G}':F')$ can stand, for they were conditioned on the fact that F' actually obtained. When, however, we turn to $Q(E':G')$ and $Q(E':\bar{G}')$, we find that we must take care. In particular, when we calculate

$$Q(E':G') = \log\{[1 - P(E'|\bar{G}')]/[1 - P(E'|G')]\}$$

we must stipulate that $P(E'|\bar{G}') = 0$, for, given that G' obtains and $\bar{G}'$ does not, no E' can result from $\bar{G}'$. Similarly, when we calculate

$$Q(E':\bar{G}') = \log\{[1 - P(E'|G')]/[1 - P(E'|\bar{G}')]\}$$

we must set $P(E'|G') = 0$ for the analogous reason.[9] With this understanding we can properly calculate the strength of the causal chains of the Bloggs cousins. For Jane we have

$$Q(G':F') = 1.39; \qquad R(G':F') = 0.286$$
$$Q(E':G') = 2.08; \qquad R(E':G') = 0.133$$

which yields

$$S(F' \rightarrow G' \rightarrow E') = 1.07$$

For Amy we have

$$Q(\bar{G}':F') = 0.288; \qquad R(\bar{G}':F') = 1.39$$
$$Q(E':\bar{G}') = 0.981; \qquad R(E':\bar{G}') = 0.470$$

which yields

$$S(F' \rightarrow \bar{G}' \rightarrow E') = 0.169$$

Notice that positive relevance now obtains within this chain. Clearly, conditioning on the state of the universe at the time $\bar{G}'$ occurs does not affect the value of $P(E'|\bar{G}')$, but it does change the value of $P(E'|G')$, which occurs explicitly in the calculation of $Q(E':\bar{G}')$. There might be some temptation to add together the strengths of the two chains in order to calculate the strength of the entire net containing F', G', $\bar{G}'$, and E', but this temptation should be resisted, inasmuch as the two chains, being mutually exclusive, are not independent. It would be highly counterintuitive, to my mind, to assign the same strengths to the causal nets for Jane and Amy.

The method just outlined goes far in handling the problem of negative relevance for a number of problematic cases. However, as I pointed out in essay 14 and Salmon (1984b, pp. 200–202), a putative counterexample from atomic physics apparently cannot be handled successfully by that method. Good (1980, p. 303) has denied the pertinence of this example on the ground that it is noncausal. I disagree with that claim (Salmon, 1984b, pp. 201–202), but I shall not say more about it here.

If we want to pursue the analogy with circuit theory, it seems crucial to distinguish two types of causal nets. Good defines a causal net as a collection of events, some actual and others perhaps only possible, in which is embedded at least one causal chain of actual events connecting the initial event F to the final event E. In some cases the net may contain a single actual chain—e.g., the Bloggs and Doakes cases, my tetra-card games, Good's pinball example—where there are other possible chains that are actualized (one at a time) in other similar situations, but they are all mutually exclusive. In other cases—e.g., Good's firing squad examples—there are several independent chains composed of actual or possible events connecting the initial event (the order "Fire!") to the final event (the death of the condemned person). The Russian roulette firing squad example, in which more than one chain of actual events linking the initial and final stages is possible, is analogous to the circuit in which all three channels are open simultaneously. Good's skin cancer example illustrates the same sort of situation. Examples of the first type, with mutually exclusive possible chains, are analogous to circuits containing two or more conductors connected by switches that allow only one channel to be connected at a time (see fig. 15.3b–c). The conductance in this case is not the sum of the several conductances; it is simply the conductance of the conductor that happens to be connected at any given time. Where the possible chains are mutually exclusive, the merely potential conductors have no effect whatever on the conductance of the circuit.

The electrical circuit analogy is heuristically useful, I believe, in the case of nets containing mutually exclusive chains and for those containing mutually independent chains. Other cases are, of course, possible. One chain may be positively or negatively relevant to another; in such cases it would presumably be necessary to add strength or resistance, respectively, to the alternative possible chains. As we shall see, the psychotherapy example may be a case in point. I do not have a concrete proposal for dealing with them, and it is not clear to me whether the circuit analogy is helpful in these instances.

We must now turn to one final fundamental distinction. It can be illustrated by comparing the tetra-card games, or Good's firing squad examples, on the one hand, with the psychotherapy case on the other. In either game, the result E (a winning draw) can come about only as a result of one or the other of the causal chains specified for that game. A person who does not enter the game cannot win. In the psychotherapy example the fact that "cures" can come about by spontaneous remission is a crucial feature. Even though we know that one or the other type of therapy (directive or nondirective) has been undertaken, we still cannot attribute the "cure" unequivocally to the therapy.

One sort of strategy that might be suggested for the psychotherapy cases is to add another "cause"—namely, chance. Thus, we might say there are two initial alternatives, F (the decision to undertake psychotherapy) and $\bar{F}$ (to do nothing and hope for the best). Given $\bar{F}$, the probability of a "cure" is simply the spontaneous remission rate. When $\bar{F}$ obtains, of course, neither G nor $\bar{G}$ obtains, so perhaps we should rename them G_1 and G_2, respectively (see fig. 15.4). It would, however, be more realistic to suppose that human beings have recuperative resources, with respect to both physical and psychological ailments. In the absence of therapy, they can sometimes produce spontaneous remissions. There is no reason to suppose that, when an individual undergoes psychotherapy, these internal resources cease operation—though the therapy might enhance or diminish

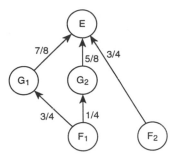

Psychotherapy - Jane Bloggs

Figure 15.4

them. Thus, instead of taking F and $\bar{F}$ as mutually exclusive initial states, we might designate them F_1 and F_2, recognizing that they are not mutually exclusive. This general approach is legitimate enough, but it does not dissolve the distinction I am discussing. The fundamental distinction is between situations in which it is possible in principle to tell which of the possible causes actually brought about the effect and those in which it is impossible to tell. By observing any winning play of either tetra-card game, it is easy to determine which of the two possible causes was operative in that play. By checking the number and trajectories of the bullets that entered the body of the condemned, it is possible to tell which members of the firing squad fired fatal shots. In contrast, by observing a person who underwent psychotherapy and experienced a "cure," it is impossible in principle to tell whether the remission was spontaneous or due to the treatment. Rather, it would quite likely be more reasonable to suppose that both causes make some contribution to the remission; our problem would be to find a way to compare them quantitatively.

Consider the case of Jane Bloggs. The probability that she would get over her frigidity if she undertook psychotherapy was $^{13}/_{16}$, while the probability that she would get over it without treatment was $^3/_4$. She underwent psychotherapy and the problem disappeared. Did the treatment cause the remission of that symptom, or would it have vanished without psychotherapy? We could ask her whether the treatment effected the cure, but her answer should not carry much weight. The chances are that the patient wants to believe that the treatment was efficacious, and is apt to give an affirmative response as a result of wishful thinking. As I understand the situation, there is no reliable way of ascertaining which individuals who undertake psychotherapy in order to deal with frigidity (or any other problem) would have experienced remission without treatment and which would not. There is no evidential basis for asserting, counterfactually, that she would not have overcome the sexual dysfunction if she had not undergone psychotherapy, and none for asserting the contrary counterfactual. Thus, I take it, there is no reliable answer to the question, Did the treatment bring about the cure? All we can say is that she had treatment and her frigidity disappeared. We cannot have adequate evidence to say that she would not have been cured if she had not had psychotherapy. The case of Doakes is similar, except for the fact that there is a large discrepancy between the spontaneous remission rate and the rate of cure among those who undergo psychotherapy for problems like his. Note that when Amy Bloggs undertook nondirective therapy, she reduced her chances of remission. In her case the therapy interfered with the process of recovery. Although these examples are totally fictitious, with probability values arbitrarily stipulated, I know of no reason to suppose that such interference is impossible in principle. Under these circumstances the therapy would be, in Paul Humphreys's (1981) terminology, a counteracting rather than a contributing cause.

In the cases of Jane and Amy Bloggs and Joe Doakes we have stipulated certain probability values, and from these we can compute the Q-value—the tendency or propensity of psychotherapy to produce a cure. The Q-values differ in these two cases: for the Bloggs cousins $Q(E':F') = 0.288$; for Doakes $Q(E:F) = 0.971$. If I understand Good's thinking on this point, he is claiming that the tendency or propensity[10] to effect a cure in either type of case is a function of the statistical relevance of the treatment to the cure. Thus, the Q-value is determined in part by the probability of spontaneous remission. This seems correct. In the case of Doakes the statistical relevance is high; in the case of Bloggs it is rather small. The tendency is, roughly speaking, the difference the treatment makes to the chance of a cure.

The problem of understanding $\chi(E:F)$—the degree to which F caused E or the contribution to the causation of E provided by F—is rather more difficult. Intuitions apparently do simply conflict. Consider again the Russian roulette firing squad. It will be recalled that the command to fire received less than the maximal score on the degree to which it caused the death of the condemned person. Many philosophers would make a different assignment. Those who hold a *sine qua non* conception of causation would point to the fact that, were it not for the commissar's command, the condemned person would not have died. It is true that the command is not a sufficient condition of death, but that does not make the command less of a cause when death results. Similarly, an inept assassin, who misses many of the shots fired, is totally responsible for the deaths of the victims actually killed. This contrasts sharply with the psychotherapy examples. In those cases we cannot say that but for the treatment, the problem would not have vanished. It is in cases of precisely this sort that a quantitative theory of probabilistic causality is most needed. I hope that an airing of conflicting intuitions will help us determine whether a satisfactory theory is available, and if not, help pave the way to finding one.[11]

Notes

1. I much appreciate the fact that Suppes (1984) gives the title "Conflicting Intuitions" to a major section of the chapter "Causality and Randomness."

2. These values are derived as follows. For Doakes,

$$P(E|F) = {}^{10}\!/_{16}; \qquad P(E|\bar{F}) = {}^{1}\!/_{100}$$
$$Q(E:F) = \log\{[1 - P(E|\bar{F})]/[1 - P(E|F)]\}$$
$$= \log\{({}^{99}\!/_{100})/({}^{6}\!/_{16})\} = 0.971$$

For Bloggs,

$$P(E'|F') = {}^{13}\!/_{16}; \qquad P(E'|\bar{F}') = {}^{3}\!/_{4}$$
$$Q(E':F') = \log\{[1 - P(E'|\bar{F}')]/[1 - P(E'|F')]\}$$

$$= \log\{(\tfrac{1}{4})/(\tfrac{3}{16})\} = 0.288$$

In numerical calculations in this essay I use natural logarithms and give results to three significant figures.

3. Another example of a causal net, which is similar in principle but a bit more complicated, involves the occurrence of skin cancer as a result of exposure to sunlight.

4. The relation is

$$e^{R} + e^{S} = 1.$$

5. It should be recalled, in computing the value of $Q(E{:}G)$, that the probability $P(E|G)$ is conditioned on the state of the universe just prior to G (or $\bar{G}$), at which point the decision to undertake psychotherapy has already been made. Consequently, we need not take into account—indeed, we must not take into account—the probability of spontaneous remission in the absence of therapy.

6. Good has kindly reminded me that in his "Errata and Corrigenda" (1962) to "A Causal Calculus" (1961–62), he added the observation that "a 'cut' chain can often be uncut by filling it in in more detail." This *may* provide a way to handle the case of Amy Bloggs. Good's (1961–62) is reprinted, with corrections inserted, in Good (1983).

7. Rosen's example has often been said to raise a problem concerning small probabilities. I do not think there is such a problem. In "Probabilistic Causality" I show that the real problem is one of negative relevance in causal chains.

8. In essay 14 and Salmon (1984b) I tried to show that the phenomenon of "making it the hard way" poses very pervasive problems for some theories of probabilistic causality.

9. Some people may be perturbed at my use of probabilities that appear to be undefined. In essay 14 I showed how to make them innocuous.

10. It should be noted that—as Good is perfectly aware—his usage diverges sharply from that customary in philosophy. Philosophers usually use the term "propensity" to refer to a probability relation, not a statistical relevance relation, especially in the context of the so-called propensity interpretation of probability.

11. [Important progress has been made by Hitchcock (1993).]

16

Causality without Counterfactuals

This essay presents a drastically revised version of the theory of causality, based on analyses of causal processes and causal interactions, advocated in Salmon (1984b). Relying heavily on modified versions of proposals by Phil Dowe, this article answers penetrating objections by Dowe and Philip Kitcher to the earlier theory. It shows how the new theory circumvents a host of difficulties that have been raised in the literature. The result is, I hope, a more satisfactory analysis of physical causality.

1. Background

In 1984 I offered an account of causality involving *causal processes* as the means by which causal influence is transmitted (Salmon, 1984b, chap. 5), and *causal forks,* as the means by which causal structure is generated and modified (ibid., chap. 6). Causal forks come in two main varieties, interactive forks and conjunctive forks. (Perfect forks are a limiting case of both of these types.) Interactive forks are used to define *causal interactions.* Causal processes and causal interactions are the basic causal mechanisms according to this approach. Although causal interactions are more fundamental than causal processes on this view, for various heuristic reasons, I introduced causal processes before causal interactions. (I fear that the heuristic strategy was counterproductive.) The idea was to present a "process theory" of causality that could resolve the fundamental problem raised by Hume regarding causal connections. The main point is that causal processes, as characterized by this theory, constitute precisely the objective physical causal connections which Hume sought in vain. The so-called *at-at theory* of causal propagation enables us to account for the transmission of causal influence in a manner that does not conflict with (what I take to be) Humean empirical strictures.

To implement this program it is necessary to distinguish genuine causal processes from pseudo-processes. The notion of a process (causal or pseudo-) can reasonably be regarded as a primitive concept that can be made sufficiently clear in terms of examples and informal descriptions, such as what Bertrand Russell (1948) called "causal lines"; however, even though Russell used the word "causal," he did not make a careful distinction between causal processes and pseudo-processes. Prior to this distinction, the concept of process carries no causal involvements. If one thinks in terms of relativity theory and Minkowski spacetime diagrams, processes can be identified as spacetime paths that exhibit continuity and some degree of constancy of character. These spacetime paths and their parts may be timelike, lightlike or spacelike.

Processes, whether causal or pseudo-, often intersect one another in spacetime; in and of itself spacetime intersection is *not* a causal concept. Looking at intersections, we need criteria to distinguish genuine causal interactions from mere spacetime intersections. The basic theses are (1) that causal processes could be distinguished from pseudo-processes in terms of their capacity to transmit marks, and (2) that causal interactions could be distinguished from mere spacetime intersections in terms of mutual modifications— changes that originate at the locus of the intersecting processes and persist beyond that place. In order to explain what is meant by transmitting a mark, it is necessary to explain what is involved in introducing a mark. Introducing a mark *is* a causal concept, so it needs to be explicated; this is done in terms of the notion of a causal interaction. Causal interactions are explicated without recourse to other causal concepts. Contrary to the heuristic order, causal interactions are logically more basic than causal processes.

This account of causality has certain strong points and certain defects, and it has been subjected to severe criticism by a number of philosophers. Some of the criticisms are well founded; some are based on misinterpretations. In this essay I address these difficulties. First, I try to clear away the misinterpretations. Second, I attempt to show how the account can be modified so as to remove the genuine shortcomings. In this latter endeavor I rely heavily on the work of Dowe (1992c). The result is, I believe, a tenable (more tenable?) theory of physical causation.

2. The Circularity Charge

Dowe (1992c) claims that the foregoing account is circular and discusses similar criticisms made by several authors. As a basis for his discussion of this and other criticisms he advances, Dowe formulates six propositions to characterize my position:

D-I   A process is something that displays consistency of characteristics.

D-II   A causal process is a process that can transmit a mark.

D-III A mark is transmitted over an interval when it appears at each spacetime point of that interval, in the absence of interactions.

D-IV   A mark is an alteration to a characteristic, introduced by a single local interaction.

D-V   An interaction is an intersection of two processes.

D-VI   A causal interaction is an interaction where both processes are marked. (ibid., p. 200: "D" stands for Dowe)

In order to evaluate various criticisms, we must examine the foregoing propositions. Dowe's concern with circularity focuses on D-IV and D-VI: taken together, he argues, they contain a circularity. Statements D-I through D-IV are acceptable just about as they stand; only D-IV requires a bit of modification, namely, the substitution of "intersection" for "interaction." Propositions D-V and D-VI require more serious revision; in fact, D-V should be deleted, while D-VI should be modified to read, "A causal interaction is an *intersection* in which both processes are marked *and the mark in each process is transmitted beyond the locus of the intersection.*" There are two crucial points. First, in my terminology "causal interaction" and "interaction" are synonymous; there are no such things as noncausal interactions. There are, of course, noncausal intersections. Second, for an intersection to qualify as a causal interaction, the modifications that originate in the intersection must persist beyond the place at which the intersection occurs.

Let us rewrite the foregoing propositions, taking the required modifications into account and rearranging the order. For the sake of further clarity I substitute a different proposition for D-V. Let "S" stand for Salmon; in each case Dowe's counterpart is indicated parenthetically:

S-I   A process is something that displays consistency of characteristics (D-I).

S-II   A mark is an alteration to a characteristic that occurs in a single local intersection (D-IV).

S-III   A mark is transmitted over an interval when it appears at each spacetime point of that interval, in the absence of interactions (D-III).

S-IV   A causal interaction is an intersection in which both processes are marked (altered) and the mark in each process is transmitted beyond the locus of the intersection (D-VI).

S-V   In a causal interaction a mark is introduced into each of the intersecting processes. (This substitute for D-V can be construed as a definition of "introduction of a mark.")

S-VI   A causal process is a process that can transmit a mark (D-II).

This revised list of propositions involves certain problems to which I will return, but it does not suffer from circularity. We assume that spatiotemporal concepts such as intersection and duration are clear. We assume (for the moment, but see §3) that we know what it means to say that a property of a process changes, or that two characteristics of processes differ from each other. Proposition S-I indicates what counts as a *process*. Material particles in motion, light pulses, and sound waves are paradigm examples. Proposition S-II introduces the concept of a *mark*. Notice that a mark is simply a modification of some kind; it need not persist. When the shadow of an automobile traveling along a road with a smooth berm encounters a signpost, its shape is altered, but it regains its former shape as soon as it passes beyond the post. It was marked at the point of intersection, but the mark vanishes immediately. Notice also that a pair of *causal processes* can intersect without constituting a causal interaction; for example, light waves that intersect are said to interfere in the region of intersection, but they proceed beyond as if nothing had happened.

Statement S-III is a key proposition; it characterizes the notion of transmission. Even if we give up the capacity for mark transmission as a fundamental explication of causal process—as I will do—the concept of transmission remains crucial (see §7, def. 3).

Statement S-IV is also a key proposition because it introduces the most basic notion— causal interaction. It says that a causal interaction is an intersection of processes in which mutual modifications occur that persist beyond the locus of intersection.

Proposition S-V, in contrast, is trivial; it defines "introduction of a mark," a concept we will not need. Proposition S-VI is one of the central theses of my 1984 theory; it is one I am prepared to abandon in the light of Dowe's alternative proposal. I return to this issue in §7.

3. The Problem of Vagueness

Dowe (1992c, pp. 201–204) justifiably complains that in my discussions of marking and mark transmission (Salmon, 1984b, chap. 5) I used terms such as "characteristic" and "structure" without specifying their meanings. He suggests that introduction of the concept of a *nonrelational property* might have clarified the situation. He is right. Somewhat ironically, in an earlier chapter of the same book (ibid., pp. 60–72), I worked hard to precisely characterize the concept of objective homogeneity of reference classes, and dealt with the kind of problem that comes up in the discussion of marks. Unfortunately, I neglected to carry the same type of consideration explicitly into the context of marking. The key concept is that of an *objectively codefined* class (ibid., p. 82, def. 2), which is explicated in terms of physically possible detectors attached to appropriate kinds of computers that receive carefully specified types of information. It is possible to ascertain, on the basis of local observations—detections—whether an entity possesses a given property at a particular time. Since, in scientific contexts, we often detect one property by observing another, it must be possible in principle to construct a computer to make the determination. For example, when we measure temperature by using a thermocouple, we actually read a potentiometer to detect an electromotive force (emf). The computer to which the explication refers must be able to translate the potentiometer reading into a temperature determination on the basis of laws concerning the electrical outputs of thermocouples, but without receiving information from other physical detectors. Notice that this explication is physical, not epistemic. This kind of definition would easily suffice to rule out properties such as *being the shadow of a scratched car* (Kitcher, 1989, p. 463) or *being a shadow that is closer to the Harbour Bridge than to the Sydney Opera House* (Dowe, 1992c, p. 201), as well as properties such as grue (Goodman, 1955).

No basic problems concerning the nature of marks arise in connection with the distinction between causal processes and pseudo-processes that cannot be handled through the use of the techniques involved in explicating objectively codefined classes. As Dowe (1992c, p. 203) notes, I made a remark to this effect in Salmon (1985a), but regrettably (owing to severe space limitations) I neglected to give details. However, since I am about to abandon the mark criterion altogether, there is no need to pursue the question here.

4. Statistical Characterization of Causal Concepts

In an illuminating discussion of the possibility of characterizing causal concepts in statistical terms, Dowe (1992c, pp. 204–207) voices the opinion that this enterprise is hopeless. He quotes my remark, "I now think that the statistical characterization is inadvisable" (Salmon, 1984b, p. 174, n. 12), correctly noting that it expresses agreement with his thesis. Citing the paucity of reasons given in my note, he offers reasons of his own. In essay 13 I attempt briefly to spell out reasons of my own. As nearly as I can tell, Dowe and I have no basic disagreement on this issue.

5. Counterfactuals

I have frequently used the example of a rotating spotlight in the center of a circular building to illustrate the difference between causal processes and pseudo-processes. A brief pulse of light traveling from the beacon to the wall is a causal process. If you place a red filter in its path, the light pulse becomes red and remains red from the point of insertion to the wall without any further intervention. The spot of light that travels around the wall is a pseudo-process. You can make the white spot red by intervening at the wall where the light strikes it, but without further local intervention it will not remain red as it passes beyond the point of intervention. Thus, causal processes transmit marks but pseudo-processes do not.

The untenability of this characterization was shown forcefully by Nancy Cartwright (in conversation) by means of a simple example. Suppose that a few nanoseconds before a red filter at the wall turns the moving spot red, someone places a red lens on the rotating beacon so that, as the spot moves, it remains red because of the new lens on the beacon. In such a case the spot *turns* red owing to a local interaction and *remains* red *without any additional local interactions.* With or without the intervention at the wall, the spot of light moving around the wall would have been red from that point on. Consideration of such cases required a counterfactual formulation of the principle of mark transmission. I had to stipulate, in effect, that the spot would have remained white from that point on if there had been no local marking (Salmon, 1984b, p. 148). In Cartwright's example, the spot would have turned red anyhow, regardless of whether any marking had occurred at the wall.

In an extended and detailed discussion of scientific explanation, Kitcher articulates a penetrating critique of my causal theory (1989, §6), making heavy weather over the appeal to counterfactuals. He summarizes this aspect of his critique as follows:

> I suggest that we can have causation without linking causal processes. . . . What is critical to the causal claims seems to be the truth of the counterfactuals, not the existence of the processes and the interactions. If this is correct then it is not just that Salmon's account of the causal structure of the world needs supplementing through the introduction of more counterfactuals. The counterfactuals are the heart of the theory, while the claims about the existence of processes and interactions are, in principle, dispensable. Perhaps these notions may prove useful in protecting a basically counter-factual theory of causation against certain familiar forms of difficulty (problems of preemption, overdetermination, epiphenomena, and so forth).* But, instead of viewing Salmon's account as based on his explications of process and interaction, it might be more revealing to see him as developing a particular kind of counterfactual theory of causation, one that has some extra machinery for avoiding the usual difficulties that beset such proposals. [*Kitcher's note: See Lewis (1973), both for an elegant statement

of a counterfactual theory of causation and for a survey of difficult cases. Loeb (1974) endeavors to cope with the problem of overdetermination.] (Kitcher, 1989, p. 472)

When Hans Reichenbach proposed his mark method, he thought it could be used to determine a time direction ([1928] 1957, pp. 136–137). This was a mistake, as Adolf Grünbaum (1963, pp. 180–186) has shown. However, drawing on suggestions offered in Reichenbach (1956, §23) concerning the mark method and causal relevance, I concluded that the mark method provided a criterion for distinguishing between causal processes and pseudo-processes, without any commitment to time direction (earlier/later). That is a separate problem (see Dowe, 1992b). It has always been clear that a process is causal if it is *capable* of transmitting a mark, whether or not it is *actually* transmitting one. The fact that it has the capacity to transmit a mark is merely a *symptom* of the fact that it is actually transmitting something else. That other something I described as information, structure, and causal influence (1984b, pp. 154–157).

When the mark criterion was clearly in trouble because of counterfactual involvement, it should have been obvious that the mark method ought to be regarded only as a useful experimental method for tracing or identifying causal processes (e.g., the use of radioactive tracers) but that it should not be used to explicate the very concept of a causal process. Dowe took the crucial step. He pointed out that causal processes transmit conserved quantities; and by virtue of this fact, they are causal. I had come close to this point by mentioning the applicability of conservation laws to causal interactions, but did not take the crucial additional step (ibid., pp. 169–170). Dowe's theory is not counterfactual.

6. Dowe's Conserved Quantity Theory

Dowe's proposed *conserved quantity* theory is beautiful for its simplicity. It is based on just two definitions (1992c, p. 210):

DEFINITION 1.   *A causal interaction is an intersection of world-lines which involves exchange of a conserved quantity.*

This definition is a substitute for my much more complex and contorted principle CI (for causal interaction), which was heavily laden with counterfactuals (Salmon, 1984b, p. 171).

In any discussion of interactions it is essential to keep in mind the fact that we are dealing with conserved quantities. In an interaction involving an exchange of momentum, for example, the total momentum of the outgoing processes must be equal to that of the incoming processes. This point is important in dealing with certain kinds of interactions in which three or more processes intersect in virtually the same spacetime region. For example, a solidly hit baseball and an atmospheric molecule, say, nitrogen, strike a glass window almost simultaneously. It may be tempting to say that the baseball caused the window to shatter, not the nitrogen molecule, because the window would not have shattered if it had not been struck by the baseball. But this analysis is unacceptable if we want to avoid counterfactuals.

We should say instead that, in the interaction constituted by the nitrogen molecule and the shattering window, momentum is not conserved. Take the window to be at rest; its linear momentum is zero. The linear momentum of the nitrogen molecule when it strikes the window is not zero but is fairly small. The total linear momentum of the pieces of the shattered window after the collision is enormously greater than that of the incoming molecule. In contrast, the total linear momentum of the baseball as it strikes the window is about equal to the momentum of the pieces of glass and the baseball after the collision. So if we talk about causes and effects, we are justified in saying that the window was broken by the collision with the baseball, not by the collision with the nitrogen molecule. With these considerations in mind, I think we can say that Dowe's definition (1) is free of counterfactuals, and is acceptable as it stands.

DEFINITION 2_a.   *A causal process is a world-line of an object which manifests a conserved quantity.*

As we will see, definition (2) requires some further work—hence the designation 2_a.

In his elaboration of the foregoing definitions, Dowe mentions mass-energy, linear momentum, angular momentum, and electric charge as examples of conserved quantities. He explains the meanings of other terms:

> A *exchange* means at least one incoming and at least one outgoing process manifest a change in the value of the conserved quantity. "Outgoing" and "incoming" are delineated on the spacetime diagram by the forward and backward light cones, but are essentially interchangeable. The exchange is governed by the conservation law. The intersection can therefore be of the form X, Y, λ or of a more complicated form. An *object* can be anything found in the ontology of science (such as particles, waves or fields), or common sense. (1992c, p. 210)

Dowe offers several concrete examples of causal interactions and causal processes involving electric charge and kinetic energy, and a pseudo-process not involving any conserved quantity (ibid., pp. 211–212). I made passing mention of two sorts of interaction which, to my great frustration, I did not know how to handle (1984b, pp. 181–182). A Y-type interaction occurs when a single process splits in two, such as radioactive decay of a nucleus or a hen laying an egg. A λ-type interaction occurs when two separate processes merge, such as the absorption of a photon by an atom or the consumption of a mouse by a snake. Dowe points out that his conserved quantity theory handles interactions of these two kinds.

7. Conserved Quantities and Invariants

A curious ambiguity arises near the conclusion of Dowe (1992c). David Fair (1979) had proposed a theory of causality in terms of transmission of energy which Dowe criticizes on the basis of several considerations: "Another advantage [of Dowe's theory] concerns Fair's admission that energy is not an invariant and therefore will vary according to the frame of reference. . . . On our account, however, cause is related to conserved quantities and these *are* invariant, for example, energy-mass, energy-momentum, and charge" (Dowe 1992c, p. 214). Up to this point Dowe has formulated and discussed his theory entirely in terms of conserved quantities, and the concept of an *invariant* has not entered. The terms "conserved quantity" and "invariant" are not synonymous. To say that a quantity is *conserved* (within a given physical system) means that its value does not change over time; it is constant with respect to time translation. To say that a quantity is *invariant* (within a given physical system) means that it remains constant with respect to change of frame of reference.

Consider linear momentum, which Dowe identifies as a conserved quantity (ibid., p. 210). We have a law of conservation of linear momentum; it applies to any interaction described with respect to any particular frame of reference, for instance, the 'lab frame' in which an experiment is conducted. Within any closed system the total quantity of linear momentum is constant over time. If you switch to a different frame of reference to describe the same physical system, the quantity of linear momentum will again be constant over time, but not necessarily the *same* constant as in the lab frame. On Einstein's famous train, for instance, the linear momentum of the train is zero, but in the frame of the ground observer it has a great deal of linear momentum. Linear momentum is a conserved quantity, but *not* an invariant. Its value differs from one frame to another. Electric charge *is* an invariant; the electric charge of the electron has the same value in any frame of reference. It is also a conserved quantity. Kinetic energy—which Dowe mentions in one of his examples (ibid., p. 212, example 3)—is neither a conserved quantity nor an invariant. In inelastic collisions it is not conserved, and its value changes with changes of reference frame. This example is easily repaired, however, by referring to linear momentum instead of kinetic energy.

The question arises as to whether we should require causal processes to possess invariant quantities, or whether conserved quantities will suffice. At first blush it would seem that conserved quantities will do. We should note, however, that causality is an invariant notion. In special relativity the spacetime interval is invariant; if two events are causally connectable in one frame of reference, they are causally connectable in every frame. Spacelike, lightlike, and timelike separations are invariants. If two events are causally connected in one frame of reference, they are causally connected in all frames. Since we are attempting to explicate frame-independent causal concepts, it seems reasonable to insist that the explicans be formulated in frame-independent terms (see Mühlhölzer, 1994).

If, however, we rewrite Dowe's definition (2$_a$) as follows, substituting "invariant" for "conserved,"

DEFINITION 2$_b$.   *A causal process is a world-line of an object that manifests an invariant quantity,*

we find ourselves in immediate trouble. Consider, for example, a shadow cast by a moving cat in an otherwise darkened room when a light is turned on for a limited period. This shadow is represented by a world-line with an initial point and a final point. The spacetime interval between these two endpoints is an invariant quantity that is manifested by a pseudo-process. Any pseudo-process of finite duration manifests such an invariant quantity. Definition (2$_b$) is patently unacceptable.

The main trouble with definition (2$_b$) may lie with the term "manifests," for with its use we seem to have abandoned one of the most fundamental ideas about causal processes, namely, that they *transmit* something (e.g., marks, information, causal influence, energy, electric charge, momentum). A process, causal or pseudo-, cannot be said to transmit its invariant spacetime length. A necessary condition for a quantity to be transmitted in a process is that it can meaningfully be said to characterize or be possessed by that process at any given moment in its history. A proton, for example, has a fixed positive electric charge—which, as already noted, is both conserved and invariant—and it has this charge at every moment in its history. Thus, it makes sense to say that the charge of a particular proton changes or stays the same over a period of time. Perhaps, then, we could reformulate definition (2_b) as follows:

DEFINITION 2_c. *A causal process is a world-line of an object that manifests an invariant quantity at each moment of its history (each spacetime point of its trajectory).*

We should also note that "manifests" contains a possibly serious ambiguity. A photon, for example, has an electric charge equal to zero. Do we want to say that it manifests that particular quantity of electric charge?

Consider first the claim that a neutral hydrogen atom manifests an electric charge of zero. This seems unproblematic because the atom is composed of two parts, a proton and an electron, each of which has a nonzero charge. Since the atom can be ionized, we can separate the two charged particles from each other. The neutron is also unproblematic for two reasons. First, it is thought to be composed of three quarks, each of which has a nonzero charge, but separating them is extremely difficult if not impossible. Second, a free neutron has a half-life of a few minutes, and when it decays, it yields two charged particles, a proton and an electron (plus an uncharged antinutrino); unlike the hydrogen atom, however, the neutron is not composed of a proton and an electron. The photon is more difficult. Under suitable circumstances (e.g., near a heavy atom) a high-energy photon will vanish, yielding an electron-positron pair, each member of which has a nonzero charge. Thus energetic photons may be said to have zero electric charge, an attribution that can be extended to less energetic photons. Any entity that can yield products with nonzero electric charge can be said to *manifest* an electric charge of zero. However, this kind of principle might not hold for all invariants that a process might manifest. The important point is that we must block the assertion that a shadow is an entity that manifests an electric charge (whose value is zero) and similar claims. Let us make an additional modification:

DEFINITION 2_d. *A causal process is a world-line of an object that manifests a nonzero amount of an invariant quantity at each moment of its history (each spacetime point of its trajectory).*

We need not fear for the causal status of photons on this definition; they manifest the invariant speed c.

When we speak in definition (2_d) of a nonzero amount of a given quantity, it must be understood that this refers to a 'natural zero' if the quantity has one. Although temperature is neither a conserved quantity nor an invariant, it furnishes the easiest exemplification of what is meant by a 'natural zero'. The choice of the zero point in the Fahrenheit and Celsius scales is arbitrary. The fact that water freezes at 0° C and boils at 100° C does not remove the arbitrariness, since the scale is referred to a particular substance as a matter of convenience. In contrast 0° K on the absolute scale is a natural zero, because it is the greatest lower bound for temperatures of any substance under any physical conditions. Applying similar considerations, we can argue that electric charge has a 'natural zero' even though it can assume negative values. An entity that has a charge of zero esu (electrostatic units) it not attracted or repelled electrostatically by any object that has any amount of electric charge. When it is brought into contact with an electroscope, the leaves do not separate. It would be possible (as an anonymous referee pointed out) to define a quantity electric-charge-plus-seventeen, which is possessed by photons, shadows, neutrons, and so on in a nonzero amount. This would be an invariant quantity, but it lacks a 'natural zero'. Given the 'natural zero' from which it departs, it should be considered inadmissible in the foregoing definition. I believe that all of the quantities we customarily take as conserved or invariant have 'natural zeros', but I do not have a general proof of this conjecture.

Definition (2_d) brings us back, regrettably, to Fair's (1979) account of causation in terms of energy transfer, which is open to the objection that it gives us no basis for distinguishing cases in which there is genuine *transmission* of energy from those in which the energy just happens to show up at the appropriate place and time. Making reference to the rotating beacon in the Astrodome, I argued that uniform amounts of radiant energy show up along the pathway of the spot moving along the wall, and that therefore the fact that the world-line of this spot manifests energy in an appropriately regular way cannot be taken to show that the moving spot is a causal process (Salmon, 1984b, pp. 145–146). Dowe (1992c, p. 214) complains that it is the wall rather than the spot that possesses the energy; however, we can take the world-line of the part of the wall surface that is absorbing energy as a result of being illuminated. This world-line manifests energy throughout the period during which the spot travels around the wall, but it is not the world-line of a causal process because the energy is not being transmitted; it is being received from an exterior source. (If Dowe's objection to this example is not overcome by the foregoing considerations, other examples could be supplied.) For this reason, I would propose a further emendation of Dowe's definition:

DEFINITION 2_c. *A causal process is a world-line of an object that transmits a nonzero amount of an invariant quantity at each moment of its history (each spacetime point of its trajectory).*

This definition introduces the term "transmits," which is clearly a causal notion, and which requires explication in this context. I offer the following modification of my mark transmission principle (MT) (Salmon 1984b, p. 148):

DEFINITION 3. *A process transmits an invariant (or conserved) quantity from A to B (A ≠ B) if it possesses this quantity at A and at B and at every stage of the process between A and B without any interactions in the half-open interval (A, B] that involve an exchange of that particular invariant (or conserved) quantity.*

The interval is specified as half open to allow for the possibility that there is an interaction at A that determines the amount of the quantity involved in the transmission. This definition embodies *the at-at theory of causal transmission* (ibid., pp. 147–157), which still seems to be fundamental to our understanding of physical causality. Definition (3) does not involve counterfactuals.

Speaking literally, the foregoing definitions imply that a causal process does not enter into any causal interactions. For example, a gas molecule constitutes a causal process between its collisions with other molecules or the walls of its container. When it collides with another molecule, it becomes another causal process which endures until the next collision. A typical value for the mean free path is 10^{-7} m, which, though small, is much greater than the size of the molecule. If we consider the life history of such a molecule during an hour, it consists of a very large number of causal processes each enduring between two successive collisions. Each collision is a causal interaction in which momentum is exchanged. When we understand this technical detail, there is no harm in referring to the history of a molecule over a considerable period of time as a single (composite) causal process that enters into many interactions. The history of a Brownian particle suspended in the gas is an even more extreme case, for it is undergoing virtually continuous bombardment by the molecules of a gas, but I think it should be conceived in essentially the same manner.

In many practical situations definition (3) should be considered an idealization. As an anonymous referee remarked, "You'd want to say that the speeding bullet transmits energy-momentum from the gun to the victim, but what about its incessant, negligible interactions with ambient air and radiation?" Of course. In this and many similar sorts of situations, we would simply ignore such interactions because the energy-momentum exchanges are too small to matter. Pragmatic considerations determine whether a given 'process' is to be regarded as a single process or a complex network of processes and interactions. In the case of the 'speeding bullet', we are not usually concerned with the interactions among the atoms that make up the bullet. In dealing with television displays, we may well be interested in the flight paths of individual electrons. In geophysics we might take the collision of a comet with the earth to be an interaction between just two separate processes. It all depends on the domain of science and the nature of the question under investigation. Idealizations of the sort just exemplified are not unfamiliar in science.

There is, however, another source of concern. According to Dowe, "A *conserved quantity* is any quantity universally conserved according to current scientific theories" (1992c, p. 210). This formulation cannot be accepted. Parity, for example, was a conserved quantity according to theories current prior to the early 1950s, but in 1956 it was shown by T. D. Lee and C. N. Yang that parity is not conserved in weak interactions. According to more recent theories, parity is not a conserved quantity. What we should say is that we look to currently accepted theories to tell us what quantities we can reasonably regard as conserved. We had good reason to regard parity as a conserved quantity prior to 1956; subsequently, we have had good reason to exclude it from the class of conserved quantities. So our current theories tell us what quantities to think of as conserved; whether or not they are conserved is another question.

We might be tempted to say that conserved quantities are those quantities governed by conservation laws, where by "law" we mean either a true lawlike statement or a lawful regularity in nature. If we were to take this tack, however, we would free ourselves from the curse of counterfactuals only at the price of taking on the problem of laws. An approach of this sort is given by F. J. Clendinnen (1992), in which he proposes a "nomic dependence" account of causation as an alternative to David Lewis's counterfactual theory. This may not take us out of the frying pan into the fire, but it does seem to offer another hot skillet in exchange for the frying pan. The hazard can be avoided, I think, by saying that a conserved quantity is a quantity that does not change. I am prepared to assert that the charge of the electron is 4.803×10^{-10} esu, and that the value is constant. Obviously I could be wrong about its value or about its constancy, but what I said in the foregoing sentence is true to the best of my knowledge. If the statement about the electron charge is true, then there is a true generalization about the charge of the electron. However, *it makes no difference whether or not that true generalization is lawful; only its truth is at stake.* The problem of laws is the problem of distinguishing true lawlike generalizations from other true generalizations. That is a problem we do *not* have to face.

In discussing the relationship between conserved quantities and laws, I deliberately chose as an example a quantity that is also an invariant. Thus, in fact, I want to stick to the formulation of definition (2_e) in terms of invariants. I have a further reason for this choice. When we ask about the ontological implications of a theory, one reasonable response is to look for its invariants. Since these do not change with the selection of different frames of reference—different perspectives or points of view—they possess a kind of objective status that seems more fundamental than that of noninvariants.

Apparently, although Dowe's conserved quantity (CQ) theory of causality embodies important improvements over my mark transmission theory, it is not fully satisfactory as he has presented it. In definitions (1), (2_e), and (3) I think we have made considerable progress toward an adequate theory of causality. This is a result, to a large extent, of Dowe's efforts in developing a process theory of causality that avoids the problems of counterfactuals with which my former theory was involved. We have, I believe, clean definitions of causal interaction, causal transmission, and causal processes on which to found a process theory of physical causality.

8. Kitcher's Objections

Among the many critiques of my account of causality, those of Dowe (1992c) and Kitcher (1989) seem the most penetrating and significant. Dowe's discussion is motivated by a desire to provide an account of process causality that is more satisfactory than mine. As I have indicated, I think he has succeeded in very large measure, and in the preceding section I tried to improve on his version. Kitcher's motivation is essentially the opposite; he supports an altogether different account of causality. His thesis is that "the 'because' of causation is always derivative from the 'because' of explanation" (ibid., p. 477). My view is roughly the opposite, and I think Dowe would agree. Dowe (1992a) assesses the difficulties posed by Kitcher and offers his own defense against them.

Kitcher does not claim to have refuted the causal account of explanation: "The aim of this section has been to identify the problems that they will have to overcome, not to close the books on the causal approach" (1989, p. 476). While I do not think that the theory I advocated in Salmon (1984b) contained adequate resources to overcome the difficulties he pointed out, I am inclined to believe that the version developed herein—

leaning heavily on Dowe's work—does have the capacity to do so. For example, Kitcher considers the problem of counterfactual entanglement "the most serious trouble of Salmon's project and [one] which I take to threaten any program that tries to use causal concepts to ground the notion of explanation while remaining faithful to an empiricist theory of knowledge" (1989, p. 470). Dowe's primary contribution is to free the concept of causality from its dependence on counterfactuals. This I consider a major part of the answer to Kitcher's challenges.

Another cluster of problems involves the concept of a mark (ibid., pp. 463–464). Inasmuch as I have now abandoned the mark transmission approach and substituted the invariant (or conserved) quantity transmission view, the difficulties concerning marks have been bypassed. In addition, Kitcher points to the problem of sorting out in complex situations which interactions are relevant and which are not pertinent (ibid., p. 463). Definition (3) goes some distance in responding to this problem inasmuch as it identifies a particular invariant (or conserved) quantity that is involved in the transmission. So Kitcher's misgivings—well taken regarding my (1984b) treatment—have been circumvented.

[9. Postscript

"Causality without Counterfactuals" evoked two penetrating critiques in *Philosophy of Science,* namely, Dowe (1995) and Hitchcock (1995). This is the journal in which Dowe (1992c) and Salmon (1994) appeared. Dowe focuses on residual differences between us regarding the process theory of causality, challenging primarily my treatment of causal transmission. Hitchcock raises broader questions about the causal theory of scientific explanation articulated in Salmon (1984b) and, in modified form, in "Causality without Counterfactuals," where I accept a view of causality close to Dowe's conserved quantity theory. My reply to both of these discussions is contained in "Causality and Explanation: A Reply to Two Critiques" (Salmon, 1997) in the same journal. In this reply, in order to deal with a counterexample posed by Hitchcock, I abandon the appeal to invariant quantities, and in partial agreement with Dowe, I opt for my version of the conserved quantity theory. I strongly urge any reader interested in the issues raised in "Causality without Counterfactuals," as well as the other essays in this book, to pursue the developments in this ongoing discussion, because I feel that we are making genuine progress in understanding causality and scientific explanation. The fact that such able philosophers as Dowe, Hitchcock, and Kitcher are seriously engaged in these topics is a source of enormous gratification to me.]

Note

I should like to express my sincere thanks to Philip Kitcher, Allen Janis, and an anonymous referee for extremely valuable comments on this essay.

17

Indeterminacy, Indeterminism, and Quantum Mechanics

In several of the preceding essays I mention radioactive decay of unstable atoms as a type of occurrence that, according to quantum mechanics, is essentially undetermined. For example, tritium is an isotope of hydrogen whose nucleus contains two neutrons and one proton. Its half-life is 12.32 years; in any sample containing a large number of tritium atoms, about one half will decay in 12.32 years and the rest will remain intact. (Each tritium atom that decays is transformed into an atom of helium-3 by emitting an electron.) Prior to the beginning of such a period there is no way even in principle to distinguish those atoms that will decay within that period from those that will not. However, it is tempting to maintain that there *must* be some physical characteristic of an atom that determines whether it will decay at a particular time or remain intact. In this essay I examine some of the issues involved in denying this claim and affirming that our universe is ineluctably indeterministic. At present I am much less confident than before that quantum mechanics is intrinsically indeterministic, and I believe that the problem is far more complex than I previously realized.[1]

1. Quantum Indeterminacy and the Wave-Particle Duality

Many attempts to explain quantum indeterminacy invoke Werner Heisenberg's famous *indeterminacy principle,* which is usually called the *uncertainty principle.* It is often stated in terms of our inability to know with complete precision both the position and the momentum of a particle such as an electron. The reference to knowledge, ignorance, or uncertainty is regrettable. It would be far better to stick with the word *indeterminacy* to reinforce the idea that we cannot know exact values for position and momentum because such exact values simply do not exist. In this way, one can more readily understand that the future behavior of a particle such as an electron is not just unpredictable but also

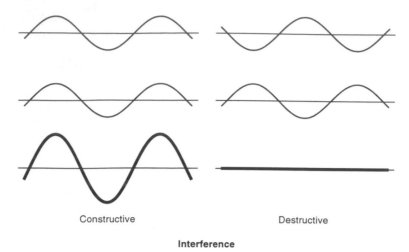

Constructive   Destructive

Interference

Figure 17.1

indeterminate. This fundamental indeterminacy pervades the quantum domain; it applies to phenomena such as radioactive decay, the swerving of atoms in a Stern-Gerlach apparatus (see fig. 17.8), and the motions of other particles in addition to electrons. In this essay I try to present, in the most elementary terms possible, the basis for the claim of indeterminacy; my approach is to sketch the development of the quantum theory in the first quarter of the twentieth century.

To set the stage for this explanation, let us go back to the situation in optics at the beginning of the nineteenth century. At that time there were two competing theories of light—the wave theory and the corpuscular theory. Early in that century the wave theory gained strong support from various interference phenomena. Interference is the hallmark of waves. If a stone is dropped into a placid pool of water, concentric waves will spread across the surface from the point of entry. If two stones are dropped simultaneously at different places in such a pond, two sets of concentric waves will spread from the respective points of entry, and when they meet, interference will occur. Where the crest of one wave meets the crest of another, they will reinforce each other, yielding a higher crest. This is *constructive* interference. Where the crest of one wave meets the trough of another, they will tend to cancel each other out. This is *destructive* interference. You can observe such interference phenomena in your own sink or bathtub (see fig. 17.1).

Two famous phenomena, the Young two-slit experiment and the Poisson bright spot, illustrate the application of these considerations to light. The setup for the former consists of a light source and two screens. One of the screens contains two narrow slits. Light from the source passes through the two slits in the first screen and falls on the second screen, where a series of light and dark bands alternate with one another. If you wish, the second screen can be a photographic film on which the interference pattern is recorded (see fig. 17.2). The same type of experiment can be conducted in a different way. Instead of having both slits open at the same time, each of the slits can be blocked off while the other is left open. The same amount of light passes through the two slits in

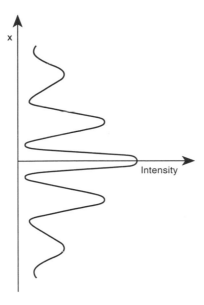

Two-Slit Interference Pattern

Figure 17.2

each experiment, but in the first version the light passes through the two slits simultaneously, while in the second version the light passes through the two slits successively. The results of the second version can also be recorded on a photographic film. When results of the two versions are compared, we find a radical difference. The superposition of the successive one-slit exposures is not the same as the exposure of the two slits open simultaneously (see fig. 17.3). The explanation is that waves coming from the two slits when both are open at the same time interfere with each other; this cannot happen when the slits are open at different times.

The mathematician S.-D. Poisson, who was no friend of the wave theory of light, showed that, according to that theory, the shadow of an illuminated disk would have a bright spot at its center.[2] He regarded this as a reductio ad absurdum of the wave theory, but lo and behold, when the experiment was conducted the bright spot appeared. The corpuscular theory of light was incapable of explaining either of these two experimental results. The wave theory of light received further theoretical support when James Clerk Maxwell identified light as a type of electromagnetic radiation. The electromagnetic spectrum contains, in addition to visible light, gamma rays, X rays, ultraviolet radiation, infrared radiation, microwaves, and radio waves. At the close of the nineteenth century the wavelike character of light and other forms of electromagnetic radiation seemed securely established.

The classical theory of electromagnetic radiation was not without problems as the nineteenth century drew to a close. In particular, the treatment of blackbody radiation presented acute difficulties. A blackbody is an object that absorbs all radiation that

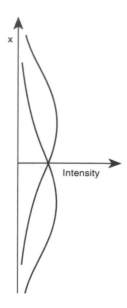

Two One-Slit Interference Patterns

Figure 17.3

impinges upon it; none of the incident radiation is reflected. A dull black iron fireplace poker is a good example—it does not reflect light that falls upon it. The mouth of a blast furnace in a steel mill is another. A body that absorbs radiant energy must also emit energy; the energy radiated by a blackbody has a characteristic distribution over a range of frequencies (see fig. 17.4). This distribution depends on the temperature of the black-body. If the poker is heated, it radiates energy in the infrared part of the spectrum, but as it gets hotter it begins to glow dark red. As it reaches higher temperatures it becomes bright red, then orange, then yellow, then white.

Classical physics was unable to give a theoretical account of the observed blackbody spectrum. Its theoretical prediction called for unlimited amounts of energy to be radiated at shorter and shorter wavelengths; this result, which obviously conflicted with the observed facts, was known as the "ultraviolet catastrophe." Max Planck's quantum theory of blackbody radiation, which resolved this problem, appeared in December 1900, the final month of the nineteenth century. According to this theory, electromagnetic radiation is *emitted* in quanta whose energy E satisfies the formula $h\nu$, where h is Planck's constant and ν (Greek nu) is the frequency of the radiation.

Another problem for classical physics involved a phenomenon known as the photo-electric effect. Under certain circumstances, when light (including ultraviolet) impinges on metal surfaces, electrons are emitted. According to Albert Einstein's 1905 theory of the photoelectric effect, electromagnetic radiation is *absorbed* as quanta whose energy

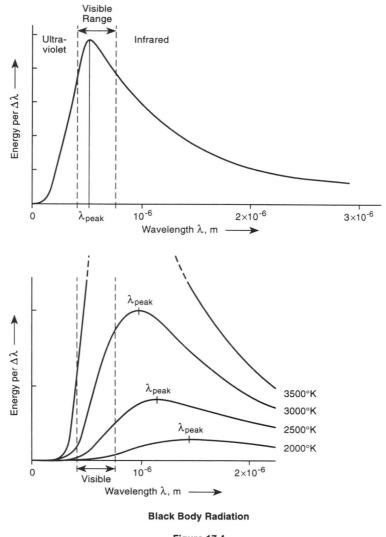

Black Body Radiation

Figure 17.4

satisfies Planck's formula. These little packets of radiant energy were later called photons. Individual photons 'kick' electrons out of the metal.

In 1923 Arthur H. Compton investigated the interactions between X rays and electrons. When an X-ray photon having energy $E = h\nu$ collides with an electron at rest, he found, a photon emerges with energy $E_1 = h\nu_1$ and the electron is set in motion with kinetic energy E_2 ($= \frac{1}{2}mv^2$). (The English letter v stands for velocity; it must not be confused with the Greek letter *nu*, which stands for frequency.) In this situation $E_1 + E_2 = E$, thus satisfying the law of conservation of energy. Moreover, the momentum of the electron is mv. If we take the momentum of the incident photon to be $h\nu/c$, where c is the speed of light, and the momentum of the scattered photon to be $h\nu_1/c$, momentum is also conserved. (Momentum is a vector having direction as well as magnitude; it is conserved as a vector.) The collision between the photon and the electron is just like the collision of two classical particles; it has no analogue in the interaction of a wave with a particle. In a collision between an X-ray photon and an electron, energy and momentum are conserved just as when two billiard balls collide. Thus, Compton showed, electromagnetic radiation is *transmitted* in quanta that also satisfy Planck's formula. In sum, then, the results of Planck, Einstein, and Compton show that electromagnetic radiation is emitted, transmitted, and absorbed as quanta—that is, as localized packets of energy.

None of these developments undermined the strong evidence for the wavelike character of light and other forms of electromagnetic radiation that emerged during the nineteenth century. Young's two-slit experiment and the Poisson bright spot still exhibit the interference phenomena that are characteristic of waves. The conclusion forced on us is that light and the other types of electromagnetic radiation have both wavelike and particlelike aspects. The wavelike character is manifest in some physical situations; the particlelike character shows itself in others. This is the *wave-particle duality for light.*

Shortly after Compton conducted his experiment with X rays and electrons, Louis de Broglie made the bold hypothesis that the same wave-particle duality holds for particles, such as electrons, that have nonzero rest mass. (Photons have rest mass equal to zero.) According to classical physics, the momentum p of a massive particle is $m\mathrm{v}$, where m is its mass and v its velocity. (Be careful not to confuse v for velocity with nu for frequency.) As Compton's experiment showed, we may take the momentum p of a photon to be $h\nu/c$, where c is the speed of light. The speed of a photon is the product of its wavelength λ and its frequency ν; that is, $c = \lambda\nu$, from which it follows that $\nu/c = 1/\lambda$. Substituting in the expression for the momentum of a photon we can write

$$h\nu/c = p = h/\lambda \qquad (1)$$

Since h is a constant, each value of momentum p is associated with a unique precise wavelength λ. If we suppose that the same relationship holds for massive particles, we have

$$m\mathrm{v} = p = h/\lambda \qquad (2)$$

For the wavelength of the particle we can write

$$\lambda = h/m\mathrm{v} \qquad (3)$$

(Notice that the wavelength of a particle depends on its velocity as well as its mass; electrons traveling at different speeds have different wavelengths.)

De Broglie's hypothesis of wave-particle duality for massive particles received support from two sources, one experimental, the other theoretical. At the very time that de Broglie was propounding his symmetry argument, Clinton J. Davisson and Lester H. Germer were performing experiments in which they bombarded nickel crystals with electrons. They found the distribution of electrons scattered from the crystals incomprehensible. As it turned out, they were observing the effects of interference of electron waves. Their interference experiment with electrons was the counterpart of Young's two-slit experiment with light. This result lent considerable credibility to the wavelike aspect of material particles.

The theoretical support came from a different quarter. Experiments conducted by J. J. Thomson and Ernest Rutherford around the turn of the twentieth century offered strong support for a nuclear model of the atom.[3] According to this account, any atom consists of an extremely small and massive nucleus with one or more electrons traveling in much larger orbits around it. In the case of ordinary hydrogen, the nucleus consists of one proton and the atom contains one electron. The problem with this model of the hydrogen atom is that, according to classical electrodynamics, the electron would radiate its energy very rapidly and would spiral into the nucleus in about a nanosecond (a billionth of a second). Hydrogen atoms would be radically unstable—a result that flagrantly contradicts well-established fact.

Moreover, when the passage of an electric current causes hydrogen gas to glow, its spectrum is not continuous but rather consists of a number of discrete lines. The discreteness of the spectrum also posed a conundrum for classical physics. Before the turn of the century, Johann Jakob Balmer had worked out a formula that made it possible to calculate the wavelengths of the light in one series (the Balmer series) of lines of the spectrum. A direct generalization of Balmer's formula made it possible to calculate the wavelengths of lines in other series of the hydrogen spectrum.[4] But even though these formulas were available for purposes of calculation, the physical mechanism that produced the spectrum was an utter mystery.

In 1913 Neils Bohr suggested a revolutionary new mechanism that egregiously violated the laws of classical electrodynamics but that gave the correct results for the spectrum of hydrogen. (It was not successful in accounting for the spectra of other elements.) Bohr postulated that the single electron of the hydrogen atom can occupy only a certain discrete set of orbits and that orbits of any other sizes are prohibited. According to his hypothesis, as long as the electron remains in any particular orbit, it does not radiate any energy (in direct contradiction to classical physics). Each orbit has a particular energy value associated with it. If the atom absorbs a photon possessing a suitable amount of radiant energy, the electron jumps from a lower-energy orbit to a higher-energy orbit. If the electron jumps from a higher-energy orbit to a lower-energy orbit, it emits a photon possessing the energy given up by the atom. In either case, whether the photon is absorbed or emitted, its energy is equal to the difference between the energy of an electron in one of the orbits and the energy of an electron in the other orbit.

Bohr was unable to offer any rationale for his scheme beyond the fact that it yielded the correct numbers for the wavelengths in the hydrogen spectrum. De Broglie's hypothesis of electron waves offered some insight. It is plausible to suppose that the wave of an electron occupying any given orbit (not making a transition from one orbit to another) is a standing wave. We are familiar with the types of standing waves generated in stringed instruments. A guitar string is fixed at both ends. Plucking the string produces waves in it whose nodes remain fixed, a node being a point of the string that does not oscillate. The nodes do not travel back and forth along the string. A variety of standing waves having different wavelengths can occur. The fundamental wave has a length twice as large as the distance between the endpoints of the string, that is, one half of a wave fits between the two fixed endpoints. Another has a length equal to the distance between the the two

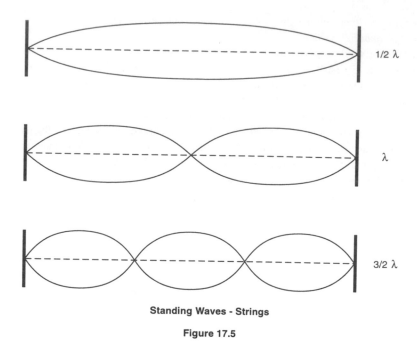

1/2 λ

λ

3/2 λ

Standing Waves - Strings

Figure 17.5

endpoints, that is, a full wave (two halves) fits between the endpoints. The next possibility is a length that allows three halves to fit between the endpoints. And so on (see fig. 17.5). For any positive integer n we can have a standing wave that puts $n/2$ wavelengths between the endpoints.

The guitar string has two distinct ends and it is fixed at both ends. If the two ends were joined together to form a ring (unattached to an external fixed object such as the guitar), the only possible standing waves would be those that fit an integral number of waves into the circumference of the ring (see fig. 17.6). This is analogous to the orbit of the electron in the hydrogen atom as Bohr conceived it. It turns out that the sizes of the Bohr orbits coincide with the de Broglie wavelengths of the orbiting electrons, ensuring that the orbits are of the correct sizes to accommodate the standing waves. Thus, the wavelike character of material particles was supported by both experimental and theoretical considerations. The *wave-particle duality for massive particles* remains secure to this day.

In order to see the parallel between the wave-particle duality for light and the wave-particle duality for material particles, let us return to the two-slit experiment. Although it is impossible to construct a screen with slits of appropriate sizes to conduct an exact analogue of the Young two-slit experiment with electrons, the diffraction of electrons by crystals is a suitable analogue.[5] For many years it has been possible to purchase off the shelf an apparatus that exhibits such diffraction patterns. Because electrons show their wave aspect so clearly, physicists have not hesitated to present a two-slit electron *thought experiment.* The setup consists of an electron source, a screen with two slits of a

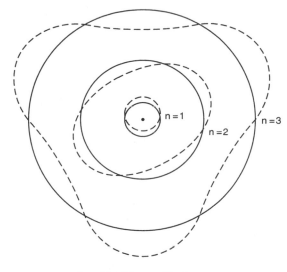

Standing Waves - Electrons

Figure 17.6

width appropriate to the electron wavelength, and another screen of some sort that can serve as a detector of electrons that pass beyond the screen with slits. The result of such an experiment would be just like that of the Young experiment.

If both slits were open at the same time, we would get a diffraction pattern like the diffraction pattern for light. If the slits are opened successively, one at a time, a different pattern results. When the two one-slit patterns are superimposed on each other, the combination does not resemble the two-slit pattern yielded by the setup in which the two slits are open simultaneously. Both light and electrons clearly exhibit wave behavior. The same result occurs even with very weak sources. If the light experiment is done with a source so weak that only one photon is present at a time—that is, a subsequent photon is not emitted until its predecessor has been absorbed at the photosensitive screen— the same patterns are built up over an extended period of time. Similarly with electrons, the same patterns emerge even if only one electron is traveling from the source to the detecting screen at any one time. Thus, the wavelike character of light and material particles is exhibited by these two experiments—the real experiment with light and the thought experiment with electrons.

The particlelike behavior of both light and material particles is exhibited by the same pair of experiments, especially when the weak sources are used. With a suitable array of detectors at the target screen, the completely localized arrival of the photons and electrons can be seen. Particles are localized entities; waves are spread out in space. Think about the surf arriving at a long smooth beach. Any individual wave arrives along an extended portion, spread out across many yards or even many miles of beach. Such a wave carries a considerable amount of energy, spread out along the beach. Any small patch of beach receives only a small part of that energy. Imagine what would happen if suddenly all of the energy of the wave were deposited at one tiny region; pity the swimmer who happened to be at that spot when the wave arrived.[6] This is just what happens in both versions of the two-slit experiment. Interference occurs when the photons or electrons pass through the slits, but they arrive at the detecting screen as compact particles.

We are now prepared to look at the connection between the wave-particle duality and quantum mechanical indeterminacy. The fundamental point to keep in mind is that any precise value of momentum p corresponds to an exact wavelength λ. However, such a wave is not spatially localized; it is completely spread out and has no precise or even approximate position. Thus, the wave of an electron having a precise momentum does not have a determinate position. If, however, we superimpose waves of diverse wavelengths upon one another, interference will occur; in some places the waves will reinforce one another while in other places they will cancel one another out as a result of constructive and destructive interference. It turns out that by suitable superposition of waves of different wavelengths, it is possible to produce a compact wave packet, one that is quite precisely localized. However, the price paid for this localization is that the wave packet involves waves of many different wavelengths that have been combined, and this gives us a range of different values of the momentum. We can obtain a precise value for the position or a precise value for the momentum but not both. Alternatively, we can settle for some indeterminacy of both, obtaining a somewhat localized position and a somewhat restricted range of values of momentum. The Heisenberg indeterminacy relation establishes the maximal joint precision possible for position and momentum taken together at any particular time. It also yields a similar relation for other pairs of quantities, for example, energy and time.

2. Indeterminism and Quantum Mechanics

De Broglie's idea of waves associated with matter was developed into a full-blown nonrelativistic quantum theory by Erwin Schrödinger in 1925; it is often referred to as Schrödinger *wave mechanics*. In classical physics, if the position and momentum of a particle are given, as well as the forces to which it will be subject, its position and momentum at a later time can be calculated. Because the electron does not possess a determinate position and momentum, it is not possible to calculate a precise position and momentum for it at a later time. This should not surprise us; the wave-particle duality has already shown that the electron is not a classical particle. The same consideration applies to other particles as well, e.g., protons, neutrons, atomic nuclei, and atoms.

According to Schrödinger's theory, a quantum mechanical system, such as an electron, is completely described by a wave function $\psi(q,t)$, where q designates position and t designates time. Schrödinger introduced a wave equation to characterize the evolution of a quantum mechanical system. With this equation it is possible to calculate the state of the system at a later time, given its state at an earlier time. Schrödinger's equation establishes a *deterministic* relationship between the state of the system at one time and its state at a later time. Of course, this later state does not embody precise values for both position and momentum. But since the electron does not possess a precise position and a precise momentum, the fact that precise values for position and momentum cannot be deduced does not necessarily undermine determinism. Determinism—as applied to a closed system—requires only that, from a complete description of the system at a given time, it is possible to deduce a complete description of the system at a later time. Given that the ψ-function provides a complete description, Schrödinger's equation does exactly what determinism requires.

In 1925, the year in which Schrödinger announced his quantum theory, Heisenberg also enunciated a quantum theory. Using mathematical techniques quite different from Schrödinger's, he offered a quantum theory known as *matrix mechanics;* it contained his indeterminacy principle. Rather soon thereafter, Schrödinger showed that his wave mechanics is logically equivalent to Heisenberg's matrix mechanics. The indeterminacy principle is therefore an integral part of Schrödinger's theory.

It is essential to insert a word of caution at this point. Whereas de Broglie's waves were waves in ordinary three-dimensional physical space, Schrödinger's waves exist in an abstract mathematical space. For a system consisting of two particles, the ψ-function evolves in a six-dimensional configuration space.[7] Given this feature of the situation, the crucial question is how such a theory can be applied to physical reality. In 1926 Max Born provided an answer. Roughly speaking, according to the Born interpretation, the square of the ψ-function gives the *probability* that the electron will be found in a given region if a measurement takes place. From the standpoint of Schrödinger's equation and the ψ-function, we may be tempted to say that an electron is a smeared-out kind of thing. However, when it is detected by measurement, it becomes a precisely localized entity, as we saw in connection with the two-slit experiment.

Consider a standard basic example. In classical mechanics we talk about kinetic energy and potential energy. A ball rolling without friction along a flat surface has a certain amount of kinetic energy. Suppose it is approaching a hill. If it were situated on top of the hill, it would have a certain amount of potential energy, depending on the height of the hill. If its kinetic energy is greater than or equal to the potential energy it *would have* if it were at the top of the hill, it *can* reach the top; in so doing it converts some or all of its kinetic energy into potential energy. If its kinetic energy is not that great, it will not reach the top of the hill; its kinetic energy will be completely converted into potential energy before it gets to the top. Lacking the requisite energy to reach the top, it will roll back down. We may say that the hill presents a *potential barrier* to the rolling ball.[8] If the ball has enough energy, it can surmount the barrier, i.e., get to the top of the hill; if its energy is insufficient, it will not. In classical mechanics this relationship is precise. If we know the kinetic energy of the ball (before it reaches the hill) and the height of the hill, we can deduce a definite answer to the question whether the ball will get to the top of the hill.

In quantum mechanics the situation is quite different. In the Heisenberg-Schrödinger theory there is an indeterminacy relationship between time and energy quite analogous to the indeterminacy of position and momentum that I have already discussed. For any given quantum mechanical system, if we specify an exact time, its energy at that moment is indeterminate. Consider an electron; it has a negative electrical charge. In quantum mechanics, as in classical electrodynamics, opposite electric charges attract each other and like electric charges repel each other. Suppose that an electron is approaching a

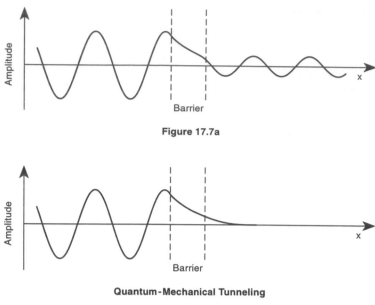

Figure 17.7a

Quantum-Mechanical Tunneling

Figure 17.7b

region of negative charge. Because of the electron's negative charge, this region presents a potential barrier to it in much the same way that the hill presents a potential barrier to the rolling ball. Consider an electron whose energy prior to encountering the potential barrier would, classically speaking, be insufficient to surmount the barrier; assuming that the kinetic energy of the electron is constant as it makes its approach to the barrier, we do not need to specify the time with any great precision. Therefore, the energy is quite precise. Nevertheless, at its precise time of encountering the barrier, the electron's kinetic energy is indeterminate, giving the electron a chance of surmounting the barrier. We can write the Schrödinger equation for the electron in this situation, specifying its wave function explicitly (see fig. 17.7a). When we do, we see that the wave extends beyond the boundary of the potential barrier.[9]

When the Born interpretation is invoked, we find that there is a certain probability that the electron has been reflected back by the barrier and a certain probability that it has passed beyond. In this latter case we say that the electron *tunneled through* the barrier. By suitable choice of the kinetic energy of the electron and of the height of the potential barrier, it can be arranged for the electron to have a fifty-fifty chance of being reflected back from whence it came and the same chance that it will tunnel through the barrier and continue moving in the same direction. If we perform precisely the same experiment repeatedly, we find that the electron is reflected back from the barrier in about half of the cases, and that it tunnels through the barrier in the remaining cases.

Similar considerations apply to an alpha-particle in a uranium-238 nucleus. The 238 protons and neutrons that make up the nucleus are bound together by the so-called strong force, but the 92 protons, having like charges, repel one another. In the nucleus, an alpha-particle, which contains two protons and two neutrons, encounters a potential barrier as it approaches the boundary of the nucleus. It is repelled by the positive charges in the nucleus, but is held back by the even stronger nuclear force. However, because of the indeterminacy relation, the alpha-particle sometimes penetrates the potential barrier, tunnels through, and escapes from the nucleus (see fig. 17.7b). When this happens we have an instance of radioactive decay; the result is an atom of thorium-234.

If the Schrödinger ψ-function provides a complete description of whatever quantum mechanical system we happen to be investigating, and if the Born interpretation affords the correct way to apply Schrödinger's wave equation to the physical world, then quantum mechanics implies that the world is objectively indeterministic. The best we can do, even in principle, is to assign probabilities to various happenings. This results not from an incompleteness in our knowledge but from a genuine lack of determinacy in the world. The antecedent state of a physical system does not determine a unique later outcome; instead, the initial state may lead to any of a number of alternative later results. As we will see, this situation intimately involves the problem of quantum mechanical measurement.

3. The Completeness of Quantum Mechanics

The preceding discussion obviously raises the question whether quantum mechanics provides a complete description of physical reality.[10] This issue was profoundly broached by Albert Einstein, Boris Podolsky, and Nathan Rosen in a classic 1935 article, "Can Quantum Mechanical Description of Reality Be Considered Complete?" (hereafter EPR), and it has continued to be a topic of concern right down to the present. Their answer was negative; they argued that certain aspects of physical reality are not captured by the quantum mechanical description. Various authors have sought to 'complete' quantum theory by the addition of "hidden variables." The addition of these variables to the standard Heisenberg-Schrödinger-Born theory would, it is hoped, render quantum mechanics complete and deterministic. Such variables would, for example, determine precisely when a given unstable nucleus would decay. In other words, according to such theorists, our impression of indeterminism regarding quantum mechanics is only a result of our ignorance of the values of the hidden variables; the physical world is not actually indeterministic after all (shades of Laplace).

The essential features of the EPR argument can be stated quite simply. Suppose we have a physical system S consisting of two particles S_1 and S_2 that initially interact with each other, but are subsequently separated and no longer interact physically. Standard quantum mechanics tells us that we can measure precisely the momentum of S_1, and from that value, along with the state of S, we can calculate the precise value of the momentum of S_2. EPR conclude that S_2 must have a precise momentum. The momentum of S_2 has thus been determined without directly measuring it—that is, without interacting with S_2 at all. However, if, instead of measuring its momentum, we had chosen to measure the position of S_1, we could have ascertained its position precisely. Standard quantum mechanics tells us that from that value, along with the state of S, we can calculate the precise value of the position of S_2. Thus, S_2 must have a definite position.

The decision on which measurement to make is our own free choice; therefore, S_2 has to be ready for both alternatives. The conclusion is—the indeterminacy principle notwithstanding—that S_2 has *both* a definite momentum *and* a definite position at the same time, even though we cannot ascertain both of them simultaneously by direct measurement of S_2. Thus, EPR argue, indeterminacy applies only to the results of measurement, not to the physical properties of the system itself. If the momentum or position of S_2 were indefinite before the measurement is made on S_1, then we would have, to use Einstein's own phrase, "spooky action-at-a-distance." The measurement of either momentum or position on S_1 would have to produce the definite value of momentum or position for S_2. Since the special theory of relativity precludes sending signals from one place to another instantaneously, we are left with a profound mystery.

Subsequent writers have elaborated and clarified the EPR argument. For example, David Bohm proposed an experiment in which S consists of a diatomic molecule. The two atoms, as well as the entire molecule, have a certain amount of angular momentum or spin; it can be measured by means of a Stern-Gerlach apparatus. This device consists of a magnet with a strongly inhomogeneous magnetic field between its two poles. An atom or molecule with a spin of zero will not be deflected as it passes through the magnetic field between the poles, but an atom or molecule with a nonzero spin will be deflected in one direction or another depending on the orientation of its spin.

Let us begin by orienting the magnet so that the field between its poles is vertical; we can label the positive direction "up" and the negative direction "down." An atom or molecule with positive spin will be deflected upward; one with a negative spin will be deflected downward (see fig. 17.8). Suppose that the total spin of our molecule is zero. Let the two component atoms be gently separated from each other in a manner that transfers no angular momentum to either. If these are sent through two Stern-Gerlach apparatuses, and one of the atoms has spin up, then the other must have spin down because of conservation of angular momentum. (Angular momentum is a vector quantity; if the direction of the one vector is up, the direction of the other is down.) Now, instead of orienting our Stern-Gerlach magnets so that the field between the poles is vertical, we could have oriented them so that the field is horizontal and perpendicular to the path of the atom. (This axis is perpendicular to the page in fig. 17.8.) We can designate the two directions of orientation as "in" and "out." If we had chosen the in-out orientation, the spin directions of the two atoms would have to be opposite. If one has spin in, the other must have spin out.

Every atom or molecule with nonzero spin that passes through a Stern-Gerlach setup comes out with a spin completely aligned with the magnetic field. In performing the Bohm version of the EPR experiment, we may choose either the horizontal orientation or the vertical one, making the measurement on one of the atoms. If we measure spin with respect to the vertical orientation, and if the one atom has spin up, the other atom must have spin down, whether it is measured or not. If we measure spin with respect to the horizontal orientation, and the one atom has spin in, the other atom must have spin out, whether it is measured or not. However, it is impossible for this other atom to have both spin down and spin out. If its spin is completely oriented in the vertical direction, then the horizontal component of the spin is zero; if the spin is completely oriented in the horizontal direction, then the vertical component of the spin is zero. Since the measure-

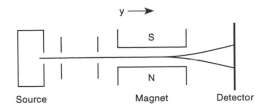

Stern-Gerlach Apparatus

Figure 17.8

ment on the one atom is remote from the other atom, it seems that the other atom must 'know' which measurement is made on the first atom. However, it is possible to choose the orientation of the Stern-Gerlach magnet after the molecule has been split and the two atoms have started on their respective ways. Spooky action-at-a-distance threatens even more ominously than before.

Notice, however, that the main thrust of the argument has changed. In the EPR version the conclusion seems to be that a particle has a precise position and a precise momentum, even though the quantum theory does not capture them. Although it violates the Heisenberg indeterminacy principle, there is no logical contradiction in supposing that a particle has both precise position and precise momentum. In the Bohm version, in contrast, the atom cannot have spin down and spin out; these are logically incompatible. It may have one or the other, but it cannot have both. There is a deterministic relationship between the spin of one atom and the spin of the other; the problem is to find a mechanism that produces the perfect correlation between the spins of the two atoms. As we will see, it cannot be any normal causal connection of the sort familiar from other areas of natural science and everyday life.

To bring the problem of determinism back into focus, consider an atom whose vertical spin orientation we are going to measure. It need not be one that has been extracted from a spin-zero molecule in the manner specified by Bohm. When it leaves the Stern-Gerlach apparatus, it will have a precise value of spin, say, spin up. Our primary question is whether that value is determined by conditions obtaining prior to its entry into the apparatus.

Consider three possibilities. Suppose (1) that the atom has just exited with spin up from another Stern-Gerlach apparatus having a vertical orientation. After passing through our apparatus, it is bound to exit with spin up. We can say that it was *prepared* in that spin state. If this experiment is repeated, it will always have the same outcome. Suppose (2) that this atom has just exited from a Stern-Gerlach apparatus having a horizontal orientation. In this case it has a fifty-fifty chance of coming out of our apparatus with spin up. If this experiment is repeated many times, we find a random sequence in which about half of the atoms come out with spin up and half with spin down. Suppose (3) that this atom escaped through a hole in a container of gas. Passing through our Stern-Gerlach apparatus, it exits with spin up. If we repeat this experiment many times, we find a random sequence of atoms with spin up and spin down.

In cases (2) and (3) the primary question for determinism is whether there is some prior physical condition that determines for every single atom whether its spin will be up or down. Standard quantum mechanics says that there is not; we can only say that it has a probability of one half for each of the two spin orientations. A hidden-variable determinist would maintain that there are as yet unknown antecedent factors that determine the response of each atom to our Stern-Gerlach device.

4. Causality and Quantum Mechanics

In "The Importance of Scientific Understanding" (essay 5) I touch briefly on the issues raised by the Bell inequalities and the Aspect experiments. In the present context it is important to reconsider these issues and their bearing on determinism. This discussion follows naturally from consideration of Bohm's experiment. The main differences between the Bohm setup and the Bell-Aspect setup are (1) that pairs of elementary particles (or photons) are used in place of Bohm's atoms; (2) that three possible orientations of the Stern-Gerlach apparatus are involved; and (3) that for any given pair of particles, the detectors may have any given combination of settings.[11]

Consider an 'atom' of 'positronium', consisting of an electron and a positron (positive electron) in orbit around each other. It is analogous to a hydrogen atom, except that it has a positron in place of hyrdogen's proton. The 'positronium atom' has spin equal to zero. The electron and positron are both spin-one-half particles; the spins can be ascertained by Stern-Gerlach devices. With respect to any orientation of the Stern-Gerlach magnets, the spin may be plus one half or minus one half. If the positron and electron are gently separated without imparting spin, then there is a perfect negative correlation between the spins of the two particles. If one spin is plus one half, the other must be minus one half.

To perform the Aspect-type experiment we need a source of particle pairs situated between two detectors. The source might be a device for separating the two components of the positronium atom, sending each component to one of the two detectors. The detectors are Stern-Gerlach devices, each of which can assume three distinct orientations, e.g., 0° (vertical), 120°, and 240°. They can be designated "left-hand detector" and "right-hand detector." There is no physical connection between the source and the detectors except for the above-mentioned pairs of particles, and no channel for direct communication between the two detectors. The experiment employs a large number of particle pairs, and the orientations of the magnets vary randomly as new pairs are emitted. For any given pair of particles, the orientations of the two magnets need not be the same. Moreover, the orientations of the detectors are selected too late for news of the setting to travel (at a speed no greater than that of light) from one of the detectors to the other detector before the arrival of the particle it detects. Therefore, the setting of one detector cannot influence the outcome of the measurement at the other detector.

To understand the import of the Aspect experiment, let us imagine that each detector has a dial that shows "1," "2," or "3" to indicate the orientation of the Stern-Gerlach magnet. In addition, each detector has two lights, one red and one green, one and only one of which lights up whenever a particle is detected. The color indicates the direction of its spin. For ease of exposition we wire the lights at the two detectors oppositely: a particle with *positive spin* illuminates the red light on the left-hand detector while a particle with *negative spin* illuminates the red light on the right-hand detector. The purpose of this cross-wiring is to make the same color show whenever the two spin orientations are opposite. We can record the result for a given particle pair by giving the setting and the color of the light that is illuminated on each detector. For example, 21RG would mean that the left-hand detector was set with the second orientation and the red light went on while the right-hand detector was set with the first orientation and the green light went on.

If we now run the experiment and record the results for a large number of pairs, we notice two features:

1. Whenever the dials on the two detectors show the same setting, the lights on the two detectors show the same color.
2. Ignoring the indications on the dials of the detectors, we find that red and green occur randomly and with probability ½, and that the color showing on one detector is statistically independent of the color showing on the other.

These results may look innocent enough at first blush, but they rapidly become disconcerting when we try to devise a mechanism to accomplish them. Since the only connection between the two detectors is the series of particle pairs coming from the source, let us consider the kinds of messages these particles might carry.

First, since the color is always the same when the detector settings are the same, the two particles in any pair must carry the same message. This message must tell the detector how to respond to that particle in each of its possible settings. We could encode such a message as RGG, which would mean that the red light goes on if the dial shows "1" and the green light goes on if the dial shows "2" or "3." There are, in total, eight such messages:

RRR, RRG, RGR, GRR, RGG, GRG, GGR, GGG

In addition, there are nine possible combinations of detector settings:

11, 12, 13, 21, 22, 23, 31, 32, 33

They occur with equal frequency. Now, consider the message RRG, for example. Out of the nine possible combinations of settings, this message gives the same color for five, namely, 11, 12, 21, 22, 33. Thus, whenever this message is sent, we get the same color $\frac{5}{9}$ of the time. The same result occurs for any message that contains one color twice and the other color once, namely—in addition to RRG—RGR, GRR, RGG, GRG, and GGR. You can easily satisfy yourself that the same result occurs for all of these combinations. Moreover, the messages RRR and GGG necessarily yield the same colors for all settings. It follows that this system of sending messages cannot reproduce our observed result. Our observations have established that, ignoring detector settings, we get RR in ¼ of all trials and GG in ¼ of all trials, that is, the same color on half of all trials and different colors on the other half of the trials. The proposed system of messages produces more than one half same-color results whatever the settings on the detectors. Notice that, although each possible combination of settings occurs equally frequently, no such assumption is required regarding the messages sent. No matter with what frequency the different possible messages are sent, the result is the same—too many same-color pairs. This system of message sending would produce a positive correlation between the results on the detectors that does not exist in fact.

The moral of the Bell inequalities and the foregoing experiment is that, under very mild assumptions about causality, there cannot be a causal explanation of the empirical results. Standard quantum mechanics, however, correctly predicts the observed outcomes. We see, then, that the quantum domain does not operate in conformity to normal causality. However, this conclusion does not address our main question, namely, is quantum mechanics unavoidably indeterministic? To answer this question we must focus on another facet of the Aspect-type experiment, as we did for the Bohm experiment.

Consider an electron (or positron) that has been emitted by the source and is approaching one of the detectors. According to standard quantum mechanics it has no definite spin orientation. It has spin of one half, but the direction of the spin vector is indeterminate. It passes through the Stern-Gerlach apparatus, and is later detected in either the plus path or the minus path. At this point we can assign a definite spin orientation—say, plus. The question is whether that particle had some characteristic, prior to entering the Stern-Gerlach magnetic field, that determines that it will register plus rather than minus in the detector. The Aspect-type experiment does not answer this question; in fact, it does not address it.

Let us extract the fundamental question of indeterminism from the subtleties of the Aspect-type experiment. Suppose that we have a heated wire that emits electrons, and that we have a single Stern-Gerlach device, which consists of a magnet with its inhomogeneous field and a detector that records whether a given electron's spin is positive or negative. Fix the orientation in whatever direction you choose; half of the electrons will come out with positive spin and half will come out with negative spin. Quantum mechanics assigns no prior property of these electrons that determines for each one whether its spin will be plus or minus.

Next, add another Stern-Gerlach magnet, with an orientation perpendicular to that of the foregoing device, in front of it. Suppose that electrons leaving the wire must first pass through this additional magnetic field. Arrange it so that only electrons in the plus channel of the first magnet can enter the following device. Standard quantum mechanics predicts that half of the electrons coming into the second Stern-Gerlach apparatus will have positive spin and half will have negative spin. This result is borne out by experiment; in fact, quantum mechanics allows one to calculate the probability that an electron exiting from the first Stern-Gerlach field will be detected with spin positive after moving through the second Stern-Gerlach field, whatever the relative orientations of the two fields. In this case, unlike that of electrons coming directly from the wire, the probability varies with differences in relative orientation. If the two magnetic fields have the same orientation— say, vertical—all of the electrons coming out of the plus channel of the first Stern-Gerlach device will be found exiting the plus channel of the second Stern-Gerlach device.

5. The Problem of Measurement

Let us return to the EPR setup. We have a physical system S consisting of two particles, S_1 and S_2, that initially interact with each other but are subsequently separated and no longer interact physically. Nevertheless, the composite system S is still described by a quantum mechanical ψ-function. If we make a measurement on one of the particles—say, position—this involves an interaction with the entire system S. If we make a precise position measurement on S_1, it also fixes the precise value of position of S_2. It has traditionally been said that such measurement brings about an instantaneous "collapse of the wave function" of the system, locating the parts of the composite system precisely. This happens even if the two subsystems, S_1 and S_2, are very widely separated. A measurement of any part of the system affects the whole system.

I mentioned this same phenomenon in connection with the two-slit experiment for photons or electrons. Before being detected at a particular place, the particle or photon is represented by a wave function that collapses instantaneously when the position is fixed by a detection after passing through the slits in the first screen. In addition, quantum mechanical tunneling, as manifested in radioactive decay of atomic nuclei, also involves a collapse of the wave function. If the alpha-particle is detected outside the uranium atom, it becomes localized as a result of this measurement of its position. Moreover, in the Aspect-type experiment, the particle (or photon) pair constitutes a physical system, and a measurement of spin (or polarization) on any part of it collapses the wave function of the whole system.

What is this process called "collapse of the wave packet" or sometimes "reduction of the state vector"? We simply do not know. This is the *problem of measurement* in quantum mechanics. According to the Born interpretation, quantum mechanics enables us to calculate the probabilities of various possible outcomes when a particular type of measurement is made on a given system, but this possibility of calculating does not tell us anything about the physical process involved. It does not tell us, for example, whether the measurement process involves 'hidden variables' in terms of which measurement is a deterministic process. Various possibilities have been suggested. One suggestion is that both the measuring apparatus and the system being measured are governed by an extremely complicated ψ-function that evolves according to Schrödinger's equation. As noted earlier, this kind of evolution is deterministic; if this is the correct account of measurement, quantum mechanics does not imply any sort of indeterminism. Another suggestion is that human consciousness is necessarily involved in collapsing the wave packet, but how this comes about is a mystery. A third suggestion is that quantum mechanical measurement splits our world into many worlds, and each possible outcome of the measurement is realized in some of these worlds. The situation is obviously desperate.

6. Conclusion

In a highly technical chapter of *A Primer of Determinism,* namely, "Determinism in Quantum Physics," John Earman offers the following concluding remarks:

> An astounding—and frustrating—feature of the [quantum] theory lies in the contrast between the exquisite accuracy of its empirical predictions on the one hand and the zaniness of its metaphysical 'consequences' on the other. The theory has been used to 'prove' not only that determinism is false but that realism fails, that logic is non-classical, that there is a Cartesian mental-physical dualism, that the world has the

structure of Borges' garden of forking paths, etc. One is tempted to say that any theory that proves all of this proves nothing. But the temptation must be resisted. Although it is not clear what the quantum theory implies about determinism, it is clear that the implications are potentially profound. Bringing the implications into sharper focus requires a simultaneous focusing on a host of other foundation issues, most especially concerning the nature of quantum magnitudes and the nature of the quantum measurement process. By now it is no surprise that pressing the question of determinism has helped to unearth the deepest and most difficult problems that challenge our understanding of the theory. (1986, p. 233)

I cannot disagree with Earman's assessment of the situation. However, I would add that quantum mechanics raises deep questions about causality in addition to those involving determinism, and that these are not the same problems. In particular, Einstein's basic worry about "spooky action-at-a-distance" still haunts us.

As my discussion has shown, quantum mechanics poses two troubling problems, one concerning determinism, the other concerning causality. In the first place, the determinism issue remains open, pending a satisfactory resolution of the measurement problem. My hunch—and it is no more than a guess—is that the collapse of the wave function involves a noncausal quantum mechanism that is irreducibly stochastic. We do not as yet understand it. If quantum mechanics turns out to be indeterministic, it will show that the world is not glued together as tightly as many philosophers and scientists have thought.

In the second place, quantum mechanics seems to involve action-at-a-distance, but it is important to distinguish two forms it might take. Consider two distinct principles. The first is *locality*—i.e., the principle that it is impossible to interact with a remote physical system. The second is *separability*—i.e., that it is possible to act on a part of a physical system that is extended in space without affecting the rest of that system. To see this distinction, return to the EPR experiment. When the two subsystems move apart, they still constitute a single quantum mechanical system. Any measurement that is made on either of the parts occurs by contact with the whole system; the condition of locality is satisfied. However, separability is violated. It is impossible to make a measurement on one part of the system that leaves the other part untouched. It is difficult to understand how the remote parts of the system can react instantaneously to a *local interaction* with one of the parts. This aspect of quantum mechanics suggests that the world is glued together more tightly than we previously realized.

A final remark concerning indeterminacy and indeterminism: I now believe that quantum *indeterminacy* is an objective feature of the world that follows from the wave-particle duality, but that the jury is still out regarding quantum *indeterminism*. In this latter respect I have changed my mind since writing "Determinism and Indeterminism in Modern Science."

Notes

1. I.e., at the time I wrote essay 2.
2. This phenomenon can be demonstrated using either a disk (e.g., a small coin) or a sphere (e.g., a ball bearing). On one occasion when I made the demonstration for a class, a

student insisted on inspecting the ball bearing closely to make sure that I had not drilled a hole through it.

3. Thomson showed that the electron is part of an atom; however, he advocated not the planetary model but rather the "plum-pudding model," according to which the electrons are embedded in the atom much as blueberries are found scattered through a blueberry muffin. Rutherford's scattering of alpha-particles from gold foil demonstrated the presence of a compact massive nucleus.

4. The other series are known, after their discoverers, as Lyman, Pfund, Brackett, and Paschen. The Balmer series lies in the visible range of the spectrum; the Lyman series lies in the ultraviolet region; the Brackett, Paschen, and Pfund series lie in the infrared region.

5. For visible light a slit width of a few thousand angstroms suffices; for an electron at a typical velocity, the slit width would have to be on the order of one angstrom ($= 10^{-10}$ meters).

6. I am reminded of the beautiful Ninety-Mile Beach on the south coast of Australia. A prodigious amount of kinetic energy is carried by a wave extending over such a distance. The unfortunate swimmer might land in Canberra, many kilometers from the shore.

7. Many modern formulations of quantum mechanics use an abstract infinite-dimensional vector space known as Hilbert space.

8. This does not mean that the hill is potentially a barrier; it is an actual barrier in relation to potential energy.

9. As shown in figure 17.7, we take the barrier to have a finite thickness; one reason for this choice is to draw a parallel to radioactive decay of atoms.

10. In "Determinism and Indeterminism in Modern Science" (essay 2) I cited a 'proof' of the completeness of quantum mechanics that had been given by John von Neumann in *Mathematical Foundations of Quantum Mechanics* (Princeton: Princeton University Press, 1955). At the time that essay was written, this 'proof' was widely, though by no means unanimously, accepted as valid; it no longer enjoys that status.

11. In Aspect's actual experiment correlations between pairs of photons with respect to their polarizations were studied instead of correlations between pairs of material particles and their spins. Aspect chose photons for practical reasons, but since both types of experiment are equivalent from a theoretical standpoint, I will continue to deal with particles and their spins. Many other authors use the particle approach when discussing the import of the work of Bell and Aspect. See, for example, the excellent exposition in Mermin (1985).

Part IV

CONCISE OVERVIEWS

The three essays in this part present brief and somewhat sophisticated summaries of material presented at greater length elsewhere. Although they would not serve as an appropriate introduction for readers with little or no background in philosophy of science, they should be fully intelligible to anyone who has absorbed the main content of the essays in the preceding three parts of this book.

Essay 18, "Causality: Production and Propagation," was published long before my conversion to a conserved or invariant quantities theory of causality, as presented in "Causality without Counterfactuals" (essay 16); nevertheless, its fundamental approach is still sound. The basic facts about causal processes and causal forks, about their mutual relationships, and about the various types of forks are presented here in some detail. The only difference is that a new criterion for causal processes and interactive forks has subsequently been adopted; however, the same processes and intersections will be classified as causal under either criterion. This essay presents a concise but comprehensive picture of the causal structure of the world along the same lines as my account in *Scientific Explanation and the Causal Structure of the World* (1984b).

Essay 19, "Scientific Explanation: How We Got from There to Here," contains a brief history of the developments in the philosophy of scientific explanation in the forty years immediately following publication of Carl G. Hempel and Paul Oppenheim's seminal article, "Studies in the Logic of Explanation" ([1948] 1965), the fountainhead from which almost all work on scientific explanation in the second half of the twentieth century flowed. Essay 19 provides historical insight into discussions of scientific explanation around the close of the century. These historical themes are treated more thoroughly in my book *Four Decades of Scientific Explanation* (1989, 1990b).

Essay 20, "Scientific Explanation: Three Basic Conceptions," was presented at the 1984 meeting of the Philosophy of Science Association just prior to the publication of *Scientific Explanation and the Causal Structure of the World*. It contains an extremely condensed précis of some of the main discussions in that book. Whereas essay 18 focuses on causal considerations, this essay focuses on the concept of explanation.

18

Causality

Production and Propagation

A standard picture of causality has been around at least since the time of Hume. The general idea is that we have two (or more) distinct events that bear some sort of cause-effect relation to each other. There has, of course, been considerable controversy regarding the nature of both the relation and the relata. It has sometimes been maintained, for instance, that facts or propositions (rather than events) are the sorts of entities that can constitute the relata. It has long been disputed whether individual events or only classes of events can sustain cause-effect relations. The relation itself has sometimes been taken to be that of sufficient condition, sometimes necessary condition, or perhaps a combination of the two.[1] Some authors have even proposed that certain sorts of statistical relations constitute causal relations.[2]

It is my conviction that this standard view, in all of its well-known variations, is profoundly mistaken, and that a radically different notion should be developed. I shall not attempt to mount arguments against the standard conception; instead, I shall present a rather different approach for purposes of comparison. I hope that the alternative will stand on its own merits.

1. Two Basic Concepts

There are, I believe, two fundamental causal concepts that need to be explicated, and, if that can be achieved, we will be in a position to deal with the problems of causality in general. The two basic concepts are *production* and *propagation,* and both are familiar to common sense. When we say that the blow of a hammer drives a nail, we mean that the impact produces penetration of the nail into the wood. When we say that a horse pulls a cart, we mean that the force exerted by the horse produces the motion of the cart. When we say that lightning starts a forest fire, we mean that the electrical discharge produces

285

ignition. When we say that a person's embarrassment was due to a thoughtless remark, we mean that an inappropriate comment produced psychological discomfort. Such examples of causal production occur frequently in everyday contexts.

Causal propagation (or transmission) is equally familiar. Experiences that we had earlier in our lives affect our current behavior. By means of memory, the influence of these past events is transmitted to the present. A sonic boom makes us aware of the passage of a jet airplane overhead; a disturbance in the air is propagated from the upper atmosphere to our location on the ground. Signals transmitted from a broadcasting station are received by the radio in our home. News or music reaches us because electromagnetic waves are propagated from the transmitter to the receiver. In 1775 some Massachusetts farmers "fired the shot heard 'round the world." As all of these examples show, what happens at one place and time can have significant influence on what happens at other places and times. This is possible because causal influence can be propagated through time and space. Although causal production and causal propagation are intimately related, we should, I believe, resist any temptation to try to reduce one to the other.

2. Processes

One of the fundamental changes that I propose in approaching causality is to take processes rather than events as basic entities. I shall not attempt any rigorous definition of processes; rather, I shall cite examples and make some very informal remarks. The main difference between events and processes is that events are relatively localized in space and time, while processes have much greater temporal duration and, in many cases, much greater spatial extent. In spacetime diagrams, events are represented by points, while processes are represented by lines. A baseball colliding with a window would count as an event; the baseball, traveling from the bat to the window, would constitute a process. The activation of a photocell by a pulse of light would be an event; the pulse of light, traveling, perhaps from a distant star, would be a process. A sneeze is an event. The shadow of a cloud moving across the landscape is a process. Although I deny that all processes qualify as causal processes, what I mean by a process is similar to what Bertrand Russell characterized as a *causal line*: "A causal line may always be regarded as the persistence of something—a person, a table, a photon, or what not. Throughout a given causal line, there may be constancy of quality, constancy of structure, or a gradual change of either, but not sudden changes of any considerable magnitude" (1948, p. 459). Among the physically important processes are waves and material objects which persist through time. As I shall use the terms, even a material object at rest qualifies as a process.

Before attempting to develop a theory of causality in which processes rather than events are taken as fundamental, I should consider briefly the scientific legitimacy of this approach. In Newtonian mechanics both spatial extent and temporal duration were absolute quantities. The length of a rigid rod did not depend on a choice of frame of reference, nor did the duration of a process. Given two events, in Newtonian mechanics, both the spatial distance and the temporal separation between them were absolute magnitudes. As everyone knows, Einstein's special theory of relativity changed all that. Both the spatial distance and the temporal separation were relativized to frames of reference. The length of a rigid rod and the duration of a temporal process varied from one frame of reference to another. However, as Minkowski showed, there is an invariant quantity—the spacetime interval between two events. This quantity is independent of the frame of reference; for any two events it has the same value in each and every inertial frame of reference. Since there are good reasons for according a fundamental physical status to invariants, it was a natural consequence of the special theory of relativity to regard the world as a collection of events which bear spacetime relations to one another. These considerations offer support for what is sometimes called an *event ontology*.

There is, however, another way (developed originally by A. A. Robb) of approaching the special theory of relativity; it is done entirely with paths of light pulses. At any point in space-time, we can construct the Minkowski light cone—a two-sheeted cone whose surface is generated by the paths of all possible light pulses that converge on that point (past light cone) and the paths of all possible light pulses that could be emitted from that point (future light cone). When all of the light cones are given, the entire spacetime structure of the world is determined (see Winnie, 1977). But light pulses, traveling through space and time, are processes. We can therefore base special relativity on a *process ontology*. Moreover, this approach can be extended in a natural way to general relativity by taking into account the paths of freely falling material particles; these moving gravitational test particles are also processes (see Grünbaum, 1973, pp. 735–750). It therefore appears to be entirely legitimate to approach the spacetime structure of the physical world by regarding physical processes as the basic types of physical entities. The theory of relativity does not mandate an *event ontology*.

Special relativity does demand, however, that we make a distinction between what I call *causal processes* and *pseudo-processes*. It is a fundamental principle of that theory that light is a *first signal*—that is, that no signal can be transmitted at a velocity greater than the velocity of light in a vacuum. There are, however, certain processes that can transpire at arbitrarily high velocities—at velocities vastly exceeding that of light. This fact does not violate the basic relativistic principle, however, for these 'processes' are incapable of serving as signals or of transmitting information. Causal processes are those that are capable of transmitting signals; pseudo-processes are incapable of doing so.

Consider a simple example. Suppose that we have a very large circular building—a sort of super-Astrodome, if you will—with a spotlight mounted at its center. When the light is turned on in the otherwise darkened building, it casts a spot of light on the wall. If we turn the light on for a brief moment and then off again, a light pulse travels from the light to the wall. This pulse of light, traveling from the spotlight to the wall, is a paradigm of what we mean by a causal process. Suppose, further, that the spotlight is mounted on a mechanism that makes it rotate. If the light is turned on and set into rotation, the spot of light that it casts on the wall will move around the outer wall in a highly regular fashion. This 'process'—the moving spot of light—seems to fulfill the conditions Russell used to characterize causal lines, but it is not a causal process. It is a paradigm of what we mean by a pseudo-process.

The basic method for distinguishing causal processes from pseudo-processes is the criterion of mark transmission. A causal process is capable of transmitting a mark; a pseudo-process is not. Consider, first, a pulse of light that travels from the spotlight to the wall. If we place a piece of red glass in its path at any point between the spotlight and the wall, the light pulse, which was white, becomes and remains red until it reaches the wall. A single intervention at one point in the process transforms it in a way that persists from that point on. If we had not intervened, the light pulse would have remained white during its entire journey from the spotlight to the wall. If we do intervene locally at a single place, we can produce a change that is transmitted from the point of intervention onward. We say, therefore, that the light pulse constitutes a causal process, whether it is modified or not, since in either case it is capable of transmitting a mark. Clearly, light pulses can serve as signals and can transmit messages.

Now, let us consider the spot of light that moves around the wall as the spotlight rotates. There are a number of ways in which we can intervene to change the spot at some point; for example, we can place a red filter at the wall with the result that the spot of light becomes red at that point. But if we make such a modification in the traveling spot, it will not be transmitted beyond the point of interaction. As soon as the light spot moves beyond the point at which the red filter was placed, it will become white again. The mark can be made, but it will not be transmitted. We have a 'process' which, in the absence of any intervention, consists of a white spot moving regularly along the wall of the building. If we intervene at some point, the 'process' will be modified *at that point,* but it will continue beyond that point just as if no intervention had occurred. We can, of course, make the spot red at other places if we wish. We can install a red lens in the spotlight, but that does not constitute a *local* intervention at an isolated point in the process itself. We can put red filters at many places along the wall, but that would involve *many* interventions rather than a single one. We could get someone to run around the wall holding a red filter in front of the spot continuously, but that would not constitute an intervention *at a single point* in the 'process.'

This last suggestion brings us back to the subject of velocity. If the spot of light is moving rapidly, no runner could keep up with it, but perhaps a mechanical device could be set up. If, however, the spot moves too rapidly, it would be physically impossible to make the filter travel fast enough to keep up. No material object, such as the filter, can travel at a velocity greater than that of light, but no such limitation is placed upon the spot on the wall. This can easily be seen as follows. If the spotlight rotates at a fixed rate, then it takes the spot of light a fixed amount of time to make one entire circuit around the wall. If the spotlight rotates once per second, the spot of light will travel around the wall in one second. This fact is independent of the size of the building. We can imagine that, without any change in the spotlight or its rate of rotation, the outer walls are expanded indefinitely. At a certain point, when the radius of the building reaches about 50,000 kilometers, the spot will be traveling at the speed of light (300,000 km/sec). As the walls are moved still farther out, the velocity of the spot exceeds the speed of light.

To make this point more vivid, consider an actual example that is quite analogous to the rotating spotlight. There is a pulsar in the Crab nebula which is about 6500 light years away. This pulsar is thought to be a rapidly rotating neutron star that sends out a beam of radiation. When the beam is directed toward us, it sends out radiation that we detect later as a pulse. The pulses arrive at the rate of 30 per second; that is the rate at which the neutron star rotates. Now, imagine a circle drawn with the pulsar at its center, and with a radius equal to the distance from the pulsar to the earth. The electromagnetic radiation from the pulsar (which travels at the speed of light) takes 6500 years to traverse the radius of this circle, but the 'spot' of radiation sweeps around the circumference of this circle in $1/30$ of a second. There is no upper limit on the speed of pseudo-processes.

A given process, whether it be causal or pseudo-, has a certain degree of uniformity; we may say, somewhat loosely, that it exhibits a certain structure. The difference between a causal process and a pseudo-process, I am suggesting, is that the causal process transmits its own structure, while the pseudo-process does not. The distinction between processes that do and those that do not transmit their own structures is revealed by the mark criterion. If a process—a causal process—is transmitting its own structure, then it will be capable of transmitting modifications in that structure. Radio broadcasting presents a clear example. The transmitting station sends a carrier wave that has a certain structure—characterized by amplitude and frequency, among other things—and modifications of this wave, in the form of modulations of amplitude (AM) or frequency (FM), are imposed for the purpose of broadcasting. Processes that transmit their own structure are capable of transmitting marks, signals, information, energy, and causal influence. Such processes are the means by which causal influence is propagated in our world. Causal influences, transmitted by radio, may set your foot to tapping, or induce someone to purchase a different brand of soap, or point a television camera aboard a spacecraft toward the rings of Saturn. A causal influence transmitted by a flying arrow can pierce an apple on the head of William Tell's son. A causal influence transmitted by sound waves can make your dog come running. A causal influence transmitted by ink marks on a piece of paper can gladden one's day or break someone's heart.

It is evident, I think, that the propagation or transmission of causal influence from one place and time to another must play a fundamental role in the causal structure of the world. As I shall argue, causal processes constitute precisely the causal connections that Hume sought but was unable to find.[3]

3. Conjunctive Forks

In order to approach the second basic causal concept, *production,* it is necessary to consider the nature of causal forks. There are three types with which we must deal— conjunctive, interactive, and perfect forks. All three types are concerned with situations in which a common cause gives rise to two or more effects that are somehow correlated with one another. The point of departure for this discussion is Reichenbach's *principle of the common cause* and his statistical characterization of the conjunctive fork as a device to elaborate that fundamental causal principle (1956, §19).

The principle of the common cause states, roughly, that when improbable coincidences recur too frequently to attribute them to chance, they can be explained by reference to a common causal antecedent. Consider some familiar examples. If two students in a class turn in identical term papers, and if we can rule out the possibility that either copied directly from the other, then we search for a common cause—for example, a paper in a sorority or fraternity file from which both of them copied independently of each other. If two friends, who have spent a pleasant day in the country together, both suffer acute gastrointestinal distress in the evening, we may find that their illnesses can be traced to poisonous mushrooms they collected and consumed. Many such examples have been mentioned in the literature, and others come readily to mind. A 1979 astronomical discovery, which has considerable scientific significance, furnishes a particularly fine example. The twin quasars 0975+561 A and B are separated by an angular width of 5.7 seconds of arc. Two quasars in such apparent proximity would be a rather improbable occurrence given simply the observed distribution of quasars. Examination of their spectra indicates equal red shifts and, hence, equal distances. Thus, these objects are close together in space, as well as appearing close together as seen from Earth. Moreover, close examination of their spectra reveals a striking similarity—indeed, they are indistinguishable. This situation is in sharp contrast to the relations between the spectra of any two quasars picked at random. Astronomers immediately recognized the need to explain this astonishing coincidence in terms of some sort of common cause. One hypothesis that was entertained quite early was that twin quasars had somehow (no one really had the slightest idea how this could happen in reality) developed from a common ancestor. Another hypothesis was the gravitational lens effect—that is, that there are not in fact two distinct quasars, but that the two images are produced from a single body by the gravitational effect on the light by an intervening massive object. This result might be produced by a massive black hole, it was theorized, or by a very large elliptical galaxy. Further observation, under fortuitously excellent viewing conditions, subsequently revealed the presence of a galaxy which would be adequate to produce the gravitational splitting of the image. This explanation is now, to the best of my knowledge, accepted virtually universally by the experts (Chaffee, 1980).

In an attempt to characterize the structure of such examples of common causes, Reichenbach (1956, §19) introduced the notion of a *conjunctive fork,* defined in terms of the following four conditions:[4]

$$P(A.B|C) = P(A|C) \times P(B|C) \tag{1}$$

$$P(A.B|\bar{C}) = P(A|\bar{C}) \times P(B|\bar{C}) \tag{2}$$

$$P(A|C) > P(A|\bar{C}) \tag{3}$$

$$P(B|C) > P(B|\bar{C}) \tag{4}$$

For reasons that will be made clear, we shall stipulate that none of the probabilities occurring in these relations is equal to zero or one. Although it is not immediately obvious, conditions (1)–(4) entail

$$P(A.B) > P(A) \times P(B)^5 \tag{5}$$

These relations apply quite straightforwardly in concrete situations. Given two effects *A* and *B,* which occur together more frequently than they would if they were statistically independent of each other, there is some prior event *C* that is a cause of *A* and is also a cause of *B,* and that explains the lack of independence between *A* and *B.* In the case of plagiarism, the cause *C* is the presence of the term paper in the file to which both students had access. In the case of simultaneous illness, the cause *C* is the common meal that included the poisonous mushrooms. In the case of the twin quasar image, the case *C* is the emission of radiation in two slightly different directions by a single luminous body.

To say of two events X and Y that they occurred independently of each other means that they occur together with a probability equal to the product of the probabilities of their separate occurrences; i.e.,

$$P(X.Y) = P(X) \times P(Y) \qquad (6)$$

Thus, in the examples we have considered, as relation (5) states, the two effects A and B arc not independent. However, given the occurrence of the common cause C, A and B do occur independently, as the relationship among the conditional probabilities in equation (1) shows. Thus, in the case of illness, the fact that the probability of both individuals being ill at the same time is greater than the product of the probabilities of their individual illnesses is explained by the common meal. In this example we are assuming that the fact that one person is afflicted does not have any direct causal influence on the illness of the other. Moreover, let us assume for the sake of simplicity that, in this situation, there are no other potential common causes of severe gastrointestinal illness.[6] Then, in the absence of the common cause C—that is, when C obtains—A and B are also independent of each other, as the relationship among the conditional probabilities in equation (2) states. Relations (3) and (4) simply assert that C is a positive cause of A and B, since the probability of each is greater in the presence of C than in the absence of C.

There is another useful way to look at equations (1) and (2). Recalling that, according to the multiplication theorem,

$$P(A.B|C) = P(A|C) \times P(B|A.C) \qquad (7)$$

we see that, provided $P(A|C) \neq 0$, equation (1) entails

$$P(B|C) = P(B|A.C) \qquad (8)$$

In Reichenbach's terminology, this says that C screens off A from B. A similar argument shows that C screens off B from A. To screen off *means* to make statistically irrelevant. Thus, according to equation (1), the common cause C makes each of the two effects A and B statistically irrelevant to each other. By applying the same argument to equation (2), we can easily see that it entails that the absence of the common cause also screens off A from B.

To make quite clear the nature of the conjunctive fork, I should like to use an example deliberately contrived to exhibit the relationships involved. Suppose we have a pair of dice that are rolled together. If the first die comes to rest with side 6 on top, that is an event of the type A; if the second die comes to rest with side 6 uppermost, that is an event of type B. These dice are like standard dice except for the fact that each one has a tiny magnet embedded in it. In addition, the table on which they are thrown has a powerful electromagnet embedded in its surface. This magnet can be turned on or off with a concealed switch. If the dice are rolled when the electromagnet is on, it is considered an instance of the common cause C; if the magnet is off when the dice are tossed, the event is designated as $\bar{C}$. Let us further assume that, when the electromagnet is turned off, these dice behave exactly like standard dice. The probability of getting 6 with either die is $\frac{1}{6}$, and the probability of getting double 6 is $\frac{1}{36}$.[7] If the electromagnet is turned on, let us assume, the chance of getting 6 with either die is $\frac{1}{2}$, and the probability of double 6 is $\frac{1}{4}$. It is easily seen that conditions (1)–(4) are fulfilled. Let us make a further stipulation, which will simplify the arithmetic, but which has no other bearing on the essential features of the example—namely, that half of the tosses of this pair of dice are made with the electromagnet turned on, and half are made with it turned off. We might imagine some sort of random device that controls the switch, and that realizes this equiprobability condition. We can readily see that the overall probability of 6 on each die, regardless of whether the electromagnet is on or off, is $\frac{1}{3}$. In addition, the overall probability of double 6 is the arithmetical average of $\frac{1}{4}$ and $\frac{1}{36}$, which equals $\frac{5}{36}$. If the occurrence of 6 on one die were independent of 6 occurring on the other, the overall probability of 6 would be $\frac{1}{3} \times \frac{1}{3} = \frac{1}{9} \neq \frac{5}{36}$. Thus, the example satisfies relation (5), as of course it must, in addition to relations (1)–(4).

It may initially seem counterintuitive to say that the results on the two dice are statistically independent if the electromagnet is off, and they are statistically independent if it is on, but that overall they are not independent. Nevertheless, they are, indeed, nonindependent, and this nonindependence arises from a clustering of sixes which is due simply to the fact that in a subset of the class of all tosses, the probability of 6 is enhanced for each die. The dependency arises not because of any physical interaction between the dice, but because of special background conditions that obtain on certain of the tosses. The same consideration applies to the earlier, less contrived, cases. When the two students each copy from a paper in a sorority or fraternity file, there is no direct physical interaction between the process by which one of the papers is produced and that by which the other is produced; in fact, if either student had been aware that the other was using that source, the unhappy coincidence might have been avoided. Likewise, as explicitly mentioned in the mushroom poisoning example, the illness of one friend had no effect on the illness of the other. The coincidence resulted from the fact that a common set of background conditions obtained, namely, a common food supply from which both ate. Similarly, in the twin quasar example, the two images are formed by two separate radiation processes that come from a common source but that do not directly interact with each other anywhere along the line.

Reichenbach claimed—correctly, I believe—that conjunctive forks possess an important asymmetry. Just as we can have two effects that arise out of a given common cause, so also may we find a common effect resulting from two distinct causes. For example, by getting results on two dice that add up to seven, one may win a prize. Reichenbach distinguished three situations (see fig. 18.1): (i) a common cause C giving rise to two separate effects, A and B, without any common effect arising from A and B conjointly; (ii) two events A and B which, in the absence of a common cause C, jointly produce a common effect E; and (iii) a combination of (i) and (ii) in which the events A and B have both a common cause C and a common effect E. He characterized situations (i) and (ii) as "open forks," while (iii) is closed on both ends. Reichenbach's asymmetry thesis was that situations of type (ii) never represent conjunctive forks; conjunctive forks that are open are always open to the future and never to the past. Since the statistical relations found in conjunctive forks are said to explain otherwise improbable coincidences, it follows that such coincidences are explained only in terms of common causes, never common effects. I believe that an even stronger claim is warranted—though I shall not try to argue it here—namely, that conjunctive forks, whether open or closed by a fourth event, always point in the same temporal direction. Reichenbach allowed that in

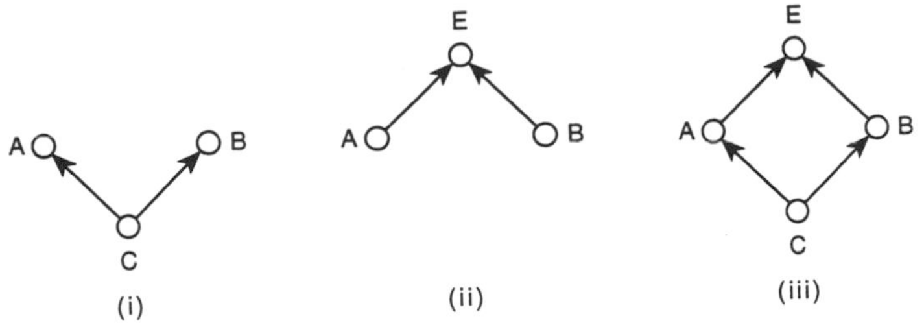

Causal Forks

Figure 18.1

situations of type (iii), the two events *A* and *B*, along with their common effect *E*, could form a conjunctive fork. Here, of course, there must also be a common cause *C*, and it is *C* rather than *E* that explains the coincidental occurrence of *A* and *B*. I doubt that, even in these circumstances, *A*, *B*, and *E* can form a conjunctive fork.

It would be a mistake to suppose that the statistical relations given in conditions (1)–(4) are sufficient to characterize common causes in their role as explanations of corre-lated effects, as an example, owed to Ellis Crasnow, clearly demonstrates.[8] Consider a woman who usually arrives at her office about 9:00 A.M., makes a cup of coffee, and settles down to read the morning paper. On some occasions, however, she arrives promptly at 8:00 A.M., and on these very same mornings her secretary has arrived somewhat earlier and prepared a fresh pot of coffee. Moreover, on just these mornings, she is met at her office by one of her associates who normally works at a different location. How, if we consider the fact that the coffee is already made when she arrives *(A)* and the fact that her associate shows up on that morning *(B)* as the coincidence to be explained, then it might be noted that on such mornings she always catches the 7:00 A.M. bus *(C)*, while on other mornings she usually takes the 8:00 A.M. bus *(C̄)*. In this example, it is plausible enough to suppose that *A*, *B*, and *C* form a conjunctive fork satisfying (1)–(4), but obviously *C* cannot be considered a cause either of *A* or of *B*. The actual common cause is an entirely different event *C'*, namely, a telephone appointment made the day before by her secretary. *C'* is, in fact, the common cause of *A*, *B*, and *C*.

In order to distinguish the cases in which the event *C* in a conjunctive fork constitutes a bona fide common cause from those in which it does not, let us add the condition that there must be a suitable causal process connecting *C* with *A* and another connecting *C* with *B*. These causal processes constitute the mechanisms by which causal influence is transmitted from the cause to each of the effects. These causal connections are an essential part of the causal fork, and without them the event *C* at the vertex of a conjunctive fork cannot qualify as a common cause.

4. Interactive Forks

There is another, basically different, type of common cause that violates the statistical conditions used to define the conjunctive fork. Consider a simple example. Two pool balls, the cue ball and the 8-ball, lie on a pool table. A relative novice attempts a shot that is intended to put the 8-ball into one of the far corner pockets; but given the positions of the balls, if the 8-ball falls into one corner pocket, the cue ball is almost certain to go into the other far corner pocket, resulting in a 'scratch.' Let A stand for the 8-ball dropping into the one corner pocket, let B stand for the cue ball dropping into the other corner pocket, and let C stand for the collision between the cue ball and the 8-ball that occurs when the player executes the shot. Assume that the probability of the 8-ball going into the pocket is $\frac{1}{2}$ if the player tries the shot, and that the probability of the cue ball going into the pocket is also about $\frac{1}{2}$. It is immediately evident that A, B, and C do not constitute a conjunctive fork, for C does not screen A and B from each other. Given that the shot is attempted, the probability that the cue ball will fall into the pocket (appx. $\frac{1}{2}$) is *not* equal to the probability that the cue ball will go into the pocket given that the shot has been attempted and that the 8-ball has dropped into the other far corner pocket (appx. 1).

In discussing the conjunctive fork, I took some pains to point out that forks of that sort occur in situations in which separate and distinct processes, which do not directly interact, arise out of special background conditions. In the example of the pool balls, however, there is a direct interaction—a collision—between the two causal processes which consist in portions of the histories of the two balls. For this reason, I have suggested that forks that are exemplified by such cases be called *interactive forks*. Since the common cause C does not statistically screen the two effects A and B from each other, interactive forks violate condition (1) in the definition of conjunctive forks.

The best way to look at interactive forks is in terms of spatiotemporal intersections of processes. In some cases two processes may intersect without producing any lasting modification in either. This will happen, for example, when both processes are pseudo-processes. If the paths of two airplanes, flying in different directions at different altitudes on a clear day, cross, the shadows on the ground may coincide momentarily. But as soon as the shadows have passed the intersection, both move on as if no such intersection had ever occurred. In the case of the two pool balls, however, the intersection of their paths results in a change in the motion of each that would not have occurred if they had not collided. Energy and momentum are transferred from one to the other; their respective states of motion are altered. Such modifications occur, I maintain, only when two causal processes intersect. If either or both of the intersecting processes are pseudo-processes, no such mutual modification occurs. However, it is entirely possible for two causal processes to intersect without any subsequent modification in either. Barring the extremely improbable occurrence of a particle-particle type collision between two photons, light rays normally pass right through each other without any lasting effect on either one of them. The fact that two intersecting processes are both causal is a necessary but not sufficient condition of the production of lasting changes in them.

When two causal processes intersect and suffer lasting modifications after the intersection, there is some correlation between the changes that occur in them. In many cases—and perhaps all—energy and/or momentum transfer occurs, and the correlations between the modifications are direct consequences of the respective conservation laws.[9] This is nicely illustrated by the Compton scattering of an energetic photon off of an electron that can be considered, for practical purposes, initially at rest. The difference in energy between the incoming photon hν and the scattered photon hν' is equal to the kinetic energy of the recoiling electron. Similarly, the momentum change in the photon is exactly compensated by the momentum change in the electron.[10]

When two processes intersect, and they undergo correlated modifications that persist after the intersection, I shall say that the intersection constitutes a *causal interaction*. This is the basic idea behind what I take as a fundamental causal concept. Let C stand for the event consisting of the intersection of two processes. Let A stand for a modification in one and B for a modification in the other. Then, in many cases, we find a relation analogous to equation (1) in the definition of the conjunctive fork, except that the equality is replaced by an inequality:[11]

$$P(A.B|C) > P(A|C) \times P(B|C) \qquad (9)$$

Moreover, given a causal interaction of the foregoing sort, I shall say that the change in each process is *produced* by the interaction with the other process.

I have now characterized, at least partially, the two fundamental causal concepts mentioned at the outset. Causal processes are the means by which causal influence is *propagated,* and changes in processes are *produced* by causal interactions. We are now in a position to see the close relationship between these basic notions. The distinction between causal processes and pseudo-processes was formulated in terms of the criterion of mark transmission. A mark is a modification in a process, and if that modification persists, the mark is transmitted. Modifications in processes occur when they intersect with other processes; if the modifications persist beyond the point of intersection, then the intersection constitutes a causal interaction, and the interaction has produced marks that are transmitted. For example, a pulse of white light is a process, and a piece of red glass is another process. If these two processes intersect—i.e., if the light pulse goes through the red glass—then the light pulse becomes and remains red, while the filter undergoes an increase in energy as a result of absorbing some of the light that impinges upon it. Although the newly acquired energy may soon be dissipated into the surrounding environment, the glass retains some of the added energy for some time beyond the actual moment of interaction.

We may, therefore, turn the presentation around in the following way. We live in a world which is full of processes (causal or pseudo-), and these processes undergo frequent intersections with one another. Some of these intersections constitute causal interactions; others do not. If an intersection occurs that does not qualify as an interaction, we can draw no conclusion as to whether the processes involved are causal or pseudo-. If two processes intersect in a manner that does qualify as a causal interaction, then we may conclude that both processes are causal, for each has been marked (i.e., modified) in the intersection with the other, and each process transmits the mark beyond the point of intersection. Thus, each process shows itself capable of transmitting marks, since each one has transmitted a mark generated in the intersection. Indeed, the operation of marking a process is accomplished by means of a causal interaction with another process. Although we may often take an active role in producing a mark in order to ascertain whether a process is causal (or for some other purpose), it should be obvious that human agency plays no essential part in the characterization of causal processes or causal interactions. We have every reason to believe that the world abounded in causal processes and causal interactions long before there were any human agents to perform experiments.

5. Relations between Conjunctive and Interactive Forks

Suppose that we have a shooting gallery with a number of targets. The famous sharp-shooter, Annie Oakley, comes to this gallery, but it presents no challenge to her, for she can invariably hit the bull's-eye of any target at which she aims. So, to make the situation interesting, a hardened steel knife-edge is installed in such a position that a *direct* hit on the knife-edge will sever the bullet in a way that makes one fragment hit the bull's-eye of target A while the other fragment hits the bull's-eye of target B. If we let A stand for a fragment striking the bull's-eye of target A, B for a fragment striking the bull's-eye of target B, and C for the severing of the bullet by the knife-edge, we have an interactive fork quite analogous to the example of the pool balls. Indeed, we may use the same probability values, setting $P(A|C) = P(B|C) = \frac{1}{2}$, while $P(A|C.B) = P(B|C.A) \simeq 1$. Statistical screening off obviously fails.

We might, however, consider another event C^*. To make the situation concrete, imagine that we have installed between the knife-edge and the targets a steel plate with two holes in it. If the shot at the knife-edge is good, then the two fragments of the bullet will go through the two holes, and each fragment will strike its respective bull's-eye with probability virtually equal to 1. Let C^* be the event of the two fragments going through their respective holes. Then, we may say, A, B, and C^* will form a conjunctive fork. That happens because C^* refers to a situation which is subsequent to the physical interaction between the parts of the bullet. By the time we get to C^*, the bullet has been cut into two separate pieces, and each is going its way independently of the other. Even if we should decide to vaporize one of the fragments with a powerful laser, that would have no effect on the probability of the other fragment's finding its target. This example makes quite vivid, I believe, the distinction between the interactive fork, which characterizes direct physical interactions, and the conjunctive fork, which characterizes independent processes arising under special background conditions.

There is a further important point of contrast between conjunctive and interactive forks. Conjunctive forks possess a kind of temporal asymmetry which was described earlier. Interactive forks do not exhibit the same sort of temporal asymmetry. This is easily seen by considering a simple collision between two billiard balls. A collision of this type can occur in reverse: if a collision C precedes states of motion A and B in the two balls, then a collision C can occur in which states of motion just like A and B, except that the direction of motion is reversed, precede the collision. Causal interactions and causal processes do not, in and of themselves, provide a basis for temporal asymmetry.

Our ordinary causal language is infused with temporal asymmetry, but we should be careful in applying it to *basic* causal concepts. If, for example, we say that two processes are modified as a result of their interaction, the words suggest that we have already determined which are the states of the processes prior to the interaction and which are the subsequent states. To avoid begging temporal questions, we should say that two processes intersect, and each of the processes has different characteristics on the two sides of the intersection. We do not try to say which part of the process comes earlier and which later. The same is true when we speak of marking. To erase a mark is the exact temporal reverse of imposing a mark; to speak of imposing or erasing is to presuppose a temporal direction. In many cases, of course, we know on other grounds that certain kinds of interactions are irreversible. Light filters absorb some frequencies, so that they transform white light into red. Filters do not furnish missing frequencies to turn red light into white. But until we have gone into the details of the physics of irreversible processes, it is best to think of causal interactions in temporally symmetric terms, and to take the causal connections furnished by causal processes as symmetric connections. Causal processes and causal interactions do not furnish temporal asymmetry; conjunctive forks fulfill that function.

6. Perfect Forks

In dealing with conjunctive and interactive forks, it is advisable to restrict our attention to the cases in which $P(A|C)$ and $P(B|C)$ do not assume either of the extreme values zero or one. The main reason is that the relation

$$P(A.B|C) = P(A|C) \times P(B|C) = 1 \tag{10}$$

may represent a limiting case of either a conjunctive or an interactive fork, even though (10) is a special case of equation (1) and it violates relation (9).

Consider the Annie Oakley example once more. Suppose that she returns to the special shooting gallery time after time. Given that practice makes perfect (at least in her case), she improves her skill until she can invariably hit the knife-edge in the manner that results in the two fragments finding their respective bull's-eyes. Up until the moment that she has perfected her technique, the results of her trials exemplified interactive forks. It would be absurd to claim that, when she achieves perfection, the splitting of the bullet no longer constitutes a causal interaction but must now be regarded as a conjunctive fork. The essence of the interactive fork is to achieve a high correlation between two results; if the correlation is perfect, we can ask no more. It therefore seems best to treat this special case as a third type of fork—the *perfect fork*.

Conjunctive forks also yield perfect forks in the limit. Consider the example of illness resulting from consumption of poisonous mushrooms. If we assume—what is by no means always the case—that anyone who consumes a significant amount of the mushroom in question is certain to become violently ill, then we have another instance of a perfect fork. Even when these limiting values obtain, however, there is still no direct interaction between the processes leading respectively to the two cases of severe gastrointestinal distress.

The main point to be made concerning perfect forks is that, when the probabilities take on the limiting values, it is impossible to tell from the statistical relationships alone whether the fork should be considered interactive or conjunctive. The fact that relations (1)–(4), which are used in the characterization of conjunctive forks, are satisfied does not constitute a sufficient basis for making a judgment about the temporal orientation of the fork. Only if we can establish, on separate grounds, that the perfect fork is a limiting case of a conjunctive (rather than an interactive) fork can we conclude that the event at the vertex is a common cause rather than a common effect. Perfect forks need to be distinguished from the other two types mainly to guard against this possible source of confusion.

7. The Causal Structure of the World

In everyday life, when we talk about cause-effect relations, we think typically (though not necessarily invariably) of situations in which one event (which we call the cause) is linked to another event (which we call the effect) by means of a causal process. Each of the two events which stands in this relation is an interaction between two (or more) intersecting processes. We say, for example, that the window was broken by boys playing baseball. In this situation there is a collision of a bat with a ball (an interactive fork), the motion of the ball through space (a causal process), and a collision of the ball with the window (an interactive fork). For another example, we say that turning a switch makes the light go on. In this case an interaction between a switching mechanism and an electrical circuit leads to a process consisting of a motion of electric charges in some wires, which in turn leads to emission of light from a filament. Homicide by shooting provides still another example. An interaction between a gun and a cartridge propels a bullet (a causal process) from the gun to the victim, where the bullet then interacts with the body of the victim.

The foregoing characterization of causal processes and various kinds of causal forks provides, I believe, a basis for understanding three fundamental aspects of causality:

1. *Causal processes* are the means by which structure and order are *propagated* or transmitted from one spacetime region of the universe to other times and places.
2. *Causal interactions,* as explicated in terms of interactive forks, constitute the means by which *modifications in structure* (which are propagated by causal processes) are *produced.*
3. Conjunctive *common causes*—as characterized in terms of conjunctive forks—play a vital role in the *production* of structure and order. In the conjunctive fork, it will be recalled, two or more processes, which are physically independent of one another and which do not interact directly, arise out of some special set of background conditions. The fact that such special background conditions exist is the source of a correlation among the various effects which would be utterly improbable in the absence of the common causal background.

There is a striking difference between conjunctive common causes on the one hand and causal processes and interactions on the other. Causal processes and causal interactions seem to be governed by basic laws of nature in ways that do not apply to conjunctive forks. Consider two paradigms of causal processes, namely, an electromagnetic wave propagating through a vacuum and a material particle moving without any net external forces acting upon it. Barring any causal interactions in both cases, the electromagnetic wave is governed by Maxwell's equations, and the material particle is governed by Newton's first law of motion (or its counterpart in relativity theory). Causal interactions are typified by various sorts of collisions. The correlations between the changes that occur in the processes involved are governed—in most, if not all, cases—by fundamental physical conservation laws. Although I am not prepared to argue the case in detail, it seems plausible to suppose that *all fundamental physical interactions* can be regarded as exemplifications of the interactive fork.[12]

Conjunctive common causes are not nearly as closely tied to the laws of nature. It should hardly require mention that, to the extent that conjunctive forks involve causal processes and causal interactions, the laws of nature apply as sketched in the preceding paragraph. However, in contrast to causal processes and causal interactions, conjunctive forks depend crucially on de facto background conditions. Recall some of the examples mentioned earlier. In the plagiarism example, it is a non-lawful fact that two members of the same class happen to have access to the same file of term papers. In the mushroom poisoning example, it is a non-lawful fact that the two participants sup together out of a common pot. In the twin quasar example, it is a de facto condition that the quasar and the elliptic galaxy are situated in such a way that light coming to us from two different directions arises from a source which radiates quite uniformly from extended portions of its surface.

There is a close parallel between what has just been said about conjunctive forks and what philosophers such as Reichenbach (1956, chap. 3) and Grünbaum (1973, chap. 8) have said about entropy and the second law of thermodynamics. Consider the simplest sort of example. Suppose that we have a box with two compartments connected by a window which can be opened or closed. The box contains equal numbers of nitrogen (N_2) and oxygen (O_2) molecules. The window is open, and all of the N_2 molecules are in the left-hand compartment, while all of the O_2 molecules are in the right-hand compartment. Suppose that there are 2 molecules of each type. If they are distributed randomly, there is a probability of $2^{-4} = \frac{1}{16}$ that they would be segregated in just that way—a somewhat improbable coincidence.[13] If there are 5 molecules of each type, the chance of finding all of the N_2 molecules in the left compartment and all of the O_2 molecules in the right is a bit less than $\frac{1}{1000}$—fairly improbable. If the box contains 50 molecules of each type, the probability of the same sort of segregation would be $2^{-100} \simeq 10^{-30}$—extremely improbable. If the box contains Avogadro's number of molecules—forget it! In a case of this sort we would conclude without hesitation that the system had been prepared by closing the window that separates the two compartments, and by filling each compartment separately with its respective gas. The window must have been opened just prior to our examination of the box. What would be a hopelessly improbable coincidence if attributed to chance is explained straightforwardly on the supposition that separate supplies of each of the gases are available beforehand. The explanation depends on an antecedent state of the world that displays de facto orderliness.

Reichenbach generalized this point in his "hypothesis of the branch structure" (1956, §16). It articulates the manner in which new sorts of order arise from preexisting states of order. In the thermodynamic context, we say that low entropy states (highly ordered states) do not emerge spontaneously in isolated systems but rather are produced through the exploitation of the available energy in the immediate environment. Given the fundamentality and ubiquity of entropy considerations, the foregoing parallel suggests that the conjunctive fork also has basic physical significance. If we wonder about the original source of order in the world, which makes possible both the kind of order we find in systems in states of low entropy and the kind of order we get from conjunctive forks, we must ask the cosmologist how and why the universe evolved into a state which is characterized by vast supplies of available energy. It does not seem plausible to suppose that order can emerge except from de facto prior order.[14]

8. Concluding Remarks

There has been considerable controversy since Hume's time regarding the question whether causes must precede their effects, or whether causes and effects might be simultaneous with each other. It seems to me that the foregoing discussion provides a reasonable resolution of this controversy. If we are talking about the typical cause-effect situation, which I characterized in terms of a causal process joining two distinct interactions, then we are dealing with cases in which the cause must precede the effect, for causal propagation over a finite time interval is an essential feature of cases of this type. If, however, we are dealing simply with a causal interaction—an intersection of two or more processes that produces lasting changes in each of them—then we have simultaneity, since each process intersects the other at the same time. Thus, the intersection of the white light pulse with the red filter produces the red light, and the light becomes red at the very time of its passage through the filter. Basically, propagation involves lapse of time, while interaction exhibits the relation of simultaneity.

Another traditional dispute has centered on the question whether statements about causal relations pertain to individual events, or whether they hold properly only with respect to classes of events. Again, I believe, the foregoing account furnishes a straightforward answer. I have argued that causal processes, in many instances, constitute the causal connections between cause and effect. A causal process is an individual entity, and such entities transmit causal influence. An individual process can sustain a causal connection between an individual cause and an individual effect. Statements about such relations need not be construed as disguised generalizations. At the same time, it should be noted, we have used statistical relations to characterize conjunctive forks. Thus, strictly speaking, when we invoke something like the principle of the common cause, we are implicitly making assertions that involve statistical generalizations. Causal relations, it seems to me, have both particular and general aspects.

Throughout this discussion of causality, I have laid particular stress upon the role of causal processes, and I have even suggested the abandonment of the so-called event ontology. It might be asked whether it would not be possible to carry through the same analysis, within the framework of an event ontology, by considering processes as continuous series of events. I see no reason for supposing that this program could not be carried through, but I would be inclined to ask why we should bother to do so. One important source of difficulty for Hume, if I understand him, is that he tried to account for causal connections between noncontiguous events by interpolating intervening events. This approach seemed only to raise precisely the same questions about causal connections between events, for one had to ask how the causal influence is transmitted from one intervening event to another along the chain. The difficulty is circumvented if we look to processes to provide the causal connections (see "An 'At-At' Theory of Causal Influence" [essay 12]). Focusing on processes rather than events has, in my opinion, enormous heuristic (if not systematic) value. As John Venn said in 1866, "Substitute for the time honoured 'chain of causation', so often introduced into discussions upon this subject, the phrase a 'rope of causation', and see what a very different aspect the question will wear" (p. 320).

Notes

This material is based on work supported by the National Science Foundation under Grant no. SES-7809146.

1. See Mackie (1974) for an excellent historical and systematic survey of the various approaches.

2. See "Probabilistic Causality" (essay 14) for a survey of statistical approaches.

3. In "An 'At-At' Theory of Causal Influence" (essay 12) I have attempted to provide a detailed analysis of the notion of transmission or propagation of causal influence by causal processes, and a justification for the claim that they legitimately qualify as causal connections.

4. The variables A, B, C which appear in the probability expressions are taken by Reichenbach to denote classes, and the probabilities themselves are understood as statistical frequencies.

5. This is demonstrated by Reichenbach (1956, pp. 160–161).

6. If other potential common causes exist, we can form a partition $C_1, C_2, C_3 \ldots$ and the corresponding relations will obtain.

7. We are assuming that the magnet in one die does not affect the behavior of the other die.

8. I had previously attributed this erroneous view to Reichenbach, but Paul Humphreys kindly pointed out that my attribution was incorrect.

9. For a valuable discussion of the role of energy and momentum transfer in causality, see Fair (1979).

10. As explained in "Why Ask, 'Why?'?" (essay 8), the example of Compton scattering has the advantage of being irreducibly statistical, and thus not analyzable, even in principle, as a perfect fork (discussed later in this essay).

11. This relation is not, however, any part of the definition of *interactive fork* or *causal interaction*.

12. [This supposition is plausible only if Y and λ interactions are included; see "A New Look at Causality" (essay 1).]

13. Strictly, each of the probabilities mentioned in this example should be doubled, for a distribution consisting of all O_2 in the left and all N_2 in the right would be just as remarkable a form of segregation as that considered in the text. However, it is obvious that a factor of 2 makes no real difference to the example.

14. [See, however, Prigogine (1984).]

19

Scientific Explanation

How We Got from There to Here

Is there a *new* consensus in philosophy of science? That is the question we must discuss. But even to pose the question in this way implies that there was an old consensus—and indeed there was, at least with respect to scientific explanation. It is with the old consensus that we should start. In order to understand the present situation we need to see how we got from there to here.[1]

I recall with some amusement a personal experience that occurred in the early 1960s. J. J. C. Smart, a distinguished Australian philosopher, visited Indiana University, where I was teaching at the time. Somehow we got into a conversation about the major unsolved problems in philosophy of science, and he mentioned the problem of scientific explanation. I was utterly astonished—literally, too astonished for words. I considered *that* problem essentially solved by the deductive-nomological (D-N) account that had been promulgated by R. B. Braithwaite (1953), Carl G. Hempel (Hempel and Oppenheim, [1948] 1965), Ernest Nagel (1961), and Karl Popper (1935, 1959), among many others— supplemented, perhaps, by Hempel's then recent account of statistical explanation (1962a). Although this general view had a few rather vocal critics, such as N. R. Hanson (1959) and Michael Scriven (1958, 1959, 1962), it was widely accepted by scientifically minded philosophers; indeed, it qualified handily as *the* received view. What is now amusing about the incident is my naïveté in thinking that a major philosophical problem had actually been solved.

1. The Received View

The cornerstone of the old consensus was the *deductive-nomological (D-N) model* of scientific explanation. The fullest and most precise early characterization of this model was given in the classic article "Studies in the Logic of Explanation" (Hempel and

Oppenheim, [1948] 1965). According to that account, a D-N explanation of a particular event is a valid deductive argument whose conclusion states that the event-to-be-explained did occur. Its premises must include essentially at least one general law. The explanation is said to subsume the fact to be explained under these laws. Hence, it is often called the *covering law model*. On the surface this account is beautiful for its clarity and simplicity, but as we shall see in §2, it contains a number of serious hidden difficulties.

Consider one of Hempel's familiar examples. Suppose someone asks why the flame of a particular Bunsen burner turned yellow at a particular moment. This why-question is a request for a scientific explanation. The answer is that a piece of rock salt was placed in the flame, rock salt is a sodium compound, and Bunsen flames always turn yellow when sodium compounds are introduced. The explanation can be laid out formally as follows:

(1)   All Bunsen flames turn yellow when sodium compounds are placed in them.
      All rock salt consists of a sodium compound.
      A piece of rock salt was placed in this Bunsen flame at a particular time.
      _____
      This Bunsen flame turned yellow at that time.

This explanation is a valid deductive argument with three premises. The first two premises are statements of natural law; the third premise formulates an initial condition in this explanation. The premises constitute the *explanans*—that which does the explaining. The conclusion is the *explanandum*—that which is explained.

From the beginning, however, Hempel and Oppenheim ([1948] 1965, pp. 250–251) recognized that not all scientific explanations are of the D-N variety. Some are probabilistic or statistical. In "Deductive-Nomological vs. Statistical Explanation" (1962a) Hempel offered his first treatment of statistical explanation, and in "Aspects of Scientific Explanation" (1965b) he provided an improved account. This theory includes two types of statistical explanation. The first of these, the inductive-statistical (I-S), explains particular occurrences by subsuming them under statistical laws, much as D-N explanations subsume particular events under universal laws. To cite another of Hempel's famous examples, if we ask why John Jones recovered rapidly from his streptococcus infection, the answer is that he was given a dose of penicillin, and almost all strep infections clear up quickly upon administration of penicillin. More formally,

(2)   Almost all cases of streptococcus infection clear up quickly after the administration of penicillin.
      Jones had a streptococcus infection.
      Jones received treatment with penicillin.
      === [r]
      Jones recovered quickly.

This explanation is an argument that has three premises (the explanans); the first premise states a statistical regularity—a statistical law—while the other two state initial conditions. The conclusion (the explanandum) states the fact to be explained. There is, however, a crucial difference between explanations (1) and (2): D-N explanations subsume the events to be explained deductively, while I-S explanations subsume them inductively. The single line separating the premises from the conclusion in (1) signifies a relation of deductive entailment between the premises and conclusion. The double line in (2) represents a relationship of inductive support, and the attached variable *r* stands for the strength of that support. This strength of support may be expressed exactly, as a numerical value of a probability, or vaguely, by means of phrases such as "very probably" or "almost certainly."

An explanation of either of these two kinds can be described as an argument to the effect that *the event to be explained was to be expected by virtue of certain explanatory facts.* In a D-N explanation the event to be explained is deductively certain, given the explanatory facts; in an I-S explanation the event to be explained has high inductive probability relative to the explanatory facts. This feature of expectability is closely related to the *explanation/prediction symmetry thesis* for explanations of particular facts. According to this thesis—which was advanced for D-N explanation in Hempel-Oppenheim ([1948] 1965, p. 249), and reiterated, with some qualifications, for D-N and I-S explanations in Hempel (1965b, §2.4, §3.5)—any acceptable explanation of a particular fact is an argument, deductive or inductive, that could have been used to predict the fact in question if the facts stated in the explanans had been available prior to its occurrence. As we shall see in §2, this symmetry thesis met with serious opposition.

Hempel was not by any means the only philosopher in the early 1960s to notice that statistical explanations play a highly significant role in modern science. He was, however, the first to present a detailed account of the nature of statistical explanation, and the first to bring out a fundamental problem concerning statistical explanations of particular facts. The case of Jones and the quick recovery can be used as an illustration. It is well known that certain strains of the streptococcus bacterium are penicillin-resistant, and if Jones's infection is of that type, the probability of the quick recovery after treatment with penicillin would be very small. We could, in fact, set up the following inductive argument:

(2')   Almost no cases of penicillin-resistant streptococcus infection clear up quickly after the administration of penicillin.
Jones had a penicillin-resistant streptococcus infection.
Jones received treatment with penicillin.
$$====================================== [q]$$
Jones did not recover quickly.

The remarkable fact about arguments (2) and (2') is that their premises are mutually compatible—they could all be true. Nevertheless, their conclusions contradict each other. This is a situation that can never occur with deductive arguments. Given two valid deductions with incompatible conclusions, their premises must also be incompatible. Thus, the problem that has arisen in connection with I-S explanations has no analogue in D-N explanations. Hempel called this *the problem of ambiguity of I-S explanation,* and he sought to resolve it by means of his *requirement of maximal specificity (RMS).*

The source of the problem of ambiguity is a simple and fundamental difference between universal laws and statistical laws. Given the proposition that all *A* are *B*, it follows immediately that all things that are both *A* and *C* are *B*. If all men are mortal, then all men who are over six feet tall are mortal. However, if almost all men who are alive now will be alive five years from now, *it does not follow* that almost all living men with advanced cases of lung cancer will be alive five years hence. There is a parallel fact about arguments. Given a valid deductive argument, the argument will remain valid if additional premises are supplied, as long as none of the original premises is taken away. Given a strong inductive argument—one that supports its conclusion with a very high degree of probability—the addition of one more premise may undermine it completely. Europeans, for example, had for many centuries a great body of inductive evidence to support the proposition that all swans are white, but one true report of a black swan in Australia completely refuted that conclusion.

There is a well-known strategy for dealing with the problem of ambiguity as it applies to inductive arguments per se: it is to impose the *requirement of total evidence*. According to this requirement, one should not rely on the conclusion of an inductive argument—for purposes of making predictions or wagers, for example—unless that argument includes among its premises all available relevant evidence. This approach is entirely unsuitable for the context of scientific explanation because normally, when we seek an explanation for some fact, we already know that it obtains. Thus, knowledge of the fact-to-be-explained is part of our body of available knowledge. We ask why Jones recovered quickly from the strep infection only after we know that the quick recovery occurred. But if we include in the explanans the statement that the quick recovery occurred, the resulting 'explanation'

(2″)   Almost all cases of streptococcus infection clear up quickly after the administration of penicillin.
Jones had a streptococcus infection.
Jones received treatment with penicillin.
Jones recovered quickly.

Jones recovered quickly.

is trivial and uninteresting. Although the conclusion follows deductively from the augmented set of premises, (2″) does not even qualify as a D-N explanation, for no law is essential for the derivation of the conclusion from the new set of premises. We could eliminate the first three premises and the resulting argument would still be valid.

Hempel (1965b, §3.4) was clearly aware of all of these considerations, and he designed his requirement of maximal specificity (RMS) to circumvent them. The purpose of this requirement is to ensure that all relevant information *of an appropriate sort* is included in any given I-S explanation. Although it is extremely tricky to say just what constitutes *appropriate information,* one could say, very roughly, that it is information that is in principle available prior to the occurrence of the event-to-be-explained.[2] Suppose that we have a putative explanation of the fact that some entity x has the property B. Suppose that this explanation appeals to a statistical law of the form "The probability that an A is a B is equal to r." Suppose, in addition, that we know that this particular x also belongs to a class C that is a subset of A. Then, if the explanation is to satisfy RMS, our body of knowledge must include the knowledge that the probability of a C being a B is equal to q, and q must be equal to r, unless the class C is *logically related* to the property B (or the class of things having the property B) in a certain way. That is, q need not equal r if the statement that the probability of a C being a B is equal to q is a theorem of the mathematical calculus of probability.

In order to clarify this rather complicated requirement, let us refer again to the example of Jones. Consider three separate cases:

(a) Suppose that we know, in addition to the facts stated in the premises of (2), that Jones's penicillin treatment began on Thursday. According to RMS that would have no bearing on the legitimacy of (2) as an explanation, for the day of the week on which the treatment is initiated has no bearing on the efficacy of the treatment. The probability of a rapid recovery after penicillin treatment that began on a Thursday is equal to the probability of rapid recovery after treatment by penicillin (regardless of the day on which it began).

(b) Suppose we were to offer argument (2) as an explanation of Jones's rapid recovery, knowing that the infection was of the penicillin-resistant type. Since we know that the probability of quick recovery from a penicillin-resistant strep infection after treatment by penicillin *is not equal to* the probability of rapid recovery from an unspecified type of strep infection after treatment with penicillin, the explanation would not be legitimate. RMS would outlaw argument (2) as an I-S explanation if the highly relevant information about the penicillin-resistant character of the infection were available.

(c) When asking for an explanation of Jones's quick recovery, we already know that Jones belongs to the class of people with strep infections. Moreover, we know that Jones belongs to the subclass of people with strep infections cured quickly by penicillin, and we know that the probability of anyone in *that* class having a quick recovery is equal to one. This knowledge does not rule out (2) as an explanation. The reason is the "unless" clause of RMS. It is a trivial consequence of mathematical probability theory that the probability of quick recovery among those who experience quick recovery is one. If Y is a proper or improper subclass of X, then the probability of Y, given X, is necessarily equal to one.[3]

Having recognized the problem of ambiguity of I-S explanation, Hempel introduced the requirement of maximal specificity. As can easily be seen, RMS makes explicit reference to our state of knowledge. Whether a given argument qualifies as an I-S explanation depends not only on the objective facts in the world but also on what knowledge the explainer happens to possess. This result led Hempel to enunciate the principle of *essential epistemic relativity of I-S explanation.* D-N explanation, in contrast, does not suffer from any such epistemic relativity. If the premises of argument (1) are true, argument (1) qualifies as a correct D-N explanation of the fact that the Bunsen flame turned yellow. The fact that it is a correct explanation does not depend in any way on our knowledge situation. Of course, whether we *think* that it is a correct explanation will surely depend on our state of knowledge. What is *considered* a correct D-N explanation at one time may be *judged* incorrect at another time because our body of knowledge changes in the meantime. But the objective correctness of the explanation does not change accordingly. By contrast, argument (2) may have true premises and correct inductive logical form, but those features do not guarantee that it is a correct I-S explanation. Relative to one knowledge situation it is legitimate; relative to another it is not. As we shall see in §2, the requirement of maximal specificity and the doctrine of essential epistemic relativization became sources of fundamental difficulty for the received view of explanations of particular facts.

On Hempel's theory it is possible to explain not only particular events but also general regularities. Within the D-N model universal generalizations are explained by deduction from more comprehensive universal generalizations. For example, the law of conservation of linear momentum can be deduced—with the aid of a little mathematics—from Newton's second and third laws of motion. In classical mechanics, consequently, the following argument constitutes an explanation of the law of conservation of linear momentum:

(3)   Newton's second law: $F = ma$.
Newton's third law: *For every action there is an equal and opposite reaction.*

Law of conservation of linear momentum: *In every physical interaction, linear momentum is conserved.*

Notice that this explanans contains only statements of law; inasmuch as no particular occurrence is being explained, no statements of particular initial conditions are required.

In the second type of statistical explanation, the deductive-statistical (D-S), statistical regularities are explained by deduction from more comprehensive statistical laws. A famous example comes from the birth of the mathematical theory of probability. A seventeenth-century gentleman, the Chevalier de Méré, wondered whether, in 24 tosses of a standard pair of dice, one has a better than fifty-fifty chance of getting double 6 ("boxcars") at least once, or whether 25 tosses are needed. He posed the question to Pascal, who proved that 25 is the correct answer. His derivation can be viewed as an explanation of this somewhat surprising fact. It can be set out as follows:

(4)   A standard die is a physically homogeneous cube whose six faces are marked with the numbers 1–6.
When a standard die is tossed in the standard manner, each side has an equal probability—namely, one sixth—of ending uppermost.
When two standard dice are tossed in the standard manner, the outcome on each die is independent of the outcome on the other.
When two standard dice are tossed repeatedly in the standard manner, the result on any given throw is independent of the results on the preceding tosses.

Twenty-five is the smallest number of standard tosses of a standard pair of dice for which the probability of double 6 occurring at least once is greater than one-half.

A small amount of arithmetic is needed to show that the conclusion of this argument follows deductively from the premises.[4] The first premise is a definition; the three remaining premises are statistical generalizations. The conclusion is also a statistical generalization.

Figure 19.1 shows the four categories of scientific explanations recognized in Hempel (1965b). However, in their explication of D-N explanation in 1948, Hempel and Oppenheim restrict their attention to explanations of particular facts, and do not attempt to provide any explication of explanations of general regularities. The reason for this

Explanada / Laws	Particular Facts	General Regularities
Universal Laws	D - N Deductive - Nomological	D - N Deductive - Nomological
Statistical Laws	I - S Inductive - Statistical	D - S Deductive - Statistical

Hempelian Models of Explanation

Figure 19.1

restriction is given in the notorious footnote 33 (Hempel and Oppenheim, [1948] 1965, p. 273):

> The precise rational reconstruction of explanation as applied to general regularities presents peculiar problems for which we can offer no solution at present. The core of the difficulty can be indicated by reference to an example: Kepler's laws, K, may be conjoined with Boyle's law, B, to [form] a stronger law $K.B$; but derivation of K from the latter would not be considered an explanation of the regularities stated in Kepler's laws; rather, it would be viewed as representing, in effect, a pointless "explanation" of Kepler's laws by themselves. The derivation of Kepler's laws from Newton's laws of motion and gravitation, on the other hand, would be recognized as a genuine explanation in terms of more comprehensive regularities, or so-called higher-level laws. The problem therefore arises of setting up clear-cut criteria for the distinction of levels of explanation or for a comparison of generalized sentences as to their comprehensiveness. The establishment of adequate criteria for this purpose is as yet an open problem.

This problem is not resolved in any of Hempel's subsequent writings, including "Aspects of Scientific Explanation." It was addressed by Michael Friedman (1974); I shall discuss his seminal article in §4. Since the same problem obviously applies to D-S explanations, it affects both sectors in the right-hand column of figure 19.1. The claim of the received view to a comprehensive theory of scientific explanation thus carries a large promissory note regarding explanations of laws.

The Hempel-Oppenheim ([1948] 1965) article marks the division between the prehistory and the history of the modern discussions of scientific explanation. Although Aristotle, John Stuart Mill (1843), and Karl Popper (1935), among many others, had previously expressed similar views about the nature of deductive explanation, the Hempel-Oppenheim essay spells out the D-N model with far greater precision and clarity. Hempel's 1965 "Aspects" article is *the* central document in the hegemony (with respect to scientific explanation) of logical empiricism. I shall refer to the account given there as *the received view*. According to the received view, *every legitimate scientific explanation* fits into one of the four compartments in figure 19.1.

2. Attacks on the Received View

The hegemony of logical empiricism regarding scientific explanation did not endure for very long. Assaults came from many directions; most of them can be presented in terms of old familiar counterexamples. Some of the counterexamples are cases that satisfy all of the criteria set forth in the received view but which clearly are not admissible as scientific explanations. These are designed to show that the requirements imposed by the received view are not *sufficient* to determine what constitutes a correct scientific explanation. Other counterexamples are cases that appear intuitively to be satisfactory scientific explanations but that fail to satisfy the criteria of the received view. They are designed to show that these requirements are not *necessary* either.

(1) One of the best known is Sylvain Bromberger's flagpole example.[5] A certain flagpole casts a shadow of a certain length at some particular time. Given the height of the flagpole, its opacity, the elevation of the sun in the sky, and the rectilinear propagation of light, it is possible to deduce the length of the shadow and, ipso facto, to provide a D-N explanation of its length. There is no puzzle about this. But given the length of the shadow, the position and opacity of the flagpole, the elevation of the sun, and the rectilinear propagation of light, we can deduce the height of the flagpole. Yet hardly anyone would allow that the length of the shadow explains the height of the flagpole.[6]

(2) It has often been noted that, given a sudden drop in the reading of a barometer, we can reliably infer the occurrence of a storm. It does not follow that the barometric reading explains the storm; rather, a drop in atmospheric pressure explains both the barometric reading and the storm.[7]

Examples (1) and (2) show something important about causality and explanation. The first shows that we explain effects in terms of their causes; we do not explain causes in terms of their effects. See "Explanatory Asymmetry" (essay 10) for a deeper analysis of this problem. The second shows that we do not explain one effect of a common cause in terms of another effect of that same cause. Our common sense has told us for a long time that to explain an event is, in many cases, to find and identify its cause. One important weakness of the received account is its failure to make explicit reference to causality— indeed, Hempel has explicitly denied that explanations must always involve causes (1965b, pp. 352–353).

(3) Many years ago Scriven (1959) noticed that we can explain the occurrence of paresis in terms of the fact that the patient had latent syphilis untreated by penicillin. However, given someone with latent untreated syphilis, the chance that he or she will develop paresis is about one fourth, and there is no known way to separate those who will develop paresis from those who won't.

(4) My favorite example is the case of the man who regularly takes his wife's birth control pills for an entire year, and who explains the fact that he did not become pregnant during the year on the basis of his consumption of oral contraceptives (Salmon, 1971, p. 34).

Examples (3) and (4) have to do with expectability, and consequently with the explanation/prediction symmetry thesis. Scriven (1959) had offered example (3) in order to show that we can have explanations of events that are improbable, and hence are not to be expected; indeed, he argued that evolutionary biology is a science containing many explanations but virtually no predictions. Example (4) shows that an argument that fully qualifies as a D-N explanation, and consequently provides expectability, can fail to be a bona fide explanation. Peter Railton has pointed out that Hempel's view can be characterized in terms of the *nomic expectability* of the event to be explained. He argues—quite correctly, I believe—that nomicity may well be a bona fide requirement for scientific explanation, but that expectability cannot be demanded.

My own particular break with the received doctrine occurred in 1963, very shortly after the aforementioned conversation with Smart. At the 1963 meeting of the AAAS, I argued that Hempel's I-S model, with its high probability requirement and its demand for expectability, is fundamentally mistaken.[8] Statistical relevance instead of high probability, I argued, is the key concept in statistical explanation.

In support of this contention I offered the following example (which, because of serious questions about the efficacy of psychotherapy, happens to have some medical importance). Suppose that Jones, instead of being afflicted with a strep infection, has a troublesome neurotic symptom. Under psychotherapy this symptom disappears. Can we explain the recovery in terms of the treatment? We could set out the following inductive argument, in analogy with argument (2):

(5)   Most people who have a neurotic symptom of type N and who undergo psychotherapy experience relief from that symptom.
Jones had a symptom of type N and underwent psychotherapy.
$$\overline{\qquad\qquad\qquad\qquad\qquad\qquad\qquad\qquad\qquad\qquad\qquad}\ [r]$$
Jones experienced relief from this symptom.

Before attempting to evaluate this proffered explanation, we should take account of the fact that there is a fairly high spontaneous remission rate—that is, many people who suffer from that sort of symptom get better regardless of treatment. No matter how large the number r, if the rate of recovery for people who undergo psychotherapy is no larger than the spontaneous remission rate, it would be a mistake to consider argument (5) a legitimate explanation. A high probability is not sufficient for a correct explanation. If, however, the number r is not very large, but is greater than the spontaneous remission rate, the fact that the patient underwent psychotherapy has at least some degree of explanatory force. A high probability is not necessary for a sound explanation.[9]

Examples (3) and (4) both pertain to the issue of relevance. In example (3) we have a factor (syphilis not treated with penicillin) that is highly relevant to the explanandum (contracting paresis) even though no high probabilities are involved. This example exhibits the explanatory force of relevance. In example (4) we have an obviously defective 'explanation' because of the patent irrelevance of the consumption of birth control pills to the non-pregnancy of a man. Furthermore, in my example of psychotherapy and relief from a neurotic symptom, the issue of whether the explanation is legitimate or not hinges on the question whether the psychotherapy was, indeed, relevant to the disappearance of the symptom. Henry Kyburg, who commented on my AAAS paper, pointed out—through an example similar in principle to example (4)—that the same sort of criticism could be leveled against the D-N model. It, too, needs to be guarded against the introduction of irrelevancies into putative explanations.

While my initial criticism of the received view centered on issues of high probability versus relevancy, other philosophers attacked the requirement of maximal specificity and the associated doctrine of essential epistemic relativization of I-S explanation. On this front the sharpest critic was J. Alberto Coffa (1974), who challenged the very intelligibility of an epistemically relativized notion of inductive explanation. His argument ran roughly as follows. Suppose someone offers a D-N explanation of some particular fact, such as we had in argument (1) above. If the premises are true and the logical form correct, then (1) is a *true* D-N explanation. In our present epistemic state we may not know for sure that (1) is a true explanation; for instance, I might not be sure that placing a sodium compound in a Bunsen flame always turns the flame yellow. Given this uncertainty, I might consult chemical textbooks, ask chemists of my acquaintance, or actually perform experiments to satisfy myself that the first premise of (1) is true. If I have doubts about any other premises, there are steps I could take to satisfy myself that they are true. In the end, although I cannot claim to be *absolutely certain* of the truth of the premises of (1), I can conclude that they are well confirmed. If I am equally confident of the logical correctness of the argument, I can then claim to have good reasons to believe that (1) is a true D-N explanation. Crucial to this conclusion is the fact that I know what sort of thing a true D-N explanation is.

The situation regarding D-N explanations is analogous to that for more commonplace entities. Suppose I see a bird in a bush but I am not sure what kind of bird it is. If I approach it more closely, listen to its song, look at it through binoculars, and perhaps ask an ornithologist, I can establish that it is a hermit thrush. I can have good reason to believe that it is a hermit thrush. It is a well-confirmed hermit thrush. But all of this makes sense only because we have objective non-epistemically relativized criteria for what an actual hermit thrush is. Without that, the concept of a well-confirmed hermit thrush would make no sense because there would be literally nothing we could have good reason to believe we are seeing.

When we turn to I-S explanations, a serious complication develops. If we ask, prior to an inquiry about a particular I-S explanation, what sort of thing constitutes a true I-S explanation, Hempel must reply that he does not know. All he can tell us about are epistemically relativized I-S explanations. He can tell us the criteria for determining that an I-S explanation is acceptable in a given knowledge situation. It appears that he is telling us the grounds for justifiably believing, in such a knowledge situation, that we have a genuine I-S explanation. But what can this mean? Since, according to Hempel's 1965 view, there is no such thing as a bona fide I-S explanation, unrelativized to any knowledge situation, what is it that we have good reason to believe that we have? We can have good reason to believe that we have a true D-N explanation because we know what sort of thing a true D-N explanation is. The main burden of Hempel and Oppenheim, ([1948] 1965) was to spell out just that. According to Hempel, it is impossible in principle to spell out any such thing for I-S explanations.

On the basis of a careful analysis of Hempel's doctrine of essential epistemic relativity, it is possible to conclude that Hempel has offered us not an independent stand-alone conception of inductive explanation of particular facts, but rather a conception of inductive explanation that is completely parasitic on D-N explanation. One is strongly tempted to draw the conclusion that an I-S explanation is essentially an enthymeme—an incomplete deductive argument. Faced with an enthymeme, we may try to improve it by supplying missing premises, and in so doing we may be more or less successful. But the moment we achieve complete success by supplying all of the missing premises, we no longer have an enthymeme—instead we have a valid deductive argument. Similarly, it appears, given an epistemically relativized I-S explanation, we may try to improve our epistemic situation by increasing our body of knowledge. With more knowledge we may be able to furnish more complete explanations. But when we finally succeed in accumulating all of the relevant knowledge and incorporating it into our explanation, we will find that we no longer have an inductive explanation—instead we have a D-N explanation.

A doctrine of inductive explanations that construes them as incomplete deductive explanations seems strongly to suggest determinism. According to the determinist every fact of nature is amenable, in principle, to complete deductive explanation. We make do with inductive explanations only because of the incompleteness of our knowledge. We appeal to probabilities only as a reflection of our ignorance. An ideal intelligence, such as Laplace's famous demon, would have no use for probabilities or inductive explanations (see Salmon, 1974a). Although Hempel has explicitly denied any commitment to determinism, his theory of I-S explanation fits all too neatly into the determinist's scheme of things. Eventually Hempel (1977) retracted his doctrine of *essential* epistemic relativization.

Careful consideration of the various difficulties in Hempel's I-S model led to the development of the statistical-relevance (S-R) model. Described concisely, an S-R explanation is an assemblage of all and only those factors relevant to the fact-to-be-explained. For instance, to explain why Albert, an American teenager, committed an act of delinquency, we cite such relevant factors as his sex, the socioeconomic status of his family, his religious background, his place of residence (urban versus suburban or rural), ethnic background, etc. (see Greeno, [1970] 1971). It would clearly be a mistake to mention such factors as the day of the week on which he was born or whether his social security number is odd or even, for they are statistically irrelevant to the commission of delinquent acts.

It should be pointed out emphatically that an assemblage of relevant factors—along with an appropriate set of probability values—is not an argument of any sort, deductive or inductive. Acceptance of the S-R model thus requires abandonment of what I called the third dogma of empiricism, namely, the general thesis that every bona fide scientific explanation is an argument (see essay 6). It was Richard Jeffrey ([1969] 1971) who first explicitly challenged that dogma.

The S-R model could not long endure as an independent conception of scientific explanation, for it embodied only statistical correlations, without appeal to causal relations. Reacting to Hempel's I-S model, I thought that statistical relevance, rather than high inductive probability, has genuine explanatory import. I no longer think so. Statistical-relevance relations are important to scientific explanation for a different reason, namely, because they constitute important evidence of causal relations. Causality, rather than statistical relevance, is what has explanatory import.

It may seem strange that the received view excised causal conceptions from its characterization of scientific explanation. Have we not known since Aristotle that explanations involve causes? It would be reasonable to think so. But putting the "cause" back into "because" is no simple matter, for Hume's searching analysis strongly suggested that to embrace physical causality might involve a rejection of empiricism. Those philosophers who have strongly insisted on the causal character of explanation—e.g., Scriven—have simply evaded Hume's critique. My own view is that the "cause" cannot be put back into "because" without a serious analysis of causality. The essays in part III of this book offer some suggestions as to how such an analysis might go (see also Salmon, 1984b, chaps. 5–7).

One of the major motivations for the received view was, I believe, the hope that scientific explanations could be characterized in a completely formal manner. (Note that the title of the Hempel-Oppenheim article is "Studies in the *Logic* of Explanation.") This makes it natural to think of explanations as arguments, for, as Carnap showed in his major treatise on probability (1950), both deductive logic and inductive logic can be explicated within a single semantic framework. Hempel and Oppenheim ([1948] 1965) offer a semantic analysis of lawlike statements.[10] This makes it possible to characterize a *potential explanation* as an argument of correct form containing at least one lawlike statement among its premises. A true explanation fulfills in addition the empirical condition that its premises and conclusion be true. A correct I-S explanation must satisfy still another condition, namely, Hempel's requirement of maximal specificity. This relevance requirement is also formulated in logical terms.

The upshot for the received view is that there are two models (three if D-S is kept separate from D-N) of scientific explanation, and that every legitimate explanation conforms to one or the other. Accordingly, any phenomenon in our universe, even in domains in which we do not yet have any scientific knowledge, must be either amenable to explanation by one of these models or else not susceptible to any sort of scientific explanation. The same would hold, it seems, for scientific explanations in any possible world.

Such universalistic ambitions strike me as misplaced. In our world, for example, we impose the demand that events be explained by their temporal antecedents, not by events that come later. But the structure of time itself is closely connected with entropic processes in our universe, and these depend on de facto conditions in our universe. In another universe the situation might be quite different—for example, time might be symmetric rather than asymmetric. In the macrocosm of our world, causal influence is apparently propagated continuously; action-at-a-distance does not seem to occur. In the microcosm of our world, what Einstein called "spooky action-at-a-distance" seems to occur. What counts as acceptable scientific explanation depends crucially on the causal and temporal structure of the world, and these are matters of fact rather than matters of logic. The moral I would draw is just this: we should not hope for formal models of scientific explanation that are universally applicable. We do best by looking at explanations in various domains of science, and by attempting adequately to characterize their structures. If it turns out—as I think it does—that very broad stretches of science employ common explanatory structures, that is an extremely interesting fact about our world.

3. The Pragmatics of Explanation

From the beginning Hempel and the other proponents of the received view recognized the obvious fact that scientific monographs, textbooks, articles, lectures, and conversation do not present scientific explanations that conform precisely to their models. They also realized that to do so would be otiose. Therefore, in the writing and speech of scientists we find partial explanations, explanation sketches, and elliptically formulated explanations. What sort of presentation is suitable is determined by factors such as the knowledge and interests of those who do the explaining and of their audiences. These are pragmatic factors.

Hempel devoted two sections of "Aspects" (1965b, §4–5) to the pragmatics of explanation, but the discussion was rather narrow. In 1965 (and a fortiori in 1948) formal pragmatics was not well developed, especially in those aspects that bear on explanation. Bromberger's path-breaking article "Why-Questions" appeared in 1966, but it dealt only with D-N explanation; the most prominent subsequent treatment of the pragmatics of explanation was provided in van Fraassen's *Scientific Image* (1980). A rather different pragmatic approach can be found in Peter Achinstein's *Nature of Explanation* (1983).

Van Fraassen adopts a straightforward conception of explanations (scientific and other) as answers to why-questions. Why-questions are posed in various contexts, and they have presuppositions. If the presuppositions are not fulfilled, the question does not arise; in such cases the question should be rejected rather than answered. If the question does arise, then the context heavily determines what constitutes an appropriate answer. Now, van Fraassen is not offering an 'anything goes as long as it satisfies the questioner' view of explanation, for there are objective criteria for the evaluation of answers. But there is a deep problem.

Van Fraassen characterizes a why-question as an ordered triple $\langle P_k, X, R \rangle$. P_k is the topic (what Hempel and most others call the "explanandum"). X is the contrast class, a set of alternatives with respect to which P_k is to be explained. In the Bunsen flame example, the contrast class might be:

$$P_1 = \text{the flame turned orange;}$$
$$P_2 = \text{the flame turned green;}$$

$$\cdot$$
$$\cdot$$
$$\cdot$$

$$P_k = \text{the flamed turned yellow;}$$

$$\cdot$$
$$\cdot$$
$$\cdot$$

$$P_s = \text{the flame did not change color.}$$

A satisfactory answer to the why-question is an explanation of the fact that P_k rather than any other member of the contrast class is true. R is the relevance relation; it relates the answer to the topic and contrast class. In the Bunsen flame example we can construe R as a causal relation; putting the rock salt into the Bunsen flame is what causes it to turn yellow. The problem is that van Fraassen places no restrictions on what sort of relation R may be. That is presumably freely chosen by the questioner. An answer A is relevant if A bears relation R to the topic P_k. In "Van Fraassen on Explanation" (essay 11) Philip Kitcher and I have shown that, without some restrictions on the relation R, any answer A can be the explanation of any topic P_k. Thus, van Fraassen needs to provide a list of types of relations that qualify as bona fide relevance relations. This is precisely the problem that philosophers who have not emphasized pragmatic aspects of explanation have been concerned to resolve. Even acknowledging this serious problem, van Fraassen and others have clearly demonstrated the importance of pragmatic features of explanation; what they have not shown is that pragmatics is the whole story.

One of the most important works on scientific explanation since Hempel's "Aspects" is Peter Railton's doctoral dissertation, "Explaining Explanation" (1980). In this work he introduces a valuable pair of concepts: *ideal explanatory texts* and *explanatory information*.[11] An ideal explanatory text for any fact to be explained is an extremely extensive and detailed account of everything that contributed to that fact—everything that is causally or lawfully relevant to it. Such texts are ideal entities; they are virtually never written out in full. To understand the fact being explained, we do not have to have the whole ideal text; what is required is that we be able to fill in the needed parts of it. Explanatory information is any information that illuminates any portion of the ideal text. Once we are clear on just what it is that we are trying to explain, the ideal explanatory text is fully objective; its correctness is determined by the objective causal and nomic features of the world. It has no pragmatic dimensions.

Pragmatic considerations arise when we decide which portions of the ideal explanatory text are to be illuminated in any given situation. This is the contextual aspect. When a why-question is posed, various aspects of the context—including the interests, knowledge, and training of the questioner—determine what explanatory information is salient. The resulting explanation must reflect objective relevance relations, but it must also honor the salience of the information it includes. Seen in this perspective, van Fraassen's account of the pragmatics of explanation fits admirably into the overall picture by furnishing guidelines that determine what sort of explanatory information is appropriate in the context.

4. The Moral of the Story

What have we learned from all of this? Several lessons, I believe. First, we must put the "cause" back into "because." Even if some types of explanation turn out not to be causal, many explanations do appeal essentially to causes. We must build into our theory of explanation the condition that causes can explain effects but effects do not explain causes. By the same token, we must take account of temporal asymmetries; we can explain later events in terms of earlier events, but not vice versa. Temporal asymmetry is closely related to causal asymmetry (see "Explanatory Asymmetry" [essay 10]).

Second, the high probability or expectedness requirement of the received view is not acceptable. High probability is neither necessary nor sufficient for scientific explanation, as examples (3) and (4) respectively show.

Third, we can dispense—as Hempel himself did (1977)—with his doctrine of *essential epistemic relativity* of I-S explanation. Many authors have found this aspect of the received view unpalatable. Coffa (1974), Fetzer (1974b), and Railton (1978) employ a propensity conception of probability to characterize types of statistical explanation that are not epistemically relativized. In my (1984b, chap. 3) I try to avoid such relativization by means of objectively homogeneous reference classes. Railton's notion of the ideal explanatory text provides the basis for a fully objective concept of statistical explanation.

Fourth, our theory of scientific explanation should have a place for a robust treatment of the pragmatics of explanation. Considerations of salience arise whenever we attempt to express or convey scientific explanations.

Fifth, we can relinquish the search for one or a small number of formal models of scientific explanation that are supposed to have universal applicability. This point has been argued with considerable care by Achinstein (1983).

Do we have the basis for a new consensus? Not quite yet, I fear. It would, of course, be silly to expect unanimous agreement among philosophers on any major topic. But leaving that impossible dream aside, there are serious issues on which fundamental disagreements exist. One of these concerns the nature of laws. Is there an objective distinction between true lawlike generalizations and generalizations that just happen to be true? Or is the distinction merely epistemic or pragmatic? The problem of laws remains unresolved, I believe, and—given the enormous influence of the covering law conception of explanation—of fundamental importance.

Another major issue concerns the question whether there are bona fide statistical explanations of particular events. Hempel's I-S model, with its high probability requirement and its essential epistemic relativization, has encountered too many difficulties. It is not likely to be resuscitated. The S-R model gives rise to results that seem strongly counterintuitive to many. For instance, on that model it is possible that factors negatively relevant to an occurrence help to explain it. Even worse, suppose (as contemporary physical theory strongly suggests) that our world is indeterministic. Under circumstances of a specified type C, an event of a given type E sometimes occurs and sometimes does not. There is, in principle, no way to explain why, on a given occasion, E *rather than* non-E occurs. Moreover, if on one occasion C explains why E occurs, then on another occasion the same kind of circumstances explain why E fails to occur. Although I do not find this consequence intolerable, I suspect that the majority of philosophers do.

One frequent response to this situation is to claim that all explanations are deductive. Where statistical explanations are concerned, they are of the kind classified by Hempel as D-S. Thus, we do not have statistical explanations of particular events; all statistical explanations are explanations of statistical generalizations. We can explain why the vast majority of tritium atoms now in existence will very probably decay within the next fifty years, for the half-life of tritium is about $12\frac{1}{4}$ years. Perhaps we can explain why a particular tritium atom has a probability of just over $^{15}/_{16}$ of decaying within the next fifty years. But we cannot, according to this line of thought, explain why a given tritium atom decayed within a given half century. The consequence of this view is that, insofar as indeterminism holds, we cannot explain what happens in the world. If we understand the stochastic mechanisms that indeterministically produce all of the various facts, we may claim to be able to explain *how the world works.* That is not the same as being able to explain *what happens.* To explain why an event has a high probability of occurring is *not* the same as explaining why it occurred. Moreover, we can explain why some event that did not occur—such as the disintegration of an atom that did not disintegrate—had a certain probability of occurring. But we cannot explain an event that did not happen.

Let me mention a third point of profound disagreement. Kitcher (1985) has suggested that there are two widely different approaches to explanation; he characterizes them as *bottom up* and *top down*. They could be described, respectively, as *local* and *global*. Both Hempel's approach and mine fall into the bottom-up or local variety. We look first to the particular causal connections or narrow empirical generalizations. We believe that there can be local explanations of particular facts. We try to work up from there to more fundamental causal mechanisms or more comprehensive theories.

Kitcher favors a top-down approach. Although many scientists and philosophers had remarked on the value of unifying our scientific knowledge, the first philosopher to provide a detailed account of explanation as unification is Friedman (1974). On his view, we increase our understanding of the world to the extent that we are able to reduce the number of independently acceptable hypotheses needed to account for the phenomena in the world. Both Kitcher (1976) and I (1989, 1990b) have found problems in the technical details of Friedman's theory; nevertheless, we both agree that Friedman's basic conception has fundamental importance. The main idea of the top-down approach is that one looks first to the most comprehensive theories and to the unification of our knowledge that they provide. To explain something is to fit it into a global pattern. What qualifies as a law or a causal relation is determined by its place in the simplest and most comprehensive theories. In his (1981) Kitcher began the development of an approach to explanatory unification along rather different lines from that of Friedman; in his (1989 and 1993) he works his proposals out in far greater detail.

Let us return, finally, to the fundamental question of this essay: Is there a new consensus concerning scientific explanation? At present, quite obviously, there is not. I do not know whether one will emerge in the foreseeable future, though I have recently come to see a basis for some hope in that direction (Salmon, 1989, 1990b, §5). However that may be, I am convinced that we have learned a great deal about this subject in the years since the publication of Hempel's magisterial "Aspects" essay. To my mind, that signifies important progress.[12]

APPENDIX

The preceding essay is a summary of material treated at much greater length in *Four Decades of Scientific Explanation* (Salmon, 1990b). Near the end I wrote:

> We have arrived, finally, at the conclusion of the saga of four decades. It has been more the story of a personal odyssey than an unbiased history. Inasmuch as I was a graduate student in philosophy in 1948 [the beginning of the first decade], my career as a philosopher spans the entire period. . . . My specific research on scientific explanation began in 1963, and I have been an active participant in the discussions and debates during the past quarter-century. Full objectivity can hardly be expected.
> . . . I know that there are . . . important pieces of work . . . that have not been mentioned. . . . My decisions about what to discuss and what to omit are, without a doubt, idiosyncratic, and I apologize to the authors of such works for my neglect. (p. 180)

One philosopher to whom such an apology was due is Adolf Grünbaum. "Explanatory Asymmetry" (essay 10) embodies my attempt to make amends.

James H. Fetzer is another worker who deserves a major apology. At the close of §3.3, which is devoted to a discussion of Alberto Coffa's dispositional theory of inductive explanation, I wrote:

> Another of the many partisans of propensities in the third decade is James H. Fetzer. Along with Coffa, he deserves mention because of the central place he accords that concept in the theory of scientific explanation. Beginning in 1971, he published a series of papers dealing with the so-called propensity interpretation of probability and its bearing on problems of scientific explanation (Fetzer 1971, 1974a, 1974b, 1975, 1976, 1977). However, because the mature version of his work on these issues is contained in his 1981 book, *Scientific Knowledge,* we shall deal with his views in the fourth decade. (p. 89)

Although these remarks are true, they fall far short of telling the whole truth. In my praise for Coffa, I said:

> In his doctoral dissertation Coffa (1973, chap. IV) argues that an appeal to the propensity interpretation of probability enables us to develop a theory of inductive explanation that is a straightforward generalization of deductive-nomological explanation, and that avoids both epistemic relativization and the reference class problem. This ingenious approach has, unfortunately, received no attention, for it was never extracted from his dissertation for publication elsewhere. (p. 83)

Without retracting my positive comments about Coffa, I must now point out that Fetzer's paper "A Single Case Propensity Theory of Explanation," published in 1974, contains a systematically developed theory of statistical explanation that has the same virtues I claimed for Coffa's approach. The issue here is not one of priority but one of complementarity. Coffa and Fetzer approach the problems in very different ways; both authors are highly deserving of our attention.

With complete justice, Fetzer has articulated his dissatisfaction in two articles (Fetzer, 1991, 1992). Those who want a more balanced view than I gave should refer to these writings as well.

I should emphasize, however, that although I find the approach to explanation via propensities valuable, I cannot agree that propensities furnish an admissible interpretation of the probability calculus. As Paul Humphreys argues cogently in his (1985), the probability calculus requires probabilities that propensities cannot furnish. For my view on the matter see Salmon (1979b).

Notes

This essay resulted from an NEH institute titled "Is There a New Consensus in Philosophy of Science?" held at the Center for Philosophy of Science, University of Minnesota. The section on scientific explanation was held in the fall term, 1985. Kitcher and Salmon (1989) contains results of this portion of the institute.

1. For a much more complete and detailed account of this development, see Salmon (1989, 1990b).

2. Neither Hempel nor I would accept this as a precise formulation, but I think it is an intuitively clear way of indicating what is at issue here.

3. Y is a *proper subclass* of X if and only if every Y is an X but some X are not Y; Y is an *improper subclass* of X if and only if X is identical to Y.

4. It goes as follows. Since the probability of 6 on each die is $\frac{1}{6}$, and the outcomes are independent, the probability of double-6 is $\frac{1}{36}$. Consequently, the probability of not getting double-6 on any given toss is $\frac{35}{36}$. Since the successive tosses are independent, the probability of not getting double-6 on n successive tosses is $(\frac{35}{36})^n$. The probability of getting double-6 at least once in n successive tosses is $1 - (\frac{35}{36})^n$. That quantity exceeds $\frac{1}{2}$ if and only if $n > 24$.

5. As far as I know, Bromberger never published this example, though he offers a similar one in his (1966).

6. In his stimulating book *The Scientific Image* (1980), Bas van Fraassen offers a charming philosophy-of-science-fiction story in which he maintains that, in the context, the length of a shadow does explain the height of a tower. Most commentators, I believe, remain skeptical on this point. See "Van Fraassen on Explanation" (essay 11).

7. This example is so old, and has been cited by so many philosophers, that I am reluctant to attribute it to any individual.

8. This essay is based on a presentation at the 1986 meeting of the American Association for the Advancement of Science.

9. I also offered another example. Around that time Linus Pauling's claims about the value of massive doses of vitamin C in the prevention of common colds was receiving a great deal of attention. To ascertain the efficacy of vitamin C in preventing colds, I suggested, it is *not* sufficient to establish that people who take large doses of vitamin C avoid colds. What is required is a double-blind controlled experiment in which the rate of avoidance for those who take vitamin C is compared with the rate of avoidance for those who receive only a placebo. If there is a significant difference in the probability of avoidance for those who take vitamin C and for those who do not, then we may conclude that vitamin C has some degree of causal efficacy in preventing colds. If, however, there is no difference between the two groups, then it would be a mistake to try to explain a person's avoidance of colds by constructing an argument analogous to (2) in which that result is attributed to treatment with vitamin C.

10. It is semantical rather than syntactical because it involves not only the characterization of a formal language but also the intended interpretation of that language.

11. These concepts are discussed more briefly and more accessibly in Railton (1981).

12. Suggestions regarding rapproachment between the unification theory and the causal theory are offered in "Scientific Explanation: Causation *and* Unification" (essay 4).

20

Scientific Explanation

Three Basic Conceptions

W hen one takes a long look at the concept (or concepts) of scientific explanation, it is possible and plausible to distinguish three fundamental philosophical views. These might be called the *epistemic, modal,* and *ontic.* They can be discerned in Aristotle's *Posterior Analytics,* and they are conspicuous in the twentieth-century literature. The classic form of the epistemic conception—the *inferential version*—takes scientific explanations to be arguments. During the period since the publication of the landmark Hempel-Oppenheim article "Studies in the Logic of Explanation" ([1948] 1965), the chairman of this symposium, Carl G. Hempel, has done more than anyone else to articulate, elaborate, and defend this basic conception and the familiar models that give it substance (Hempel, 1965b), though it has, of course, had many other champions as well. According to the modal conception, scientific explanations do their jobs by showing that what did happen had to happen. Among the proponents of this conception, Rom Harré and Edward Madden (1975), D. H. Mellor (1976), and G. H. von Wright (1971) come readily to mind. The ontic conception sees explanations as exhibitions of the ways in which what is to be explained fits into natural patterns or regularities. This view, which has been advocated by Michael Scriven (1975) and Larry Wright (1976), usually takes the patterns and regularities to be causal. It is this third conception—the ontic conception—that I support.

If these conceptions are viewed in the context of classical physics, construed in a Laplacian deterministic fashion, there seems not much point in distinguishing them or trying to choose among them; the distinctions give a strong appearance of being merely verbal. In the more contemporary context of possibly indeterministic physics, these distinctions take on a great deal of significance. Thus, a careful examination or probabilistic or statistical explanation turns out to be crucial. In that context the choice among the basic conceptions has major philosophical ramifications. The purpose of this essay is to examine each conception in its strongest form, and to consider the basic philosophical issues involved in accepting or rejecting each.[1]

1. The Epistemic Conception

Since the objections to the inferential version of the epistemic conception are quite familiar, I shall not spend much time rehearsing them. Let me simply cite one issue that has crucial importance for the present discussion, namely, the problem of explaining events that are relatively improbable. According to Hempel's most detailed account, an explanation is an argument to the effect that the event to be explained was to be expected, given the explanatory facts. D-N (deductive-nomological) explanations obviously have this character, for the explanandum follows with deductive certainty from the explanans. An I-S (inductive-statistical) explanation shares this characteristic, for the explanandum is rendered highly probable with respect to the explanans. The main problem for the present discussion is the status of events that occur with middling or low probabilities. The point is well illustrated by a familiar, though admittedly highly over-simplified, example. If two heterozygous brown-eyed parents produce a brown-eyed child, that fact can presumably be explained statistically on the basis of the 0.75 probability of such an occurrence.[2] If these same parents produce a blue-eyed child, that fact seems unexplainable because of its low probability. Nevertheless, as Richard Jeffrey ([1969] 1971) and others have argued persuasively, we understand each of these occurrences equally well. To say that we can explain the one but not the other is strangely asymmetrical.

In a more recent discussion of statistical explanation, Hempel (1977) relinquished his high probability requirement, thus apparently removing the difficulty associated with the explanation of improbable occurrences. He did not, however, pursue the ramifications of this change for his whole theory of scientific explanation. I suspect that it creates more problems than it solves—for example, violating a principle he characterized as "*a general condition of adequacy for any rationally acceptable explanation of a particular event*" (1965a, p. 367; emphasis in original). It is closely related to the principle CA-I, which I introduce shortly.

Two other versions of the epistemic conception have also been advanced. The first is an *information-theoretic version,* proposed by James Greeno ([1970] 1971) and later defended by Joseph Hanna (1978). This approach is extremely useful, but if it is supplemented, as Kenneth Sayre (1977) has suggested, with suitable consideration of the mechanisms of transmission of information, it should be reclassified under the ontic conception, which stresses physical mechanisms.

The second is an *erotetic version,* first broached by Sylvain Bromberger (1966) and later elaborated by Bas van Fraassen (1980). Bromberger's account was restricted to explanations of the D-N sort, whereas van Fraassen's theory deals explicitly with statistical explanations. Space does not permit a detailed exposition and critique of van Fraassen's theory (see Salmon, 1984b); nevertheless, I should like to make two comments. First, it is a subtle and well-worked-out theory—to my mind the most promising of any falling under the epistemic conception. Second, it presents a basic philosophical issue on which we take opposite sides.

The problem centers on van Fraassen's handling of the contrast class. On his view, each properly formulated explanation-seeking why-question presupposes a set of alternatives, one and only one of which (the topic) is true. To revert to the eye color example, we might choose a contrast class consisting of the alternatives {the child has brown eyes, the child has blue eyes}. Now, I have no objection whatever to the invocation of the contrast class; indeed, I require just such a class under the name "explanandum-partition." But van Fraassen requires the explanans to "favor" the topic—that is, to enhance the probability of the topic vis-à-vis the other alternatives. That seems to cause no problem in case we want to explain the eye color of the brown-eyed child, but it causes exactly the same kind of difficulty as Hempel's high probability requirement in the case of the blue-eyed child. The requirement of favoring has the consequences that we can explain the more probable outcome when it occurs, but not the less probable outcome when it occurs. Since, as noted before, it seems that we understand each alternative equally well, or equally poorly, the favoring requirement leads to an unsatisfactory asymmetry. I shall return to this issue later on.

2. The Modal Conception

The most obvious major consequence of indeterminism is that it appears to make the modal conception untenable. This conception seems to be impaled on the horns of a trilemma: one must either (1) make an a priori commitment to determinism; (2) admit degrees of necessity; or (3) grant that, to the extent that there are irreducibly statistical occurrences, they are inexplicable. Since, to my mind, a priori commitments to determinism are archaic and degrees of necessity unilluminating, we seem to be stuck with the third alternative. The problem is that quantum mechanics, which is arguably the most powerful explanatory theory in all of science, appears to require a statistical interpretation. Even if, in the end, it should turn out that a deterministic interpretation is tenable, we must not at present beg that question by a priori fiat.

The adherents of the modal conception have one further line of defense. It is natural—as von Wright has seen—to emphasize the strong affinity between their conception and Hempel's D-N model and to reject Hempel's I-S model. There is, however, one additional model—the D-S (deductive-statistical)—to which they might appeal. This model, it will be recalled, is used to explain a statistical regularity by showing that it follows with necessity from one or more statistical laws (and initial conditions in some cases). There is, however, no real need to treat the D-S model as a separate model, for D-S explanations (being both deductive and nomological) automatically qualify as D-N explanations, as long as we do not insist (as Hempel has not) that D-N explanations embody only universal laws. The advocate of the modal conception can accept, as legitimate D-N explanations, those of the deductive-statistical variety as well. Following this tack, the modal theorist can claim that quantum mechanics does explain statistical regularities—for example, tunneling phenomena—while denying that it explains any individual occurrences at all.

Confronted with the example of eye color, the modal theorist might say that Mendelian genetics does explain the fact that among all heterozygous brown-eyed parents, three-fourths of the children have brown eyes and one-fourth have blue eyes, though it offers no explanation at all of the eye color of any given child. The difficulty is that it also precludes explanation of distributions in large samples, for example, why among a sample of a thousand children of such parents about 750 were brown-eyed.

Many proponents of the modal conception will be totally unsatisfied by the eye color example, for it is easy to suppose that underlying Mendelian statistical distributions there are deterministic causal mechanisms, and consequently there is the possibility in principle of providing D-N explanations in all such cases. So let us take another example that involves different physical systems but the same probability distribution. Suppose that a single tritium atom is placed in a container, that the container is sealed for twenty-four and a half years, and that it is reopened at the end of that period. Since the half-life of tritium is about twelve and a quarter years, there is a probability of about three-fourths that it will have undergone spontaneous radioactive decay, transmuting it into a helium-3 atom, and a probability of about one-fourth that it will still be intact. According to this view, we can explain the fact that about three-fourths of all tritium atoms decay in any period of twenty-four and a half years, but we cannot explain the decay of any single tritium atom, the nondecay of any single tritium atom, or the percentage of decays in any restricted collection of tritium atoms in any given span of time.

There might be some plausibility in arguing that theoretical science does not contain explanations of individual events or restricted sets of events—only explanations of universal or statistical regularities—but I do not think it is true. Rutherford wanted to explain why small numbers of alpha-particles were deflected through large angles by gold foil. Hiraizumi wondered why small numbers of matings of fruit flies produced radically non-Mendelian distributions of eye colors (see Crow, 1979). But even if we grant the point about theoretical science, one can hardly doubt that applied science often tries to explain individual occurrences or limited sets of occurrences. And this is true in cases—such as the onset of a given disease in a given patient—in which D-N explanations are not available. A philosophical account of scientific explanation that excludes the explanations provided by applied science can hardly be considered adequate.

3. The Ontic Conception

There is a fundamental intuition—shared, I believe, by almost everyone who thinks seriously about the matter—according to which causality is intimately involved in explanation. Those who are familiar with Hume's critique of causality may deny the validity of that intuition by constructing noncausal theories of scientific explanation. Others may skirt the issue by claiming that the concept of causality is clear enough already, and that further analysis is unnecessary. My own view is (1) that the intuition is valid—scientific explanation does involve causality in an extremely fundamental fashion—and (2) that causal concepts do stand in serious need of further analysis.

It may be possible—though I seriously doubt it—to construct a regularity analysis of causality that would be adequate within the context of Laplacian determinism. The most promising such approach is J. L. Mackie's treatment of the problem in terms of INUS conditions (Mackie, 1974). In the contemporary scientific context, in which irreducibly statistical laws may well obtain, it seems necessary to admit that causal relations may also be irreducibly statistical. Among the authors who have tried seriously to construct theories of probabilistic causality, several—for example, I. J. Good (1961–1962), Hans Reichenbach (1956), and Patrick Suppes (1970)—have tried to found their analyses on statistical regularities. Although I do not have any knock-down argument to support my contention, my sense of the objections to such theories convinces me (at least tentatively) that no such regularity analysis of probabilistic causality will be adequate. We must, instead, look to the causal mechanisms. (See "Probabilistic Causality" [essay 14] for a detailed examination of these theories.)

Two causal mechanisms seem to me to be fundamental. First, there are spatiotem-porally continuous causal processes that transmit causal influence from one part of spacetime to another. Causal processes must be distinguished from pseudo-processes. Pseudo-processes exhibit considerable regularity, thus closely resembling causal pro-cesses. However, pseudo-processes do not possess the ability to transmit causal influ-ence. Causal processes are distinguished from pseudo-processes by the fact that causal processes can transmit marks while pseudo-processes cannot.

The second causal mechanism is the causal interaction. When two or more causal processes intersect in spacetime, they may or may not produce lasting modifications in one another. If they do, the intersection constitutes a causal interaction. Pseudo-processes do not enter into causal interactions. Pseudo-processes are produced by causal processes, and these causal processes that give rise to pseudo-processes can participate in causal interactions. Thus, a pseudo-process may be momentarily altered by intersection with another process (causal or pseudo-), but such modifications do not persist beyond the locus of the intersections.

When these causal mechanisms are deployed, it is possible to distinguish two distinct aspects of causal explanation. First, in many cases the explanation or an event tells the causal story leading up to its occurrence. To explain the presence of a worked bone, found by radiocarbon dating to be about 30,000 years old, in an archaeological site in the Yukon requires a causal account of how it got there. Since no well-authenticated New World sites of human habitation are nearly that old, one possible explanation accounts for the age of the bone by hypothesizing that the caribou died 30,000 years ago, and its carcass was preserved in ice for many millennia before it was found and worked by a human artisan (Dumond, 1980). Explanations of this sort can be called (following Larry Wright's felicitous terminology) *etiological*.

Other explanations account for a given phenomenon by providing a causal analysis of the phenomenon itself. For example, we explain the pressure exerted by a gas on the walls of a container in terms of momentum exchanges between the molecules and the walls. Such explanations may be termed *constitutive*.

In many cases, I presume, causal explanations possess both etiological and constitu-tive aspects. To explain the destruction of Hiroshima by a nuclear bomb, we need to explain the nature of a chain reaction (constitutive aspect) and how the bomb was transported by airplane, dropped, and detonated (etiological aspect).

According to the ontic conception—as I see it, at least—an explanation of an event involves exhibiting that event as it is embedded in its causal network and/or displaying its internal causal structure. The causal network, external or internal, consists of causal processes transmitting causal influence and causal interactions in which the structures of the interacting processes are modified. The whole structure is probabilistic. When two processes—such as an alpha-particle and the nucleus of a gold atom—intersect, there is a certain probability distribution for scattering of the alpha-particle at various angles. When a single process—such as an amoeba—is simply transpiring, there is a certain probability that it will split into two processes—in the case of the amoeba, undergoing mitosis that yields two daughter amoebas. In the case of the tritium atom, a neutron decays into a proton, an electron, and an antinutrino, thus transmuting hydrogen-3 into helium-3.

The ontic conception, in the causal version I have tried to elaborate, faces two major problems. First, there is the question whether adequate analyses of the basic causal concepts have been furnished. I hope that the answer is yes, but in case it is not, further work would need to be done toward achieving that goal. (See essay 16 for subsequent developments.)

The second—and far more difficult—problem concerns quantum mechanical explanation. Remote correlational phenomena of the type first treated in the famous Einstein-Podolsky-Rosen paper (1935), and widely discussed at present in connection with Bell's inequalities, suggest that there are fundamental difficulties in principle in attempting to provide causal explanations in terms of spatiotemporally continuous causal processes and localized interactions in the quantum domain. I am not inclined to dispute this claim. Rather, I should say, it appears that causal explanations of the sort just discussed are adequate and appropriate in many domains of science, but that other mechanisms—possibly of a radically noncausal sort—operate in the quantum domain. If that is true, then we need to learn what we can about those mechanisms so that we can arrive at a satisfactory characterization of quantum mechanical explanation. It may turn out that the causal conception of scientific explanation has limited applicability; nevertheless, the ontic conception could be maintained in a mechanistic version even as applied to quantum phenomena.

To the best of my knowledge all of the problematic cases involve quantum mechanical systems, each of which is in a pure state that can be described by a single wave function. In each such case the problem arises out of an interaction of that system with a measuring apparatus that results in reduction of the wave packet or collapse of the wave function. What kind of mechanism is this? I do not pretend to know, but I suspect that no one else does either. In his address on the situation in the philosophy of quantum mechanics, presented at the 1982 Philosophy of Science Association meeting, Howard Stein (1983) maintained that our lack of understanding is so profound that we do not even know whether there is in nature any such process as reduction of the wave packet. Under these circumstances, it is hardly surprising that we have no satisfactory treatment of quantum mechanical explanation.

Proponents of the epistemic conception might claim to have a viable account of quantum mechanical explanation, for there is a well-established scientific theory that correctly predicts the outcomes of the remote correlation experiments and other puzzling quantum phenomena. This is, perhaps, another instance of the principle that one person's counterexample is another person's *modus ponens*. To my mind, the fact that what quantum theory offers qualifies, under the epistemic conception, as correct scientific explanation constitutes strong evidence of the inadequacy of the epistemic conception.[3]

It is a basic principle of my approach that we cannot get very far in attempting to understand scientific explanation if we try to articulate a universally applicable logic of scientific explanation. What constitutes an adequate explanation depends crucially, I think, on the kind of world in which we live; moreover, what constitutes an adequate explanation may differ from one domain to another in the actual world. Even if a causal account of explanation cannot be extended into the quantum domain, that does not mean that its application in other domains is illegitimate. The ontic conception mandates attention to the mechanisms that actually operate in the domain in which explanation is sought.

4. A Criterion of Adequacy

One preeminent criterion of adequacy (which I shall call CA-I) has guided much of the discussion of scientific explanation for a long time. I do not know when or by whom it was first explicitly formulated; Wolfgang Stegmüller (1973) calls it "the Leibniz principle." Careful consideration of this criterion will, I think, bring out some of the basic philosophical issues separating the three general conceptions of scientific explanation I have outlined. According to CA-I,

> if, on one occasion, the fact that circumstances of type C obtained is taken as a correct explanation of the fact that an event of type E occurred, then on another occasion, the fact that circumstances of type C obtained cannot correctly explain the fact that an event of type E' (incompatible with E) occurred.

Most philosophers, I believe, have taken this criterion as an unexceptionable condition of adequacy for any theory of scientific explanation, and surely within the Laplacian deterministic context it *is* a correct criterion. It is satisfied by D-N explanations, and within the deterministic context every event is explainable by a D-N explanation. As long as the high probability requirement is maintained, I-S explanations also satisfy this criterion, for, under any given conditions, it is impossible for both the occurrence and the nonoccurrence of a given type of event to have probabilities greater than one-half. If, however, the high probability requirement is relinquished, it is not clear how the foregoing criterion can be satisfied. As noted earlier, Hempel's (1977) admission of low probability explanations could lead to violation of it.

In the case of eye color, for example, genetically identical pairs of parents produce brown-eyed children three-fourths of the time and blue-eyed children one-fourth of the time. Unless we can produce further information that would enable us to provide D-N explanations of the eye color of offspring, the adherent of the modal conception will claim that we have no explanation of either outcome. The advocate of the epistemic conception who allows for the traditional I-S model will admit the explanation of the more probable outcome but not that of the less probable. If I am correct in thinking that we understand the less probable outcome just as much or as little as we understand the more probable one, then the asymmetry in this epistemic approach becomes unacceptable. One is driven to say that we understand both or we understand neither. Because of the reasons already offered for rejecting the modal approach, I claim that we can have explanations of both. This violates the fundamental criterion CA-I. I am sure that many philosophers will maintain that, with this admission, I have just dug my own grave.

Let us examine for a moment the rationale for holding the foregoing criterion sacrosanct. Its main point, I think, is to rule out certain familiar sorts of pseudo-explanation. We do not want to allow the dormitive virtue to explain the power of opium to produce sleep. We do not want to allow an appeal to the will of God to explain whatever happens. We do not want to allow a psychoanalytic theory that is compatible with all possible behavior to have explanatory power. CA-I is not, however, needed to do that job. The dormitive virtue theory is patently too ad hoc to have legitimate scientific status. The will of God 'explanation' is scientifically unacceptable because it appeals to a supernatural agency. Psychoanalytic theories that are compatible with all possible psychological phenomena cannot be scientifically well confirmed. To rule out such cases as these, we do not need CA-I. It suffices to require that scientific explanations appeal to bona fide scientific laws and theories.

If indeterminism is true—and I think we must allow for that possibility in our theories of scientific explanation—then circumstances of a type C sometimes yield an outcome E and sometimes one or more other outcomes E′ that are incompatible with E. Heterozygous brown-eyed parents sometimes have brown-eyed offspring and sometimes blue-eyed. When the offspring is brown-eyed, the explanation is that the parents are both brown-eyed and heterozygous, and three-fourths of all children of such parents are brown-eyed. When the off-spring is blue-eyed, the explanation is that the parents are brown-eyed and heterozygous, and one-fourth of all children of such parents are blue-eyed. A tritium atom left alone in a box for twenty-four and a half years sometimes yields a tritium atom in the box and sometimes a helium-3 atom in the box. When a helium-3 atom is found, the explanation is that the tritium atom placed in the box underwent beta-decay and was transmuted to helium-3, and about three-fourths of all tritium atoms undergo such decay in that period of time. When a tritium atom is found, the explanation is that the tritium atom placed in the box remained intact, and that happens to about one-fourth of such atoms in that period of time.

Strong protest is likely to be raised at this point on the ground that none of the foregoing explanations is acceptable, for we cannot explain why the eye color is brown rather than blue or blue rather than brown. Nor can we explain why the tritium atom decayed rather than remaining intact or remained intact rather than decaying. The point can be put in terms of van Fraassen's contrast class. In the eye color example, the contrast class contains blue and brown. In the case of the brown-eyed child, I should say, we can *explain why* the topic is true, and we *know that* the only alternative is false, but we cannot explain why the other *rather than* the other obtains.

The demand that a satisfactory explanation of any occurrence must contain the "rather than" component stems most naturally from the modal conception of scientific explanation. According to this conception, an explanation explains by showing that what did happen had to happen, from which it follows that no incompatible alternative could have happened. Such an explanation would explain why the alternative did not happen because under the circumstances it could not have happened. To my mind, this demand for the "rather than" component stems from the Laplacian deterministic context in which the same circumstances always lead to the same outcome. If one holds on to the modal conception, the natural response to indeterminism is to suppose that it makes explanations of certain kinds of occurrences impossible. The Laplacian orientation strikes me as scientifically anachronistic.

If one supports the epistemic conception, there is a strong temptation to identify explanation and rational expectability. In the deterministic context, rational expectability takes the form of deductive certainty. In the indeterministic context, the type of expectability can be weakened, allowing for high inductive probability. This weakening can still allow for the "rather than" condition, for if one outcome is highly probable, the alternatives must all be rather improbable. It makes sense to expect the probable outcome rather than an improbable alternative even though we are sometimes wrong in such expectations. A 'explanation' of a low probability occurrence is likely to be rejected as any sort of genuine explanation. How, for example, can one be said to have explained the presence of a tritium atom in the container if it is reasonable to bet three-to-one against that outcome under the stated conditions?

In van Fraassen's erotetic version of the epistemic conception, the "rather than" condition is preserved through the requirement that an adequate explanation *favor* the topic of the why-question.[4] This requirement is not tantamount to the high probability requirement of the inferential version; indeed, it is even weaker than a positive-relevance requirement. Nevertheless, in the simple case of a contrast class with just two members, if the explanation favors the more probable alternative vis-à-vis the less probable one, it cannot also favor the less probable alternative vis-à-vis the more probable one. The erotetic version, like the inferential version, focuses on rational expectability. The form of the expectability requirement in the erotetic version is more subtle, but it still leads to what seems to me to be an unappealing asymmetry. This is the price paid to avoid violating CA-I.

To shift from the epistemic to the ontic conception involves a radical gestalt switch. It involves relinquishing rational expectability as a hallmark of successful scientific explanation. Instead of asking whether we have found reasons to have expected the event to be explained if the explanatory information had been available in advance, we focus on the question of physical mechanisms. Scientific understanding, according to this conception, involves laying bare the mechanisms—etiological or constitutive, causal or noncausal— that bring about the fact-to-be-explained. If there is a stochastic process that produces one outcome with high probability and another with low probability, then we have an explanation of either outcome when we cite the stochastic process and the fact that it gives rise to the outcome at hand in a certain percentage of cases. The same circumstance—the fact that this particular stochastic process was operating—explains the one outcome on one occasion and an alternative on another occasion.

The gestalt switch demanded by the ontic conception is perhaps most vividly seen in connection with criterion CA-I. In an indeterministic world this criterion is inappropriate; in such a world the same circumstances do explain one outcome on one occasion and an incompatible alternative on another occasion. If we do not know for sure whether our world is deterministic or indeterministic, and if we want to leave open the possibility of indeterminism, then we should not tie our basic intuitions about scientific explanation to a criterion of adequacy that is appropriate only to a deterministic world.

The Laplacian influence is pervasive and insidious; it gives rise, I suspect, to some of the most widely held intuitions regarding scientific explanation. To adopt the ontic conception involves the rejection of at least four significant doctrines:

1. An explanation of an occurrence must show that the fact to be explained *was to be expected.*
2. An explanation must confer upon the fact to be explained a *favored position* vis-à-vis various alternative eventualities.
3. An explanation must show why one outcome *rather than* another alternative occurred.
4. It is impossible that circumstances of type C can, on one occasion, explain the occurrence of an event of type E, and, on another occasion, explain the occurrence of an incompatible alternative E' (CA-I).

It may not be easy to abandon the foregoing intuitions, but I am inclined to think that they must be overcome if we are to come adequately to terms with scientific explanation in the contemporary scientific context.

Notes

The material in this paper is based on work supported by the National Science Foundation under Grant no. GS-42056 and Grant no. SOC-7809146. Any opinions, findings, and conclusions or recommendations expressed in this publication are those of the author and do not necessarily reflect the views of the National Science Foundation.

1. The issues raised in this paper are discussed in much greater detail in Salmon (1984b).
2. If any reader considers a probability of 0.75 insufficient to qualify as a high probability, it is a routine exercise to construct similar examples that involve probabilities as large as one wishes—that is, arbitrarily close to 1.
3. This criticism applies to the inferential version of that conception, but this version has constituted the 'received view' of scientific explanation for several decades at least.
4. It should be noted that the "rather than" condition need not be embodied in every erotetic theory. Peter Achinstein's (1983) illocutionary theory is erotetic, but it does not contain any "rather than" condition.

Part V

APPLICATIONS TO OTHER DISCIPLINES

*Archaeology and Anthropology,
Astrophysics and Cosmology, and Physics*

Although the essays in the first four parts make many references to empirical sciences and the philosophical problems arising out of them, those in this part address specific issues in particular scientific disciplines. They aim to show that the area of philosophy of science that deals explicitly with causality and explanation is not irrelevant to contemporary empirical science.

Essay 21, "Alternative Models of Scientific Explanation," written in collaboration with Merrilee H. Salmon, is addressed to archaeologists and other anthropologists interested in the nature of scientific explanation. A group called the *new archaeologists,* concerned to assure the scientific status of archaeology, had become convinced that a *sine qua non* of science is the construction of explanations conforming to Hempel's D-N model. Our aim was to show that a much wider class of models of explanation is available and that others in this set are more suitable than the D-N model for archaeology and anthropology. At the same time, we show that the so-called *systems approach,* advocated by other archaeologists, has all of the shortcomings of the D-N model without offering any improvements in exchange.

Essay 22, "Causality in Archaeological Explanation," extends the discussion of essay 21, emphasizing the causal dimensions of explanation in archaeology.

Essay 23, "Explanation in Archaeology: An Update," published 15 years after essay 21, deals with subsequent developments in the philosophical discussions of scientific explanation that have particular relevance to archaeology. These philosophical developments are treated in more technical detail in essay 19.

Essay 24, "The Formulation of Why-Questions," is a response to a criticism, posed by an archaeologist, of the S-R model of explanation. I claim that the objection does not hold; in answering, I clarify some fundamental features of the model. This response connects directly to van Fraassen's pragmatic theory, which is discussed from a different angle in essay 11.

Essay 25, "Quasars, Causality, and Geometry: A Scientific Controversy That Did Not Occur," shows how the failure of astrophysicists to honor the distinction between causal processes and pseudo-processes vitiates much of the discussion of quasars since their discovery in 1963. In this case the theory of causality presented in this book bears directly on recent developments in an exciting branch of physical science.

Essay 26, "Dreams of a Famous Physicist: An Apology for Philosophy of Science," explores in some depth the relationship between physics and philosophy of science. Here I expose misconceptions regarding philosophy of science that seem to pervade the attitudes of many physicists toward this field. I try to show that philosophy of science is not the pointless enterprise one famous physicist, Steven Weinberg, takes it to be. Because his argument depends crucially on explanation in physics, this essay is a fitting conclusion for the entire set of essays in this book.

21

Alternative Models of Scientific Explanation

With Merrilee H. Salmon

For a number of years, archaeologists have evinced considerable interest in the nature of scientific explanation. One group has even identified adoption of a particular model of scientific explanation as *the* hallmark of scientific method (Watson et al., 1971). The model in question is the deductive-nomological (D-N) model, which has been elaborated most thoroughly and precisely by Hempel (1965b). Since that model is known also as a "covering-law model," its proponents have come to be called the "law-and-order" group. Other archaeologists, finding various shortcomings in the deductive-nomological model, have sought alternative approaches. The present essay explores some of the available alternatives.

Some archaeologists—identified as the "Serutan" group[1]—have urged adoption of a "systems approach" to the problem of explanation in archaeology. The interests and purposes of archaeology are best served by adopting that strategy, they claim, and the information available to archaeologists can best be utilized within a systems framework. The systems approach is offered as a superior alternative to Hempel's D-N account of explanation, for various reasons. The D-N model is believed by some archaeologists (Tuggle et al., 1972) to be deficient primarily on the following ground: the D-N model requires laws, and the archaeologist, unlike the physical scientist, does not have a ready stock of laws with which to construct explanations.

As our analysis will show, the systems model of explanation must surely be classified, along with the D-N model, as a covering-law model. Natural systems are governed in their operations by more or less complex causal interactions, and such causal relations are instances of laws of nature. Advocates of systems theory cannot legitimately ignore such causal laws in constructing a model of scientific explanation. This fact about systems theory explanation illustrates a fundamental point that deserves emphasis: *There are many different covering-law models of scientific explanation*; the deductive-nomological model is not the only covering-law model. The remainder of this essay discusses several models that differ radically from the D-N model. All of these qualify as covering-law models. One can embrace a "covering-law conception" of scientific explanation without claiming that *all,* or even *any,* scientific explanations fit the D-N schema.

The requirement for laws in explanations, such a stumbling block for acceptance by some archaeologists, can be seen as quite natural when the nature and function of laws in explanation are understood. Scientific explanations are answers to certain kinds of why-questions: "Why did this granary collapse? It was destroyed by termites." This answer is accepted as an explanation only because it is true that, in general, when termites are permitted to feed unmolested on the wooden foundations of buildings, such buildings collapse.

Laws are crucial to explanations because they provide the *link* between the particular circumstances surrounding the event to be explained and that event itself. It is by virtue of a uniform connection—embodied in a general principle, or law of nature, or physical regularity pertaining to termite damage—that *this* granary collapsed because of a *particular instance* of termite damage.

Laws are simply true statements of such regularities. Even events that occur only once—unique events—can be explained in terms of regularities, for we may be able to say that *if* such a combination of circumstances were to occur again, an event of the same type would also occur.

That is not to say that laws are unproblematic. For example, there are serious questions about the precise characterization of laws. There are questions about how we can discover laws and confirm their truth. Statements that fulfill all the requirements for being a law except that their truth value is uncertain are called "lawlike statements"; these would be laws if they were true. Examples of lawlike statements that are believed to be laws are: "[T]he learning of the sounds of [a] language proceeds at the [same] time as the learning of the higher order units" (Olmsted, 1971, p. 36), and $PV = nRT$, the ideal gas law, which relates the pressure, volume, and temperature of gases under certain conditions. These laws are universal laws—laws affirming that all or no members of some class have certain properties.

All laws are general statements, and they cannot refer to any specific time or place. But laws may be statistical or probabilistic rather than universal generalizations, such as those just mentioned. Examples of statistical lawlike statements are "0.35 of schizophrenic patients improve after treatment regardless of the type of treatment received" (Kiev, 1964, p. 5), and "The success rate of folk curers in treating patients of their own culture is very high" (Madsen, 1964, p. 421). Schiffer (1975, p. 843), who ably defends the importance of laws for archaeologists, cites examples of statistical lawlike statements used by anthropologists:

The greater the distance between groups in time and space, the more unlikely it is that diffusion would take place between them. (Sanders and Price, 1968, p. 59)

Cultures developing in isolation will normally change less rapidly than those with more extensive cultural interrelation. (Griffin, 1956, p. 48)

[S]pecific items are more likely to be lost and more rapidly lost than broad, varied activities or large systems . . . they are also accepted more rapidly. (Wauchope, 1966, p. 26)

We shall adhere to the thesis that explanations require an appeal to laws.

A concern with providing satisfactory explanations, as well as with establishing suitable criteria for such explanations, is not unique to archaeology. This concern is shared by all subfields of anthropology. Every anthropologist is familiar with debates about whether the appearance of similar cultural items in situations spatially or temporally remote from one another is best explained by principles of diffusion or in some other way; whether synchronic or diachronic laws are better able to explain certain phenomena; whether one variety of functionalism has more explanatory power than another.

Archaeologists' interest in explanation has been unusual in one respect: they have focused attention on the *logical form* of explanations. In most anthropological writings no specific attention is paid to this aspect of the problem. Formal questions are usually ignored, although some writers have questioned the propriety of applying standards of "Western scientific explanation" to explanations offered in non-Western cultures (see Basso, 1978, and Wilson, 1970). Anthropological writings on explanations are, for the most part, concerned with the nature of theoretical principles (e.g., diffusionist or evolutionary) that are employed in explanatory frameworks.

An important difference between the two approaches is that the latter is substantive. Deciding which general principles are correct is ultimately an empirical matter, although the moves from data to theory may be extremely complex. Uncovering the logical structure of satisfactory explanations, however, is a philosophical task. A priori reasoning, such as occurs in pure mathematics, is the appropriate tool for investigating this problem. Understanding the logical form of explanations is important and useful but not sufficient to guarantee the construction of satisfactory explanations. Their forms must be "fleshed out" with correct substantive principles.

This essay addresses primarily the logical form of explanation in the behavioral sciences. Used as illustrations are examples from cultural anthropology, social anthropology, and linguistics, as well as from archaeology.[2] It is somewhat unfortunate that discussion among archaeologists of Hempel's models of explanation has focused so exclusively on his D-N model, given the prevalence of statistical laws in the behavioral sciences. In a classic 1948 paper, Hempel and Oppenheim explicitly acknowledged that not every scientific explanation could conform to a deductive model (reprint in Hempel, 1965a, pp. 250–251). The need for an inductive or statistical model was recognized, and Hempel later fully elaborated his inductive-statistical (I-S) model (1965b, pp. 381–403). This model shares with the D-N model the requirement that every explanation must incorporate at least one law statement, but in the I-S model the laws are statistical rather than universal. Statistical generalizations that meet certain conditions *can* qualify for the status of lawlike generalizations. That archaeologists sometimes appeal to such statistical laws becomes evident when we recognize that the radioactive disintegration of carbon-14 is governed by a statistical law (and there is no compelling reason to think that there are underlying universal laws to be discovered). Thus, archaeologists who doubt the possibility of finding fully deterministic explanations of all phenomena that concern them could still invoke the I-S model in support of the claim that archaeologists should employ covering-law explanations.

Another objection to the D-N model claims that it is committed to an oversimple, linear view of causality, one that does not allow for the importance of feedback processes in the causal chain (Flannery, 1973). Flannery's objection to the D-N model is a complaint often expressed by adherents of general systems theory, who criticize the physical sciences and their laws as inadequate and simplistic with respect to the complex nature of causality. The accusation of oversimplicity is ungrounded. The laws used in physical sciences can be quite complex. For example, look at Maxwell's equations:[3]

$$\nabla \cdot E = 4\pi\rho$$
$$\nabla \cdot B = 0$$
$$\nabla \times E = -\left(\frac{1}{c}\right)\left(\frac{\delta B}{\delta t}\right)$$
$$\nabla \times B = \left(\frac{1}{c}\right)\left(\frac{\delta E}{\delta t}\right) + \left(\frac{4\pi}{c}\right) J$$

These laws can be used in D-N explanations. Many physical laws, such as Maxwell's equations, are interactive laws relating several variables, e.g., electric charge, electric field, magnetic field, electric current. There is no restriction on the number of laws that can be used in a D-N explanation. Moreover, if feedback processes are important for the explanation of some event, then the laws concerning these processes, such as the physical laws regarding operation of a governor on a steam engine, can be used in D-N explanation. In general *any* multivariate causal law that is universal can serve as a covering law in a D-N explanation.

Another problem with D-N explanation, apparently unnoticed by archaeologists, makes it unsatisfactory for archaeologists who are interested in the causes of the phenomena they investigate. Not all laws are causal laws, and the D-N model does not *require* causal laws for explanation.

Causal explanations explain later events by reference to antecedent events, using "laws of succession." D-N explanations, however, may explain by using "laws of coexistence," as when the period of pendulum is explained by the length of the pendulum. The connection between period and length is lawlike, but neither *causes* the other (Hempel, 1965a, p. 352). When archaeologists try to discover why some event occurred, such as the collapse of the Mayan civilization, they are asking for causal explanation. A D-N explanation need not refer to any causes of this event.

To summarize the discussion so far: All the models discussed in this essay qualify as covering-law models. Some laws appropriate for archaeological explanation are available, and others may be discovered through empirical research. The restriction to universal laws, however, makes the D-N model less useful to archaeologists than would be a model that allows statistical laws. The D-N model can accommodate complex, as well as simple, laws. But since those laws need not be causal laws, conformity to the D-N model does not guarantee a causal explanation.

Having considered the objections that proponents of a systems approach have leveled against the D-N model, let us now turn to a closer examination of the systems approach to the problem of explanation. To do that we must say precisely what an explanation must be according to the standards required by this approach. Vague claims, such as "to explain some event is to show its place in a system," do not count as providing a model of explanation.

Whatever the shortcomings of the D-N model, it is a genuine model of explanation in the sense that it states precisely what features a proposed explanation must have for it to be counted as a genuine explanation. One thing that has made the D-N model so widely discussed, pro and con, is the clarity and force with which it is presented.

"The systems approach" is a phrase applied to many different ways of dealing with scientific problems, so it is particularly important that its use here be delineated in a careful way.

One well-known attempt to provide an alternative model of explanation within the framework of the "systems approach" is that of Meehan (1968). He, like some others attracted to systems theory, has claimed that the model of explanation used in physical science is not appropriate for the social sciences. All the same, it is interesting that much of the work done by pioneers in modern systems theory, such as Wiener (1961) and Ross-Ashby (1956), has not been directed to attempts to provide a systems model of explanation. Instead, they have tried to reformulate statements of purpose or function, using the concept of "feedback" (which, by the way, is a term coined by mathematicians and engineers, not by social scientists) in order to eliminate any objectionable features of "purpose," so that statements involving purpose can play a role in D-N explanations of the type used in physical science. Commitment to the importance of systems theory does not demand a special "systems approach" to explanation.

Meehan's model appears to be theoretically attractive to some archaeologists, although few actually try to use it. For example, Tuggle et al. (1972) merely cite it. Although we have found no reference to it in Flannery's work, nor any explanations in Flannery that conform to this model, LeBlanc (1973, p. 211) claims that Flannery is influenced by Meehan. Rice (1975) does attempt to use Meehan's model to provide an explanation of changes in Mogollon settlement-subsistence patterns, but as far as we know, that is the only such attempt.

Meehan's own reasons for rejecting the D-N model are not entirely the same as those of the archaeologists. For example, he accuses Hempel of confusing logical with empirical results, and claims that the D-N model does not emphasize prediction sufficiently. (We do not agree with those criticisms, although that is not relevant to the present discussion.)

To be fair to Meehan, we quote the passage in which he most clearly states the criteria of adequacy for explanations conforming to a systems paradigm:

> First, the phenomenon to be explained must be embedded in an empirical description that is dynamic and not static, that stipulates change as well as differences. The phenomena will be defined in terms of such changes. Second, a system, a formal calculus, is used to generate entailments or expectations with reference to a set of

symbols. Third, the symbols or variables in the calculus are "loaded," given empirical referents so that the entailments of the formal system have empirical meaning defined in terms of the concepts used to load the basic symbols. If the loaded system is isomorphic to the situation in which the phenomenon occurs, *the system provides an explanation for the event.* (Meehan, 1968, pp. 56–57; emphasis added)

We want to show that Meehan's model does not avoid the problems associated with the D-N model. That can be demonstrated by looking at an example that accords with Meehan's model. The physical example is chosen primarily for simplicity. Meehan himself does not apply his model to any examples of explanation in social science.

The event to be explained is the change in volume of helium gas at moderate temperature and pressure in a container with a close-fitting cover. The cover is equipped with a movable piston. We can apply a force to the piston, or heat or cool the container, and note changes in volume. Next, following Meehan's model, we adopt the formal calculus $PV = nRT$ and use it to generate the value of V with respect to changes in P and T. Then we "load" the system. Let P refer to pressure (newtons/m^2) on our sample of helium, V refer to volume (m^3), and T to temperature (deg. K.). R is a universal constant, and n is a measure of the quantity of gas (the number of moles). The loaded system behaves just like the gas. Predictions made with the use of the loaded system are accurate. Thus, according to Meehan, we have explained changes in the volume of gas.

But although this example fits Meehan's model, it does not provide a *causal* explanation of the change in volume. The scientifically accepted causal explanation is found in the kinetic theory of gases, in the statistical laws governing the motions of the molecules that make up the gas. The fact that the ideal gas law does not provide causal explanations can be made evident from the following consideration. Assume that we have put a certain amount of gas—n moles—into our container at the outset. Thereafter we do not remove any of the gas or add any more. We go through various processes of changing the temperature, pressure, and volume of the gas, as outlined. Then, noting the values of P, V, and T, we can calculate the value of n. If that is an explanation, it 'explains' the amount of gas put into the container in the first place in terms of the *subsequent* values of P, V, and T. Such an 'explanation' would 'explain' the cause in terms of the effect.

With respect to the problem of causal explanation, Meehan's model is no better than Hempel's. Finding a mathematical system that fits the real system, even when such a mathematical system yields accurate predictions, is not the same as finding a causal explanation. There is a mathematical system that fits the relation between the length and the period of a pendulum, and one that fits the relations between pressure, volume, and temperature of a gas. Neither system is explanatory in any causal sense. At best, those mathematical relationships can be said to encourage us to look for causal relationships; they do not, in and of themselves, constitute causal relations.

To illustrate the point further, let us imagine the following situation. Archaeologists have reason to suspect that the ratio of large, storage-size pots to small pots found at a previously inhabited site is related in some way to the length of time the site was occupied. Suppose—as a *totally fictitious* example—that on the basis of sherd analyses at many sites, the following regularity or law could be established:

$$\frac{L}{s} = \frac{1.5 + (t - 3.375)^{1/3}}{6}$$

In Meehan's terms, this formula is the mathematical system, and the system is "loaded" by letting L represent the number of large pots, s the number of small pots, and t the time in years. This mathematical relationship is plotted in figure 21.1.

Now, such a law ("mathematical system") would be very useful to archaeologists because it would allow them to determine ("predict") the duration of occupation of a site. At the same time, no archaeologist would be satisfied that the duration of occupation has been explained.

It should be obvious from what has been said about Meehan's model that laws are as important to this model as they are to the D-N model. The statements connecting variables in the formal system in a regular manner are laws, and the correlating principles that are used to "load" a system are also laws. No matter whether they are called regularities or rules, they are still laws. Moreover, since Meehan requires a relation of deductive entailment between the "regularities" and the event to be explained, he is implicitly committed to universal laws. That is because no occurrence of an individual event can be *deduced* from statistical regularities. The mere requirement for some sort of laws in explanation is not a defect, as we have already said. But archaeological laws, like other laws of behavioral science, are apt to be statistical laws, although the laws reveal their statistical nature not by the use of numerical percentages, but rather by employing terms such as "more likely," "normally" (e.g., examples given earlier), or "frequently." Consider, for example, Deetz's (1967, p. 111) account of the matrilocal residence of Pawnee and earlier Arikara. His explanation is based on the law, drawn from ethnological theory, that "such households frequently develop in situations where women produce the majority of the food and in which there is a large degree of permanence in the location of the communities." An adequate theory of explanation for archaeology should recognize the importance of statistical laws. Meehan does not bring out in a careful way, as Hempel does, the deep logical differences between explanations involving statistical

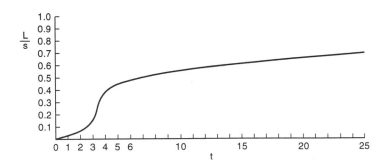

Fictitious "Law" Relating Pot Size to Duration of Occupancy

Figure 21.1

laws and those involving universal laws. Hempel provides an alternative model of explanation to handle these differences, whereas Meehan does not.

A certain fundamental philosophical difficulty, not mentioned by the critics who opt for a systems approach, infects both the D-N and I-S models of scientific explanation. Before we present that problem, it will be useful to characterize these models.

Briefly, the D-N model says that an explanation of an event is an argument to the effect that the event to be explained was to be expected in view of explanatory statements. The event to be explained (the *explanandum*) is described by the conclusion of an argument whose premises are the explanatory statements (the *explanans*). The explanans must be true, and it must include at least one universal law. Taken together, these premises must deductively entail the conclusion. In this model, explanation is presented as a logical relation (deductive implication) between the explanatory statements and the statement of the occurrence of the explanandum event.

The I-S model is different from the D-N model in the following respects: laws in the I-S model may be statistical rather than universal, and the event to be explained must follow from the explanatory premises with high probability, rather than deductive entailment. There is also a requirement of "total evidence" (automatically fulfilled in the deductive case) to guarantee that the inductive argument constituting the I-S explanation is a strong inductive argument. [This is a reference to Hempel's "requirement of maximal specificity."]

The problem with these Hempelian models can be illustrated by two simple examples.

(1) John Jones avoided becoming pregnant during the past year because he faithfully consumed his wife's birth control pills, and any man who regularly takes oral contraceptives will escape pregnancy.

This 'explanation' fulfills *every requirement* imposed upon D-N explanations.

(2) Susan Smith experienced only mild symptoms after her infection with valley fever, for she was wearing a turquoise necklace when she contracted the disease, and almost everyone who is wearing turquoise at the time of infection has a mild case.

This 'explanation' fulfills *every requirement* imposed upon I-S explanations.

These examples share a common defect. No man becomes pregnant, whether he takes oral contraceptives or not. Almost every case of valley fever is mild, regardless of the jewelry worn by the victim. In each case, the fact cited to explain the phenomenon in question was *irrelevant* to the occurrence of the fact to be explained. Hempel has, to be sure, imposed a "requirement of explanatory relevance," but that requirement does *not* block the kinds of irrelevancies that undermine the foregoing two examples. Moreover, these examples are not unique. Once you see how it is done, additional examples can be constructed in profusion (Salmon, 1971, pp. 33–35). And some may not be scientifically trivial and transparent, as we shall see in connection with the psychotherapy example.

Hempel has characterized an explanation as "an argument to the effect that the event to be explained was to be expected by virtue of certain explanatory facts" (1962b, p. 10). Both D-N and I-S explanations fulfill those conditions. A D-N explanation is a deductive argument whose conclusion follows from its premises with logical necessity (Meehan's model also has this deductive entailment requirement.) An I-S explanation is an inductive argument whose conclusion has high probability relative to the given premises. Hempel's requirement of explanatory relevance demands only that the argument in question—whether deductive or inductive—have correct logical form. As the examples show, the statement that a given event occurs can 'follow' deductively or inductively from premises describing facts that have nothing whatever to do with such occurrences.

In an effort to overcome this fundamental difficulty in the D-N and I-S models, a new model of scientific explanation has been constructed, known as the *statistical-relevance* (S-R) model (Salmon, 1971). The basic idea is not that the event to be explained should be rendered highly probable (including probability of one) by the explanatory facts but, rather, that the explanatory facts should *make a difference* to the occurrence or nonoccurrence of the fact to the explained. We might ask, for example, whether the fact that one has undergone psychoanalysis explains the fact that one no longer manifests a certain neurotic symptom. This should not be construed as the question of whether people with such symptoms are likely to get rid of them in the course of analysis; rather, it should be understood as the question of whether the probability of recovery under psychoanalytic treatment is any different from the spontaneous remission rate.

Two features of this nontrivial example deserve special notice. (1) The spontaneous remission rate may be quite high; in that case, a high probability for recovery during treatment—if it is no higher than the spontaneous remission rate—does not confer any explanatory value on the psychotherapy. In view of (1), a high probability is not a sufficient condition for explanatory import. (2) Even if the probability of recovery during psychoanalytic treatment were low—but still higher than the spontaneous remission rate—the fact that the patient has undergone treatment would explain the recovery, at least partially. In view of (2), a high probability is not a necessary condition for explanatory import.

According to the statistical-relevance model of scientific explanation, *an explanation is an assemblage of factors that are statistically relevant to the occurrence of the event to be explained,* accompanied by an associated probability distribution. Consider an example that may help clarify this formulation. Suppose that a teenager has committed an act of delinquency, and that a sociologist attempts to provide an explanation of that fact. The investigator would presumably look for certain factors that are known to be relevant to delinquent behavior—for example, sex, age, marital status of parents, religious background, socioeconomic status, residence (rural/urban/suburban), etc. Each of those factors, as well as many others, no doubt, is statistically relevant to delinquency in the sense that the probability of delinquency is (we assume) different for boys than for girls, different for urban dwellers than for rural youngsters, different for children whose parents have remained married than for those from broken homes, etc. If any of the factors cited did not make some difference in the probability for delinquency, it would be discarded as irrelevant, and hence as lacking in explanatory value.

Now, ideally, to provide an explanation of the instance of delinquency under consideration, we would take the population as a whole—American teenagers, let us say—and subdivide it into a series of subclasses in terms of such relevant factors as those already mentioned. For each subclass in the resulting partition the probability for delinquent behavior on the part of members of that subclass would be determined. Then, the teenager in question would be assigned to one such subclass in terms of his/her actual attributes—e.g., 17-year-old male, suburban dweller, middle-class undivorced parents, etc.—and the probability of delinquency cited for that subclass. As is obvious, the probability obtained in a given case may not be high. That is not crucial. What matters is that the relevant factors be taken into account in partitioning the population, and that the probabilities associated with the various subclasses be accurate.

A nice example of recognition of the importance of finding appropriate reference classes occurs in Olmsted's (1971) study of children's acquisition of language, in particular the development of the child's ability to pronounce various phones. Although the study focuses mainly on data gathering and some hypothesis testing, an overall goal of such research is to explain why some phones are more readily acquired than others. Preceding his research, Olmsted suggested that the differential acquisition of phones might be explained by some principle such as, "Assuming random masking of phones by other sounds in the environment, the more discriminable phones are more likely to be acquired and less likely to be mispronounced than less discriminable phones."

Careful observation of a sample of 100 children between 15 and 54 months of age yielded results that were in partial agreement with that principle as well as results that conflicted with it. Among the results are the following: semivowels are least likely to be mispronounced (9% rate of error), whereas retroflex vowels are mispronounced in about 58% of all attempts; both friction and nasality are detrimental to the acquisition of phones; one group of consonants was more difficult to acquire than expected; another group was less difficult (Olmsted, 1971, pp. 242–247). In light of that investigation, Olmsted recognized that although the discriminability principle was a suitable statistical lawlike statement to account for resistance to error in pronunciation, it was not adequate for explaining successful acquisition of phones. Avoidance of error and acquisition of phones are not the same thing. He suggests that other factors, such as articulatory difficulty, are important in explaining successes (p. 243). In other words, the reference class of easily discriminable phones must be further partitioned in terms of articulatory difficulty, and possibly other factors as well, before there can be a satisfactory explanation of the pattern of acquisition of phones by young children.

In many cases, like those just discussed, many different factors will be relevant. In some instances, however, a single relevant factor may provide an adequate explanation. To the question "Why are altered clay minerals present on the surface of this pot?" it would be sufficient to point out that the pot surface was exposed to high temperatures. Nothing further need be added, for no other information can have any influence on the probability that it contains altered clay minerals (Coles, 1973, p. 148). This example also shows that an S-R explanation may yield a probability of one for the event to be explained. Unobjectionable explanations that might have been classified as D-N within the Hempelian framework can qualify adequately as statistical-relevance explanations. The counterexamples—such as the man who takes birth control pills—will not qualify as S-R explanations (although, as noted earlier, they do fulfill the conditions formally stipulated for D-N explanations).

An archaeological example illustrating the importance of relevant partitioning is found in "Bone Frequencies—and Attritional Processes" (Binford and Bertram, 1977).

L. R. Binford and J. B. Bertram use the technique of partitioning a reference class in order to explain differential survival probabilities of anatomical parts. The authors observed that the relative frequencies of the anatomical parts of sheep that were available for archaeological observation differed between two Navajo sites although the sites were similar in most major respects, particularly in exposure of the dead animals to free-ranging dogs that fed on the carcasses. Those agents of attrition (the dogs) were solely responsible for destruction of the anatomical parts. The statistically relevant factor that partitions the reference class of "site with complete skeletons of dead and dying sheep exposed to free-ranging dogs" into two subclasses was discovered by the authors to be the season of occupation. One site was a winter site, while the other was used in summer. Since most lambs are born in early spring, the average age of dead animals differed between the two sites, so that certain bones of young sheep, being less dense than those bones in older sheep, have a different probability of surviving the actions of agents of attrition (pp. 148–149). This example also illustrates the importance of supplementing statistically relevant factors with causal considerations in order to achieve a satisfactory explanation. The direct causal factor in the differential destruction of parts is the differing degrees of resistance (degrees of density) that the bones offer to the agent of attrition. That factor depends on the age of the animal, which in turn depends on the season of occupation. Factors that are statistically relevant are not always causally relevant to the occurrence of an event but may be very important in guiding our search for causally relevant considerations.

It is, we believe, a distinct advantage of the S-R model of scientific explanation that it can accommodate events of high, middling, or low probability as amenable to explanation. In various domains of science we seem to meet with events that are improbable but not, we hope, inexplicable. In physics, radioactive disintegrations of certain types of atoms have extremely low probabilities (the half-life of U^{238} is measured in billions of years), yet they occur and they do not defy explanation. In biology, we can discover factors in a changing environment that are relevant to the evolution of a new species, but that does not make the emergence of such a species highly probable. In medicine, paresis as the tertiary stage of syphilis is far less probable than not, yet syphilis (which is obviously relevant to the occurrence of paresis) would normally be accepted as an explanation of paresis. In behavioral science, a particular suicide may be rather improbable, but it seems unreasonable to claim that we could not explain it if all relevant facts could be ascertained.

One example of archaeologists' attempts to explain an improbable phenomenon occurs in Coles (1973, p. 149). In most cases, ring-built and coil-built pottery can be distinguished from single-lump wheel-thrown pottery by the traces of junctions between rings in the finished pottery. In experiments performed to replicate prehistoric pottery, however, it was found that in a relatively small proportion of cases the junctions between the rings might be totally obscured during shaping and firing. Even though the shaping and firing process obliterates the junctions less often than not, the shaping and firing contributes to explanation of the absence of these traces in the same way that having syphilis contributes to explanation of the contraction of paresis.

The statistical-relevance conception of scientific explanation holds out some promise of an additional advantage over the Hempelian D-N and I-S models that may be of interest to archaeologists. The attempts of Hempel—and other defenders of the same general approach, such as Nagel (1961) and Braithwaite (1953)—to deal with functional explanations have notoriously met with severe difficulties.

Anthropologists are familiar with examples of functional explanation in, for example, the works of Radcliffe-Brown. Radcliffe-Brown suggests that social customs can be explained by considering their function or role in society, just as the presence of hemoglobin in the blood is accounted for by its function in transporting oxygen from the lungs to various parts of the body. "Every custom and belief of a primitive society plays some determining part in the social life of the community, just as every organ of a living body plays some part in the general life of the organism" (Radcliffe-Brown, [1933] 1967, p. 229).

Radcliffe-Brown's explanation of joking relationships in terms of their role in preventing conflict between a married man and his wife's relatives is well known ([1952] 1965, chap. 4). All the same, even though the joking relationship is quite pervasive, its existence in face of a threat of conflict between a man and his wife's kin cannot be deduced or claimed to be highly probable. That the joking relationship avoids conflict may well be established, but we cannot show that such a relationship must exist if the function of social stability is to be achieved, since there may be other means of achieving the same end. In fact, as Radcliffe-Brown shows ([1952] 1965, chap 5), in other similar situations an avoidance relation performs the same function. Because of problems with functional equivalents of this sort, Hempel was reduced to saying that functional 'explanations' are not real explanations at all—at best, they are partial explanations.

Functional explanations do not pose such difficulties within the S-R model. Because statistical-relevance explanations are not arguments—in contrast to the D-N and I-S types—the statement that oxygen is transported from the lungs to other parts of the body places one under no mandate to construct a deductively valid or inductively correct inference that hemoglobin is present in the blood. Even if no such argument can be given, the transport of oxygen is relevant to the presence of hemoglobin (as one of the various substances that can perform that function), and this relevance may constitute the basis for an adequate S-R explanation. Whatever one's views of functionalism as a general theory of anthropology, it seems to us that, in some cases that interest archaeologists, functional explanations are both familiar and appropriate. For example, we read that dwellings were constructed in a certain way in order to let smoke escape from hearth fires, that a particular location for a camp was selected to afford protection from enemies, that a function of barbs on a projectile point was to ensure its lodging in the target, and that the Southwestern diet of corn and beans was adopted because it provided a complete source of protein. An approach to explanation that accommodates explanations of that sort should be attractive to archaeologists.

Although we believe that the S-R model of explanation has certain virtues, we do not believe it can provide a fully adequate account of scientific explanation. In order to have any hope of achieving a satisfactory treatment of this notion, we must supplement the concept of statistical relevance with some kinds of causal considerations. The most obvious symptom of this need goes back to a fundamental shortcoming of the D-N and I-S models. To cite a paradigm of D-N explanation, one could explain the occurrence of a lunar eclipse, using well-established astronomical laws, in terms of some *previous* configuration of the sun, earth, and moon. Such an explanation would be widely regarded, intuitively, as fully acceptable. According to the requirements of D-N explanation, however, it would be possible to construct an equally satisfactory explanation of the same eclipse, using the same laws, but citing *subsequent* conditions rather than prior ones. Intuitively, nevertheless, most of us would balk at taking tonight's position of the sun, moon, and earth as an explanation of last night's eclipse. To do otherwise would bring us uncomfortably close to the awkward view that a person's death is explained by the fact that his obituary appeared in the paper on the following day. The obituary allows us to infer that the death occurred, and the obituary may *tell us why* the death occurred, but the printing of the obituary *is not why* the death occurred.

The relation of cause to effect has a distinct temporal asymmetry; causes come before effects, not after them. The relation of explanatory facts to the fact to be explained has a similar temporal asymmetry. In the D-N and I-S models, there is no demand that the explanatory relation be any kind of causal relation. Hempel has explicitly asserted that explanations that fit these models need not be causal. The same is true of S-R explanations; they are not explicitly causal, and it is not clear that they embody any temporal asymmetry requirements (Salmon, 1971, pp. 53–55, 65–76, 81). Causes are statistically relevant to effects, but the same effects have precisely the same statistical relevance to the same causes. Only by introducing causal considerations explicitly, it appears, can we impose the appropriate temporal asymmetry conditions upon our scientific explanations.

It is surprisingly difficult, for various philosophical reasons, to produce an adequate treatment of causal relations, and thus to implement the program of constructing a causal-relevance model of scientific explanation. But the time has come to put "cause" back into "because," and work is going on in that direction. It must be emphatically noted, however, that the causal relation need not be deterministic—it can be statistical.

The result, we hope, will be an account of scientific explanation that will be adequate to the needs of anthropologists. In the meantime, it seems reasonable to call the S-R model to the attention of anthropologists concerned with covering-law explanations, and to remind them that philosophers have not yet provided definitive answers to all of these problems—*but we are trying!*

Fortunately, anthropologists can take comfort in the knowledge that the construction of good explanations in anthropology need not wait upon solutions to these problems about models. The lack of an adequate model of scientific explanation is no more a problem for anthropologists than it is for physicists or chemists. Recognition of satisfactory explanations does not depend on a set of carefully detailed criteria any more than recognition that a member of Congress is behaving ethically depends on the ethical criteria that Congress proposes for its members. It is philosophically satisfying and theoretically important to have such criteria. Yet, in the last analysis, the development of an adequate theory of explanation for anthropology or any other science depends on a delicate balance between the invocation of logical principles and the considered judgments of scientists. Any philosophical theory of explanation that condemned in wholesale fashion all of the explanations accepted as sound by the practitioners of a particular branch of science would certainly have to be judged inadequate. At the same time, it does seem important to bring analytic tools to bear in an effort to discern the logical features that characterize successful explanations, in the hope that such knowledge may have value in the construction and criticism of actual explanations. Perhaps philosophers of science can play a useful role in this enterprise. As John Venn, a nineteenth-century philosopher, wrote in his epoch-making work on probability:

> No science can safely be abandoned entirely to its own devotees. Its details of course can only be studied by those who make it their special occupation, but its general principles are sure to be cramped if it is not exposed occasionally to the free criticism of those whose main culture has been of a more general character. (1866, p. ix)

The principle is reciprocal. While philosophers may supply logical criteria, scientists provide the hard cases against which philosophical models of explanation must be tested.

Notes

This essay is the result of fusion and revision of separate papers presented by each author at the forty-second annual meeting of the Society for American Archaeology. The authors are grateful to the editor and referees for valuable suggestions. Both authors thank the National Science Foundation for support of the research.

1. "Serutan"—"natures" spelled backward—was a product claimed by its manufacturer to help the human gastrointestinal system function correctly.
2. Since explanations in biological sciences may raise some special problems, we have omitted consideration of examples from physical anthropology.
3. "$\nabla \cdot$" and "$\nabla \times$" stand for the divergence and curl operators, respectively. Without this compact modern notation Maxwell's equations would look much more formidable.

22

Causality in Archaeological Explanation

If I were an archaeologist—which I am not—I imagine I would be seeking explanations for various phenomena and that the desired explanations would be causal. In asking for the explanation of the abandonment of the Grasshopper Pueblo at the end of the fourteenth century, I would be looking for the factors that caused the inhabitants to leave. In asking why a scarlet macaw was buried with the body of a child at that site, I would want to know what causal processes led to the presence of a bird of that species so far north of its current natural habitat. Additional examples come readily to mind, but the point seems almost too obvious even to be worth mentioning. The reason for raising it is that archaeologists have been widely influenced by Hempel's treatment of scientific explanation (1965b, 1966), and his theory leads to a radically different conception.

It will not be a primary aim of this essay to go into the details of Hempel's well-known deductive-nomological (D-N) or inductive-statistical (I-S) models of explanation, nor my statistical-relevance (S-R) model. Instead, I intend to examine certain general conceptions of scientific explanation that have motivated the construction of such formal models. As it turns out (see "Why Ask, 'Why?'?" [essay 8]), two basic intuitions have guided much of the discussion of scientific explanation during the last few decades. According to one of these—*the inferential conception*—to provide a scientific explanation of a phenomenon is to construct an argument that shows that the event to be explained was to be expected on the basis of the explanatory facts. This is the conception that Hempel—along with a number of other leading philosophers of science, including R. B. Braithwaite (1953), Ernest Nagel (1961), and Sir Karl Popper (1959)—has advocated and elaborated. According to the second basic intuition—*the causal conception*—to provide a scientific explanation is to identify the causal factors that produced the event to be explained. This conception has given rise to some of the sharpest criticisms of Hempel's models of explanation (e.g., Scriven, 1975; Wright,

1976), and to certain criticisms of my S-R model as well (e.g., Cartmill, 1980; King, 1976; Lehman, 1972). M. H. Salmon (1982) discusses the fundamental disagreement between Nagel and Wright on the subject of functional explanation—a difference that emerges rather patently from the fact that Nagel adheres to an inferential conception, while Wright embraces a causal conception.

The main purpose of this essay is to explore the divergence between these two conceptions in the hope of illuminating certain fundamental problems concerning scientific explanation that have, I believe, proved troublesome to archaeologists. In order to address the issue, permit me to take a short historical detour back to the beginning of the nineteenth century, the heyday of Laplacian determinism. In expounding his deterministic outlook, Laplace made reference to his famous demon, the hypothetical superintelligence that knew all of the laws of nature, knew the state of the universe in complete detail at some particular moment, and could solve any mathematical problem. For it, Laplace remarked, "[n]othing would be uncertain and the future, as the past, would be present to its eyes" ([1820] 1951, p. 4). It seems evident that this demon would be able to subsume any actual occurrence under basic universal laws in conformity to Hempel's D-N model and would be able also to discern the causal mechanisms leading to any such occurrence. Indeed, within the Laplacian framework it is not clear that there is any point in distinguishing the two conceptions, for Laplace gives the distinct impression that he views the basic laws of nature as both *universal* and *causal* (cf. "Why Ask, 'Why?'?" [essay 8]).

When we shift our framework from classical physics, with its deterministic outlook to twentieth-century science, where we must take seriously the possibility that the world is in some respects indeterministic, we find a sharp divergence between the two fundamental conceptions. Given our current state of physical knowledge and the fact that the statistical interpretation of quantum mechanics is the received interpretation, it seems to me that our philosophical theories of scientific explanation *must* take account of the *possibility* that some phenomena are in some respects undetermined, and therefore not amenable, even in principle, to D-N explanation. Consider radiocarbon dating. Radioactive decay is an ineluctably statistical phenomenon; it is, at best, highly probable that about half of any given collection of C^{14} atoms will disintegrate in 5715 years. Moreover, leaving aside altogether the current status of microphysics, we can surely see that in most sciences the only *available* explanations of some phenomena are probabilistic or statistical. In evolutionary biology, for example, we can say that a given trait raises or lowers the probability that a particular organism will survive and procreate, but physical necessity is not involved. Evolution is a statistical phenomenon.

Although the classic Hempel-Oppenheim ([1948] 1965) essay explicitly acknowledged the fact that there are legitimate scientific explanations of the statistical sort, the first serious attempt to develop a systematic theory of statistical explanation was, to the best of my knowledge, offered by Hempel (1962a). A sketch of this theory was presented in Hempel (1962b), and the theory was significantly refined by him in 1965. A sketch of the newer version was given in the elementary textbook (Hempel, 1966). The leading proponent of the inferential conception of scientific explanation recognized the existence of statistical explanations, and he provided a philosophical account of this pattern. The inductive-statistical model is the result. He *never* maintained that all acceptable scientific explanations must conform to the D-N schema.

The transition from deductive-nomological to statistical explanation brings out the fundamental differences between the inferential and causal conceptions. If one adheres to the inferential approach, it is natural (as Hempel clearly recognized) to regard statistical explanations as inductive arguments—analogous to the deductive arguments that characterize D-N explanations—and to require that the inductive argument render its conclusion highly probable in relation to the explanatory facts. The obvious result is that events can be explained only if they are highly probable with respect to some suitable explanatory conditions, and that events that are simply improbable in relation to all available information are not amenable to scientific explanation. The inexplicability of improbable occurrences leads to difficulty on two scores. First, it appears that we *do* regard improbable events as explainable. An archaeological example is given in "Alternative Models of Scientific Explanation" (essay 21). Ring-built and coil-built pottery can usually be distinguished from single-lump, wheel-thrown ceramics by traces of junctions between the rings in the finished product. However, in a small fraction of cases the junctions are totally obliterated during shaping and firing. Although this happens far less frequently than not, when it does happen, the shaping and firing explains the absence of such traces. A variety of other, nonarchaeological examples is furnished in the same place and in "A Third Dogma of Empiricism" (essay 6).

Second, it seems to me, we *should* consider some improbable occurrences just as explainable as certain highly probable occurrences. Hempel (1965a, p. 391) mentions as example taken from Mendelian genetics. In a certain population of pea plants, there is a probability of ¾ for red blossoms and a probability of ¼ for white blossoms. If we take ¾ to be a high probability (and if we don't, it is easy enough to cook up another example with a higher value), then we can explain the occurrence of a red blossom, but we cannot explain the occurrence of a white blossom in the same population. This represents a strange lack of parity, for it seems to me that we understand the occurrence of a white blossom in that population exactly as well as we understand the occurrence of a red blossom. (In a 1976 postscript, Hempel [1977, pp. 98–123] appears to have agreed with this point.) When we consider carefully the nature of statistical explanation, we find that the inferential conception of scientific explanation encounters serious difficulties; these have been more fully elaborated in "A Third Dogma of Empiricism" (essay 6) and "Comets, Pollen, and Dreams" (essay 3).

The statistical-relevance model was originally motivated by an intuitive sense of dissatisfaction concerning the high probability requirement associated with the I-S model (Salmon, 1971, pp. 10–12). In "Alternative Models of Scientific Explanation" (essay 21), examples are offered to show that putative explanations that fulfill all of the conditions for correct I-S explanations, including the high probability requirement, cannot be considered acceptable, while other examples that do not satisfy the high probability requirement are bona fide explanations. Reflection on these and a host of other similar examples convinced me that the relation of statistical relevance, not the relation of high probability of the explanandum relative to the explanans, was the *fundamental* explanatory relation. Strangely, it seemed to me, Hempel and other proponents of the inferential conception of explanation never seemed to appreciate the explanatory significance of statistical-relevance relations. It is now easy to see why. In the first place, as the arguments of Cartmill (1980), Lehman (1972), and King (1976) have shown, it is simply incorrect to suppose that the relation of statistical relevance has explanatory import in and of itself; rather, it is at best a *symptom* of a bona fide explanatory relation.

If we find a higher incidence of lung cancer among heavy cigarette smokers than we do in the population at large, this positive correlation suggests that a causal relation exists, but the relation of positive relevance in and of itself does not explain anything. If one holds the view that furnishing an explanation consists in constructing an argument that shows that the event to be explained was to be expected—either with deductive certainty or with high inductive probability—on the basis of explanatory facts, then statistical relevance relations will not seem very important from an explanatory standpoint. At best, they will appear as pale substitutes for the desired relations of high probability required in strong inductive arguments. Statistical relevance does not say much to a proponent of the inferential conception of scientific explanation. If, however, one holds the view that furnishing a scientific explanation consists in locating and displaying causal mechanisms, then relations of statistical relevance will be precisely the kinds of clues we need to ferret out the underlying causal relations that can be used in constructing scientific explanations. To the advocate of the causal conception, relations of statistical relevance are beacons guiding our way to crucial explanatory relations.

If it is acknowledged that the inferential conception encounters severe difficulties in providing a satisfactory account of statistical explanation, then we must ask how the causal conception will fare if indeterminism is admitted. If we think of causality in the usual deterministic way—sticking rigorously to the principle *same cause, same effect* (Hempel, 1966, p. 53)—then the causal conception will be incapable of coping. It seems to me, however, that we need not saddle a contemporary philosophical theory of scientific explanation with a Laplacian deterministic or Humean constant-conjunction notion of causality, and consequently we can save the causal conception, even in the face of events whose occurrences are not fully determined in every respect. To make good on this claim obviously requires the development of a theory of probabilistic causality, as attempted, for example, by Suppes (1970). I should like to say a little about probabilistic causality, partly because the notion is so unfamiliar that people are likely, I suspect, to consider it an incoherent concept. These issues are pursued in more detail in "Probabilistic Causality" (essay 14) and "Causality: Production and Propagation" (essay 18).

Almost every morning, it seems, the newspaper carries a story about some causal claim. Cigarette smoking causes lung cancer; saccharine causes bladder cancer in laboratory rats; the use of a certain brand of tampon causes toxic shock syndrome. In each case the *evidence* for the causal claim comes in the form of a relation of positive statistical relevance. In his introductory text, Giere (1984, chap. 12) spells out the details of several interesting examples. If we compare investigations of this sort with the Humean constant-conjunction conception of causality, we are immediately struck by an enormous discrepancy, for in virtually every case we find nothing remotely approaching constant conjunction. The typical example usually involves a *small* increment in a *minute* probability. For instance, there was a suspicion that substances such as epoxies and resins to which workers in the General Motors woodshop were exposed were carcinogens. The evidence that led to this suspicion was that during a 10-year period, 14 cases of cancer of the rectum or colon developed in a population of more than 1600 woodworkers (less than 1%), while the normal rate for a sample of this size would be 6 cases. For another example, among the 2235 soldiers who witnessed an atomic blast at close range in Operation Smoky in 1957, 8 (approximately ⅓ of 1%) subsequently contracted leukemia. Nevertheless, in the opinion of the medical investigator who examined the evidence, there was "no doubt whatever" that the radiation had caused the leukemia (see "Why Ask 'Why?'?" [essay 8]). The crucial issue in all such cases is not any pretense of constant conjunction, but whether a significant positive-relevance relation exists.

The standard answer to these considerations is to maintain that the discovery of positive-relevance relations, while providing *evidence* for the claim that we have located a *causal factor* in a complicated situation, does not enter into the *analysis* of the causal relation itself. In cases of the foregoing sort, it may be said, it is possible in principle to locate numerous other causal factors, and when we have collected all of them, we will find a constant conjunction between a complex set of causal conditions and a given effect. In this way we can save the general principle *same cause, same effect* and hang onto our cherished deterministic prejudices. The basic objection to this approach is, in my opinion, that there is no reason to believe that it is true.

Consider a couple of examples. Suppose that an ice cube is placed in a glass of tepid water and melts. I think we would not hesitate to say that being placed in the water is the cause of the melting, though we know theoretically that being placed in lukewarm water is neither sufficient nor necessary for the melting of the ice cube, merely rendering the result highly probable. This is true, by the way, even in classical statistical mechanics based on deterministic laws of motion for the molecules, because the initial condition—that the water be tepid—is not sufficient under that theory for the melting of the ice cube.

Or take the case of a simple type of laser. A large number of atoms are pumped up into a relatively stable excited state for which the average decay time by spontaneous emission is fairly large. If, however, radiation of the appropriate frequency impinges upon these atoms, there is rapid decay by stimulated emission, resulting in a burst of radiation from the laser. The acronym laser stands for "light amplification by stimulated emission of radiation." There is no doubt, I believe, that the impinging radiation *causes* the emission of radiation by the laser, but it is neither necessary nor sufficient for the occurrence. There is an admittedly minute nonzero probability that the atoms would all decay rapidly by spontaneous emission, and an admittedly minute nonzero probability that they would not decay rapidly even in the presence of incident radiation of a sort suitable to stimulate emission.

The preceding examples are cases in which the effect follows the cause with such a high probability that the nonoccurrence of the effect has so low a probability as to "make no odds." Let us look at an example that is less extreme. Some children are playing baseball, and one of them hits a long fly ball that shatters a neighbor's window. Suppose, for the sake of argument, that we can specify rather precisely the position and momentum of the ball as it strikes the window and that we can specify quite precisely the type of glass that shattered. Suppose further than a window pane of that particular sort will break in 95% of all cases in which it is struck by a ball of just that type traveling with the same momentum and striking the pane in the same spot. If someone were to say that it would be possible, if we knew further details about the internal structure of the glass and other relevant features of the situation, to ascertain exactly which collisions would result in breakage and which would not, I would remain skeptical.

Classical physics, from which we derive much of our deterministic inspiration, notoriously failed to provide any satisfactory theory of the structure of matter, and hence would be at a loss to deal with the preceding example. Contemporary solid-state physics, which may or may not have an adequate theory of the structure of glass (I do not happen to know), is not fundamentally deterministic. The basic point, however, is this. We may fuss as much as we like about the details of this example, and argue ad nauseam whether a deterministic account of the breaking of the window is possible in principle. It would make no difference how the argument came out. Either way, we would all readily agree, under the conditions stipulated, that the baseball *caused* the window to break. I cannot see why anyone should fear that such a judgment would become false or nonsensical if the supposed deterministic underpinning turned out to be absent. Such metaphysical baggage is completely dispensable to a satisfactory account of causality, and can play no useful role in the understanding of archaeological explanation.

My primary thesis, in the foregoing discussion, has been the inadequacy of the inferential conception of scientific explanation and the superiority of the causal conception. Implementation of the causal conception, it has been noted, requires a probabilistic concept of causality if it is to be able to deal with statistical explanations. Although there are, admittedly, serious difficulties involved in the full elaboration of a theory of probabilistic causality, the problems do not seem insuperable.

Having sketched these claims, I would now like to apply them to some archaeological examples that are not altogether trivial. In each of these cases several alternative potential explanations will be mentioned, but I shall make no attempt to decide which explanatory accounts, if any, are correct. That is the kind of question that can be answered only by the professional archaeologist who is in full command of the relevant empirical data. I shall, instead, point to general features of all of the various alternatives that deserve to be taken seriously.

Consider, first, the case of a piece of worked bone found at an Old Crow River site in the northern Yukon that has a radiocarbon date of approximately 30,000 B.P. (Irving and Harington, 1973; Dumond, 1980). This object has obvious bearing upon the problem of how early there were human inhabitants in the New World. Let us assume for the sake of argument that the radiocarbon date is correct—that the bone is indeed about 30,000 years old. The question is how to account for the presence of this object at a North American site. According to one potential explanation, there has been continuous human habitation in the region for at least 30,000 years. According to another, there was a brief period of human habitation about 30,000 B.P., followed by a long period during which no humans were there. On either of these theories, the presence of the worked bone is explained by human production (a causal interaction between the worker and the piece of bone) in the Yukon 30,000 years ago. A different potential explanation is that the bone had existed, preserved frozen in an unworked state for about 20,000 years before it was discovered and made into a human artifact about 10,000 years ago. This is a very different causal story. As noted, I have no intention of trying to say which, if any, of these potential explanations is correct, but I do want to emphasize the fact that any satisfactory explanation will involve a complex of causal processes and interactions leading to the presence of the worked bone in the Yukon.

It should be explicitly remarked that no assumptions about causal determinism need be taken to underlie any of the explanations. On the third alternative, it may have been a matter of sheer chance that an ancient artisan came across a piece of frozen bone suitable for working; indeed, it would be gratuitous to assume that it was even probable in the circumstances that the bone would be found, picked up, and worked. Furthermore, the recent presence of the worked bone in the Yukon involves causal processes that account for its preservation for 10,000 years after it had been worked. Who knows the vicissitudes such an object might have suffered or escaped, and who can say that its endurance over 10 millennia was causally determined in a nonstatistical sense or even that it was highly probable in the circumstances? These considerations seem to me to render dubious at best the claim that we could ever hope to construct a D-N or I-S explanation of the presence of the worked bone. They do not seem to militate in the least against a probabilistically causal sort of explanatory account.

As another example, let us consider the problem of Pleistocene extinction of large mammals over large regions of the earth. I am aware of two main types of explanations that have been offered (Grayson, 1980). One kind of explanation appeals to a radical climatic change, which in some accounts led to a loss of habitat to which the various species had become adapted and which in other accounts led to the birth of the young out of season. On this sort of explanation, the extinction is a result of evolutionary processes that resulted in the nonsurvival of many species. As I remarked before, evolution is a statistical affair, and, it might be added, there is no reason to presuppose a deterministic foundation. If an organism, born into a given environment, has a certain characteristic, that fact may raise or lower the probability that it will survive and procreate, but whether it does or not is a chancy affair. A baby mammoth, born out of season, may have a lessened chance of survival, but there is no basis for claiming that its failure to survive was wholly causally determined. Nevertheless, while causal determinism is no part of the story, we are all clearly aware of the kinds of causal mechanisms involved in the relationship between an organism and its environment that have a bearing on survival and procreation.

A rather different sort of explanation attributes the extinction to human overkill. The superfluousness of deterministic underpinnings are obvious in this case as well. Whether a particular animal escaped notice by a human hunter might well be a matter of chance. Given that the hunter has spotted the animal and thrusts a weapon with a Clovis point into its body, there might be a chance of only 95% that the animal will die as a result. The matter of death owing to the penetration of a Clovis point appears to me entirely parallel to the case of the window shattered by the baseball.

What I have said about the preceding two examples applies quite generally, I suspect, to explanation in archaeology. If, for example, any adequate explanation of the abandonment of Grasshopper Pueblo can ever be found, it will, I imagine, involve appeal to a complex set of factors that, taken together, account for the phenomenon. Some factors, such as a moderately severe drought, which would tend to cause people dependent on agriculture for food to move elsewhere, will be positively relevant to the occurrence. But such droughts do not always result in abandonment. Perhaps the existence of a fairly large and complex pueblo would tend to make people remain—even under circumstances of physical hardship. Factors of this kind would be negatively relevant. Again, a recent rapid growth in population, which led to the agricultural exploitation of marginal land, might be positively relevant to departure, since production on marginal land would be affected more drastically by drought than would production on land better suited to agriculture. This kind of approach to the explanation of the abandonment of Grasshopper Pueblo clearly involves a search for contributing causes (positively relevant factors) and counteracting causes (negatively relevant factors). In this way, we hope to be able to exhibit the complex causal mechanisms that produced the event we are trying to explain.

In the decades since Hempel's work began to exert a wide influence on archaeologists, there have been developments of fundamental importance within the philosophical theory of scientific explanation. It is easy to see the powerful appeal of an account of scientific explanation that demands deductive subsumption of the event to be explained under universal laws of nature—as is schematized in the D-N model. In 1960, as I mentioned previously, no systematic theory of probabilistic or statistical explanation existed. It must have been clear to many archaeologists, however, that the demand that every explanation conform to the D-N pattern was an unrealistic goal for archaeology. Since then, we have seen the emergence of at least two models of statistical explanation—the I-S model and the S-R model. Both have been subjected to severe criticisms. The major criticism of the former model is that it imposes a high probability requirement; the latter model overcomes the basic problem by shifting emphasis from high probabilities to relations of statistical relevance. The S-R model has, in turn, been criticized for failure to take adequate account of causal considerations—a criticism that applies equally to the other two models. When we attempt to repair this difficulty, we find ourselves facing another fundamental philosophical problem, because cause-effect relations have traditionally been construed as cases of constant conjunction. This conception of causality leads to the principle *same cause, same effect,* which implies that causal laws are universal laws. On this view of causality, the problem of finding causal explanations is just as insuperable as is the problem of deductive subsumption under universal laws demanded by the D-N model. It appears that we have come full circle.

The way out of this difficulty, I have been suggesting, lies in the development of a different conception of causality—a probabilistic concept, along the lines suggested by Suppes (1970) and "Probabilistic Causality" (essay 14). Relations of statistical relevance play a crucial role in any theory of probabilistic causality. Thus, the theory of causal explanation that emerges when we employ probabilistic causality is an extension or enrichment of the S-R model. However, in order to implement the probabilistic theory of causality, we must relinquish the time-honored principle *same cause, same effect.* We must be prepared to admit that a given cause may on one occasion produce one sort of effect, but the same kind of cause may on another occasion produce a different sort of effect. If, as contemporary physics suggests, indeterminism actually obtains in our world, that is exactly what we must expect. For example, in the famous Stern–Gerlach experiment, an atom of a given type may be deflected upward when it enters a certain magnetic field, while another atom *exactly similar to the first in all physical respects* may be deflected downward by the *same* magnetic field (see "Indeterminancy, Indeterminism, and Quantum Mechanics" [essay 17]). If the world has this sort of indeterminancy at the level of fundamental physics, we should not be dismayed to encounter indeterminancy in other domains as well. In order to have any hope of developing theories of explanation adequate to the contemporary sciences—including archaeology—we must be prepared to re-examine critically some of our most cherished philosophical concepts.

Note

The material in this essay is based on work supported by the National Science Foundation under Grant no. SES-7809146. I wish to express my gratitude to the NSF for this support of research, to the University of Arizona Committee on Foreign Travel for providing a travel grant that enabled me to attend the meeting of the Theoretical Archaeology Group at which this essay was first presented, and to the Theoretical Archaeology Group for its kind hospitality during the meeting.

23

Explanation in Archaeology

An Update

During the last few years I have been taking a fairly close look at some recent history of philosophy of science that has a direct bearing on archaeology.[1] The starting point is 1948, when Carl G. Hempel and Paul Oppenheim published their epoch-making article "Studies in the Logic of Explanation." It attempted to provide an explicit and precise account of the so-called *deductive-nomological* pattern of scientific explanation. In my view this essay marks the dividing line between the prehistory and the history of philosophical work on the nature of scientific explanation. To my utter astonishment I found that this article was virtually ignored for an entire decade after its publication. Then, quite suddenly, around 1957–58, it became the subject of intense critical discussion.

What came to be called the *new archaeology* has its roots in the same period. During the 1950s, Lewis Binford, one of its chief founders, was a graduate student at the University of Michigan. One of his teachers, the famous cultural anthropologist Leslie White, advised him to find out what science is all about by studying the works of philosophers of science. He took that advice (Binford, 1972, pp. 7–8). One of the key features of the new archaeology is its emphasis on the search for scientific explanations that fit the deductive-nomological model.

The Hempel–Oppenheim article is a preliminary study. Hempel provided a much more full-blown account in his monographic essay "Aspects of Scientific Explanation" (1965b). This essay offers an inductive-statistical pattern of scientific explanation that supplements the earlier deductive-nomological model.[2] Both of these patterns are "covering-law models," in that each requires that explanations incorporate the statement of a universal or statistical law of nature. A heavily watered-down treatment of scientific explanation was given in Hempel's little 1966 textbook, *Philosophy of Natural Science*. The general account—offered in full detail in "Aspects of Scientific Explanation" and superficially in *Philosophy of Natural Science*—qualified, during the 1960s and 1970s, as the received view of scientific explanation. It was this view that profoundly influenced the new archaeology. The influence can readily be seen in Watson, LeBlanc, and Redman's *Explanation in Archeology: An Explicitly Scientific Approach* (1971), the locus classicus of the new archaeology. Their commitment to the received view is reiterated in the second edition, published under the title *Archeological Explanation: The Scientific Method in Archeology* (1984).

Influential as these standpoints were, not all archaeologists were persuaded by the new archaeology, and not all philosophers of science accepted the received view of scientific explanation. I shall not try to trace the subsequent developments in archaeology, since I am not qualified to do so. Nevertheless, I would like to say a little about developments in the philosophy of science concerning scientific explanation.[3] The philosophical situation has changed markedly since the early 1970s.

The first point to emphasize is that the 'received view' is no longer received. Indeed, there is widespread (though not complete) consensus among those actively working on scientific explanation that the 'received view' of the 1960s and 1970s is basically unsound. I shall not go into the details of the philosophical arguments that have brought about this change of attitude, but I would like to say a little about their upshot for archaeology.[4] To do so I shall briefly discuss two general approaches to scientific explanation.

1. Causal Explanation

It seems evident to common sense that, in many cases, to explain some phenomenon is to find and cite its cause. For example, to explain an airplane crash, the FAA looks for causes. Similarly, to find an explanation of the abandonment of Grasshopper Pueblo, archaeologists seek the causes of the departure. Hempel and Oppenheim ([1948], 1965, p. 250), in passing, casually identified their deductive-nomological pattern of explanation with causal explanation. In his fully developed theory, however, Hempel (1965b, pp. 352–354) explicitly denies that causality is in any way essential to explanation. Many philosophers have criticized the received view in general for its neglect of causal considerations, and in "Causality in Archaeological Explanation" (essay 22) I argue at length for its importance in archaeological explanation in particular.

The causal character of archaeological explanation has to be taken in conjunction with a recognition of the basic statistical character of explanations in the sciences—especially the behavioral sciences. In archaeology, for example, one might appeal to the fact that a particular hunting strategy is, in certain specific circumstances, more likely than another to yield success. This is obviously a probabilistic relationship. People often maintain, of course, that underlying the statistical explanations are deterministic causal relationships, and that we resort to statistical considerations only because of our ignorance of these underlying causal relationships. My own view is that we need not make gratuitous metaphysical assumptions of that sort. A more straightforward and realistic approach is to try to develop a theory of probabilistic causality.

A great deal of serious effort has been devoted to the elucidation of probabilistic causality, but the problem is not simple. For example, in elaborating a version of probabilistic explanation that he calls *aleatory explanation,* Paul Humphreys (1981, 1983, 1989) has pointed out that, in the statistical context, we must make allowances for both contributory causes and counteracting causes. In attempting to explain the abandonment of Grasshopper Pueblo, for example, we must take account of contributing factors such as the occurrence of a fairly severe drought and counteracting factors such as the existence of a highly developed stable community. The situation becomes extremely complex when we realize that, in some cases, two factors that qualify individually as contributing causes may, when they occur together, constitute a counteracting cause. Although one cannot say that a satisfactory account of probabilistic causality has been developed, I think we can say that important progress has been made in that direction.[5]

The recognition of causal aspects of scientific explanation does nothing to undermine the covering-law character of explanations. The causal processes and interactions to which we appeal for purposes of giving an explanation, whether of a deterministic or a probabilistic sort, are governed by *causal laws.*

2. Explanation by Unification

One idea—implicit in many works on scientific explanation, and made explicit by Michael Friedman (1974)—is that science enhances our understanding of the world by providing unified accounts of wide ranges of phenomena. Our understanding increases as we reduce the number of independent assumptions required to explain a given body of phenomena. Friedman cites as an example the kinetic-molecular theory of gases, which gives a unified account of a number of different gas laws—Boyle's law, Charles's law, Graham's law, etc.—and connects them with other mechanical phenomena that can be explained by Newtonian physics. The search for broad unifying theories has certainly been a major driving force in the history of science, and it has met with some striking successes in the natural sciences.

Explanatory unification often involves the reduction of one domain of science to another. When, for example, it was shown that visible light consists of electromagnetic waves that occupy a small segment of the total spectrum of electromagnetic radiation, optics was reduced to electromagnetic theory. Thereafter, it was not necessary to have two separate theories—one for optics and another for electromagnetic phenomena—because Maxwell's theory of electromagnetism covered them both.

The idea of explanatory unification in the behavioral sciences is more problematic, but it has often been associated with some notion of reduction. If, for example, methodological individualism is correct, then psychology furnishes the fundamental explanatory theory for anthropology, economics, political science, and sociology. I do not intend to argue the case for or against this sort of reductionism; I mention it only because it clearly illustrates the idea of explanatory unification.

One of the most basic issues associated with the new archaeology arises from the *covering-law* conception of scientific explanation. According to this conception of explanation—which was fundamental to the received view—every bona fide explanation makes essential reference to at least one law of nature. Considerable controversy surrounded the question whether there are any archaeological laws per se. Certainly archaeologists make use of various laws of nature. In radiocarbon dating, for example, one appeals to the law of radioactive decay. Many other examples could be given, involving laws of geology, biology, chemistry—as well, perhaps, as economics, sociology, and psychology. There still remains the question whether any *distinctively archaeological* laws exist. If not, then any legitimate archaeological explanation would have to depend on laws from other scientific disciplines, and these laws would provide unifying connections between a portion of archaeology and at least one other domain. Such explanations would thus exemplify the unification of archaeological phenomena with the phenomena in other realms of science, natural and/or behavioral. This *must not,* however, be taken to imply that explanatory unification cannot exist if laws peculiar to archaeology are invoked in archaeological explanation. There might be bona fide archaeological laws that can themselves be explained by laws of other domains. The fact that optics is reducible to electromagnetic theory *does not imply* that there are no laws of optics or that they are not used in explaining optical phenomena.[6]

3. Relations between the Two Types

Let us take a moment to compare and contrast causal explanations and unifying explanations. To provide a causal explanation of any given fact, it is often necessary to get into the nitty-gritty details of the causal mechanisms that produced the fact to be explained. To explain the location and contents of a particular burial, for instance, it may be necessary to ascertain the age and gender of the individual interred, and to determine the cause of death. In constructing causal explanations it is necessary, in general, to infer or postulate the existence of causal processes that are no longer available for our direct inspection. Moreover, causal explanations often appeal to entities such as atoms, molecules, or bacteria that are not *directly* observable under any circumstances; their observation or detection requires some sort of special apparatus. The causal explanation consists, in large part, of exposing hidden mechanisms.

Unifying explanations involve reference to broad structural features of the world. Consider a 'homey' example. A parent notices that a baby, left in a carriage with the brake on, can move it some distance by bumping and rocking, whereas it cannot move the carriage any significant distance if the brake is *off*.[7] One could, in principle, calculate the effect of each of the many causal interactions between the baby and carriage when the brake is off, thereby providing a causal explanation. Much more simply, however, we can cite the law of conservation of linear momentum—a fundamental and universal law of nature—to show that, no matter what the interactions between the baby and the carriage, no significant travel will occur. The unifying explanation shows how this peculiar bit of baby/carriage behavior fits into the universal scheme of things.

Explanations in the unification style can also occur in archaeology, as well as in other branches of the behavioral and biological sciences. Though the appeal may be to laws less fundamental than conservation of linear momentum, that does not necessarily disqualify them. Such explanations might involve fundamental principles of nutrition. It has been pointed out, for example, that corn (maize) existed as a cultivated crop in Arizona as early as 2000 B.C., but that it did not become a major cultigen until 1000 to 1500 years later. The evidence suggests that it became an important crop when beans were also available.

> The dietary needs of human organisms . . . have an important effect on patterns of adoption of domesticates. Corn, for example, cannot be used as a major source of protein because it lacks lysine, a major amino acid. Beans, however, are rich in lysine. . . . Corn and beans together can form the basis of a particular population's diet in a way that neither could alone. It is not surprising that prehistoric populations in Arizona did not seem to have begun to rely heavily on corn until after beans were also present in the region. (Martin and Plog, 1973, p. 284)

This explanation is based on extremely comprehensive principles of nutrition and biochemistry, but it does not make any attempt to fill in the details of the causal story of the introduction of beans to the area, or of the first efforts at large-scale corn production. In addition, as we shall see, explanations in terms of basic principles of biological evolution qualify for membership in this category.

Explanation through unification seems to me, fundamentally, a way of providing understanding of some phenomenon by relating that phenomenon to a *Weltanschauung* or overall conception of the world. This does not mean that one may pick just any worldview he or she happens to feel comfortable with; the picture of the world must be developed on the basis of the best scientific knowledge available to us. The adequacy of any explanation by unification is *not* just a matter of psychological comfort; the adequacy must be evaluated on the basis of objective scientific knowledge. But given an adequate scientific basis, such explanations do provide at least some measure of scientific understanding of the phenomena thereby explained.

When discussing scientific explanation, it is important to avoid thinking and talking about *the* unique correct explanation of any given phenomenon. There may, in general, be several different *correct* explanations of any such phenomenon. There will normally be many different sets of explanatory facts from which to construct *a* correct explanation. This point is recognized, at least implicitly by most philosophers who deal with this topic. I want to argue, in addition, that a given fact may have correct explanations of different types, in particular, correct causal explanations *and* correct unifying explanations (see "Scientific Explanation: Causation and Unification" [essay 4]).

4. Functional Explanation

If I am right in my assessment of the relationship between these two types of explanation, we might use it to deal with a long-standing problem regarding explanation that arises not only in archaeology but in several other biological and behavioral sciences as well. The problem to which I refer is functional explanation, and the point can best be seen in the context of evolutionary biology. It should be noted, to begin with, that the Darwinian theory of evolution provides an overarching framework for understanding the development of the various forms of life on our planet. It appeals to chance mutations, the heritability of traits, the struggle for survival, adaptedness to the environment, and the survival of the fittest. It does not, however, give an account of the details of the mechanisms of inheritance or natural selection.

Consider some particular trait of some particular type of organism—for example, the well-known case of protective coloration of the peppered moths in Liverpool. These moths live on plane trees, which have naturally a light-colored bark. Prior to the industrial revolution the moths were light gray, but when the soot from the factories blackened the bark of the trees, the moths became black. We can say that the dark color has the function of providing camouflage and thus lessening the chance of a given moth's being eaten by birds. Those moths that have a lighter color have a greater chance of becoming prey for birds. Thus, on average, the darker-colored moths tend to reproduce more frequently, and the lighter ones tend to reproduce less frequently. This unbalanced color situation with respect to progeny has the effect of producing a dark color for the species as a whole. In this century, when the industrial pollution in Liverpool was cleaned up, the color of the moths reverted to light gray. The dark color no longer functioned as effective camouflage; the lighter color functioned better in the changed environment. Functional explanations of this sort are perfectly legitimate in the context of evolutionary biology, and they provide scientific understanding of the phenomena they seek to explain. Although—as Larry Wright (1976) has convincingly argued—there is an important sense in which the evolutionary explanation is causal, it does not give a fine-grained causal account of the underlying mechanisms. As long as the trait of color is heritable, it does not matter to the evolutionary account what precisely is the mechanism of inheritance.

There is, at the same time, the possibility in principle, if not in fact, of giving a full causal explanation of the color of these moths in biochemical terms. This type of explanation would appeal to the chemistry of DNA and RNA, and to the synthesis of the proteins that provide color in the surface of the organism. It is also a legitimate explanation of the color of the moth. This kind of explanation does involve the fine-grained details of the causal mechanisms involved in the production of the trait in question. Wright (1976, p. 59) clearly recognizes the compatibility of functional (or teleological) explanation and fine-grained causal explanation.

5. Conclusions

The point for which I am arguing has not been widely acknowledged. Indeed, traditionally—going back to the Hempel-Oppenheim paper ([1948] 1965)—there has been a deep and continuing dispute between those who upheld the received view and those who supported a causal conception of explanation. And the controversy was not completely ill founded, because both the received view and the causal conception have evolved considerably from their earlier forms. It makes sense to suggest that they are compatible in their present forms only because such developments have occurred.

Earlier versions of the causal conception suffered from the lack of any adequate analysis of the very concept of causality. It was often taken as a primitive concept, about which we could make accurate intuitive judgments, without any need for further clarification. On the basis of recent work on causality—including, of course, probabilistic causality—it is possible to offer a more defensible account.

As I see the situation, the received view has evolved into the view of explanation as unification. But a significant change has occurred in the course of this evolution. Whereas the received view was prepared to accept as legitimate any subsumption of a fact to be explained under a bona fide law of nature, no matter how narrow its scope, the newer view looks at the overall structure of scientific knowledge and judges explanations in terms of their ability to unify. As Friedman pointed out, his view of explanation as unification is a global view; the received view was local.

Indeed, as Philip Kitcher (1985) has remarked, we can look at the two approaches to explanation—via unification and causation—as "top down" and "bottom up," respectively. The former looks at the entire structure of scientific knowledge, with special attention to its highest-level theories, and works down from there, so to speak. The latter pays fundamental attention to the nitty-gritty details of the causal mechanisms, and builds up from there. I see these two approaches as complementary rather than mutually exclusive (see "Scientific Explanation: Causation *and* Unification" [essay 4]).

Scientific understanding is, I think, a rather complicated matter. It has many aspects. I have discussed two of them in this essay, namely, the exposing of underlying causal mechanisms and the exhibition of global structures. Perhaps there are others as well. Our minds should be open to that possibility.

We must realize, moreover, that there are many types of explanation that make no pretense of being scientific, but which are perfectly legitimate nonetheless. An archaeologist can explain to a student how to prepare a sample of material for radiocarbon dating. This is not an attempt to explain *why* some natural phenomenon occurs; it is an attempt to explain *how to do something*. An archaeologist can explain to a mechanic what is wrong with a backhoe in the hope that the mechanic can repair it. Explaining *what* is fundamentally different from explaining *why*. An archaeologist can try to explain the meaning of a particular decorative design on pottery in a particular cultural setting. The result may involve a description of various psychological responses it arouses in members of the group. This kind of explanation is closely akin to explanations of the meanings of paintings or poems in our culture. Such explanations have aesthetic or religious significance, but they do not pretend to furnish *scientific* explanations of natural phenomena. To distinguish these other types of explanation from scientific explanations is not to disparage them—quite the contrary, to confuse different sorts of explanation with one another interferes with the appreciation of all of their importances.[8]

So, how does all of this relate to the new archaeology? It does nothing to undermine the thesis that a basic aim of archaeology is to provide scientific explanations of phenomena in its domain. The moral is that the rather simplistic and rigid notions of scientific explanation furnished by philosophers in the heyday of logical empiricism and of the new archaeology should give way to more sophisticated and complex conceptions of the nature of scientific explanations.

Notes

An earlier version of this essay was presented at the First Joint Archaeological Congress, Baltimore, 5–9 January 1989.

1. The results of this work appeared in Salmon (1989, 1990b).

2. Hempel also introduces a deductive-statistical pattern, which is not of particular concern in this discussion. Moreover, explanations of this type can appropriately be considered a subspecies of deductive-nomological explanation.

3. An account of these developments can be found in "Scientific Explanation: How We Got from There to Here" (essay 19).

4. Some of these arguments are given in essays addressed to archaeological audiences, namely, essays 21 and 22.

5. Patrick Suppes (1970) offers a classic systematic treatment; "Probabilistic Causality" (essay 14) contains a survey of various theories. Humphreys (1989), Eells (1991), and Hitchcock (1993) offer important further developments.

6. The most advanced and thorough account of the unification approach is given in Kitcher (1993).

7. For purposes of this example we must assume that the rolling friction of the carriage with the brake off can be neglected.

8. The point is made in the introduction to this book and further elaborated in "The Importance of Scientific Understanding" (essay 5).

24

The Formulation of Why-Questions

For several decades archaeologists have exhibited active interest in the nature of scientific explanation and in the attempts of philosophers of science to provide a reasonably precise characterization of it. A great deal of attention has been devoted to the account provided by Hempel (1965b) and to the D-N (deductive-nomological) and I-S (inductive-statistical) models he advocated. In the meantime, other philosophers have developed competing alternate accounts. "Alternative Models of Scientific Explanation" (essay 21) discusses several of the alternatives and offers a critical comparison of my S-R (statistical-relevance) model (1971) with Hempel's D-N and I-S models. Cartmill (1980), while acknowledging the force of one particular counterexample against Hempel's D-N model, argues that it cuts equally sharply against my S-R model. In this essay I analyze Cartmill's criticism and defend the S-R model against it.[1] I argue that the issue hinges on the nature of explanation-seeking why-questions. I use this occasion to discuss the nature of such questions, and to make some remarks about a theory of scientific explanation advanced by Bas van Fraassen (1980) which accords to explanation-seeking why-questions a genuinely central role. This theory of scientific explanation is, in my view, one of the most significant contributions to the discussion of scientific explanation in recent years.

In order to deal with the fundamental issue raised by Cartmill, I must sketch a little bit of the philosophical background. For some time I have been convinced that Hempel's well-known D-N and I-S models of scientific explanation are vulnerable to a fundamental criticism. The difficulty centers on the problem of explanatory relevance. According to Hempel's theory, an event is explained by showing that it was to be expected, either with deductive certainty (D-N explanation) or with high inductive probability (I-S explanation), on the basis of the explanatory facts. According to the S-R (statistical-relevance) approach, a satisfactory explanation must rely on facts that are *statistically relevant* to the occurrence of the event to be explained, and it must not appeal to irrelevant facts.[2]

The main point is most easily seen in connection with statistical cases. Consider, for instance, the claim that the consumption of massive doses of vitamin C explains quick recovery from a cold.[3] Let us assume, for the sake of the argument, that the probability of quick recovery—say within a week—is quite high for those who take large doses of vitamin C. The question is whether taking vitamin C *makes a difference* to quick recovery. If carefully controlled double-blind experiments show that there is no difference in speed of recovery between those who are given great quantities of vitamin C and those who receive a placebo, then the explanatory import of the consumption of massive doses of vitamin C to quick recovery from the common cold is nil. It is easy to find or concoct many other examples of I-S 'explanations' that suffer from the same sort of failure of relevance.

The switch from a high probability requirement to a relevance requirement leads to a fundamental change in the concept of statistical explanation. To say that it is highly probable that quick cold recovery will occur obviously involves only one probability. To say that consumption of vitamin C is relevant to quick cold recovery requires two probabilities whose values must be compared. We must either compare the probability of quick recovery among those who take vitamin C with the probability of quick recovery among those who do not (as in the controlled experiment just mentioned) or compare the probability of quick recovery among those who take vitamin C with the prior probability of quick recovery regardless of whether vitamin C is used. For purposes of characterizing the statistical-relevance concept of explanation, I prefer the latter comparison. This means that we must give the *prior probability* of quick recovery a prominent place in our explanatory schema.

Let us see what bearing this issue has on the nature of the explanation-seeking why-question. If one wishes to apply the I-S model, the question may be phrased, "Why did this quick recovery occur?" The answer would be "Because the subject took massive doses of vitamin C, and almost all people with colds who take massive doses of vitamin C recover quickly from their colds." If, however, one wishes to apply the S-R model, then the question would be phrased, "Why did this person who had a cold get over it quickly?" The answer would be, "Because the subject took massive doses of vitamin C, and the posterior probability of quick recovery from a cold, given use of vitamin C, is greater than the prior probability of quick recovery regardless of medication." For purposes of the I-S model the question has the form, "Why does this x have the property B?" For purposes of the S-R model the question has the form, "Why does this x, which is an A, have the property of B?" As we shall see, this difference in form between the two different why-questions, which may seem rather trivial at first blush, has profound consequences.

Problems of relevancy quite similar to those associated with I-S explanation also infect D-N explanations. Consider the classic case of John Jones, whose failure to become pregnant during the past year is 'explained' by the fact that he faithfully consumed all of his wife's birth control pills. Inasmuch as men never become pregnant anyhow, his consumption of oral contraceptives has no explanatory import.

If we ask why John Jones avoided pregnancy, we have not phrased the explanation-seeking why-question in an appropriate canonical form for application of the S-R model, for we have not specified a reference class with respect to which the prior probability is to be taken. If, however, we translate the question into "Why did John Jones, who is a man, fail to get pregnant last year?" we have a properly formed why-question. We see immediately that the prior probability of pregnancy in the class of men is equal to the posterior probability of pregnancy in the class of men who consume oral contraceptives. If we partition the class of men into two subclasses, those who consume oral contraceptives and those who do not, that partition is irrelevant to nonpregnancy because the probability of nonpregnancy is equal in each of the two subclasses and in the initial reference class. In the S-R model of scientific explanation we prohibit the use of irrelevant partitions; we insist on maximal homogeneous partitions.

The criticism offered by Cartmill—which is similar to one offered by King (1976)—is based on the claim that the class of men (human males) is not the broadest reference class available. Cartmill elaborates as follows:

> Let us assume that the pills he [John Jones] has been taking are perfectly efficacious contraceptives for *Homo sapiens* but have no effect on other organisms, and let us define a pregnant organism as one which contains viable zygotes to which it contributed gametes. (By this definition, snails, hens, and male sea horses can get pregnant; amoebas, ganders, and female sea horses cannot.) The three large classes to which Mr. Jones belongs—humans, males, and pill takers—are heterogeneous with respect to likelihood of pregnancy; some humans (e.g., young women), some males (e.g., sea horses), and some pill takers (e.g., hens) are more likely to get pregnant than others. . . . The S-R model therefore does not warrant the conclusion that John Jones's pregnancy should be explained in terms of gender rather than pharmacology. If the case of John Jones is fatal to the D-N model, it is equally fatal to the S-R model. (Cartmill, 1980, pp. 383–384)

The source of Cartmill's problem lies in the fact that the explanatory question in this example is not clearly specified. We cannot proceed with the explanation until this is done. Is John Jones, for purposes of this question, a representative of the class of humans, males, pill takers, animals, living organisms, or what? Depending on how we clarify the explanation-seeking why-question, we will call forth different answers—and surely there is nothing surprising in that fact. If the question is "Why did this entity (John Jones) which is a man (A) avoid becoming pregnant (B)?" the answer "Because he is a regular consumer of oral contraceptives (C)" is not a satisfactory answer. The appropriate answer to this question is that no explanation is either possible or needed, for the class of men is homogeneous with respect to pregnancy. If, however, the question is "Why did this entity which is a pill taker (A') avoid becoming pregnant (B)?" the answer "Because he is human (C')" is suitable, for, by Cartmill's stipulation, pills work only for humans. Once the explanation-seeking why-question is rephrased in standard form, thus furnishing an initial reference class on which the prior probability is based, there is no need to hunt for a broader reference class. The S-R model demands only that, within the reference class to which the why-question refers, no irrelevant partition is made. This is the force of the requirement that the fact to be explained be referred to the broadest homogeneous subclass of the given initial reference class. I do *not* mean to deny, however, that a different explanation that might be scientifically interesting and important could result from a different why-question.

Other important features of why-questions have been emphasized in the philosophical literature. It has long been recognized that questions have presuppositions—"Have you stopped exploiting your research assistants?"—and why-questions are no exception. We do not try to explain things that did not happen. To pose an explanation-seeking why-question is to presuppose that the event to be explained actually occurred. To ask "Why did ancient visitors from outer space leave large-scale diagrams in South America?" is very apt to be an illegitimate question because its presupposition is not fulfilled. This point is almost too obvious to need stating.

Van Fraassen (1980) has also called attention to the fact that why-questions need what he calls *contrast classes*. Consider the case of the notorious bandit Willie Sutton. Asked by a journalist why he robbed banks, he answered, "Because that's where they keep the money." This answer is humorous because the journalist was obviously considering one contrast class while Sutton employed another. The journalist presumably had in mind the notion that among the vocations available to him—e.g., plumber, civil servant, college professor, writer, robber—he chose the last, and the question was why. Sutton clearly took as his contrast class the kinds of places that might be robbed—e.g., gas stations, liquor stores, post offices, factories, banks—and he robbed banks for the reason mentioned.

If an explanation-seeking why-question is to be posed in full and explicit form, the intended contrast class must be specified. This step is familiar from standard statistical practice; it amounts to the specification of an appropriate sample space. Consider an example, owing to James Greeno ([1970] 1971), of S-R explanation. Suppose that Albert has been arrested for stealing a car. If we ask why he stole the car, we are presupposing that he did, in fact, do so. As long as he maintains his innocence, he has a perfect logical right to reject the question, "Albert, why did you steal a car?" If we satisfy ourselves that he did steal a car, we must next decide what population he is to be taken to represent. If we are concerned with juvenile delinquency, we may decide to regard Albert as a member of the class of American teenagers. This specifies the population to which the prior probability of the event is to be referred. We formulate the question more explicitly. "Why," we ask, "did this American teenager steal a car?" Third, we must choose our contrast class. In presenting his example, Greeno chose to partition the class of American teenagers into those who have committed at least one major crime, those who have committed only minor offenses, and those who have not committed any crimes at all. Since stealing a car is a major crime, the question is finally well specified. Given that Albert might have belonged to any of the three subsets in the aforementioned partition of American teenagers, why does he belong to the first (those who have committed major crimes)? Presumably some sort of answer can be given in terms of various relevant sociological or psychological factors—e.g., sex, socioeconomic background, personality traits. We explain Albert's crime, in part at least, by pointing out that he is a psychopathic male from a poverty-stricken family, etc.

Notice, however, that the original question could have been explicated quite differently with quite different results. "Why," we might have asked, "did *Albert* steal a car?" Because, we might learn, among the members of his teenage gang, Albert is the most adept at hot-wiring. In construing the why-question in this way, we are taking the initial reference class to be the members of the gang, and the contrast class as {Albert steals the car, Ben steals the car, Charlie steals the car, etc.}. Then, again, we might have asked, "Why did Albert steal a *car?*" This time the initial population would be objects that might have been stolen by Albert. The contrast class might be {cars, color TVs, money, bottles of liquor, etc.}. Because he wanted to go joy-riding might be a correct explanation. Clearly, the sorts of presuppositions that accompany explanation-seeking why-questions have a great influence on the sort of explanation that constitutes a suitable answer.

Does any of this have any scientific significance? I do not know. I cannot cite any bona fide scientific case in which any actual confusion has arisen over failure to make presuppositions of why-questions explicit. These issues are, nevertheless, matters of considerable importance to those who are trying to construct precise explications of scientific explanation. What moral should we draw? I have two suggestions, one somewhat facetious and the other more serious.

The first can be put quite succinctly: Ask a philosophical question and you may get a philosophical answer. There may be an important lesson for archaeologists here.

The second involves quite a general point. The attempt to construct a philosophically sound explication of scientific explanation is a highly abstract enterprise. As we have learned from the history of mathematics, it is often impossible to tell in advance what concrete applications, if any, will result from such endeavors. That observation suggests the possibility that philosophical attempts to characterize scientific explanation may yet have significance for practitioners of archaeology that we have no way of anticipating at the present moment.

Notes

This essay was presented in a symposium, "Prospects for Philosophy of Archaeology," at the 1983 meeting of the Society for American Archaeology in Pittsburgh.

1. Although I no longer believe that the S-R model provides an adequate account of scientific explanation, I do not believe that it falls victim to the sort of objection Cartmill raises. My current view is that scientific explanations involve causal considerations, but that the S-R schema constitutes a correct basis on which acceptable explanations can be erected. The details of this approach are spelled out in Salmon (1984b). "Causality in Archaeological Explanation" (essay 22) discusses the bearing of causal considerations on archaeological explanations.

2. If a causal account of explanation is adopted, then we must insist that only *causally relevant* facts are to be taken to have explanatory force.

3. [I now believe it is much more plausible to claim that vitamin C aids in the avoidance of colds rather than in quick recovery from them (see Pauling [1970]).]

25

Quasars, Causality, and Geometry

A Scientific Controversy That Did Not Occur

Q uasars, originally called quasi-stellar radio sources, were discovered over thirty years ago; the discovery was announced to the world at large in the December 1963 issue of *Scientific American* (Greenstein, 1963). The discovery occurred at a time when radio astronomy was beginning to achieve fairly high resolution, so that radio sources could in some cases be identified with visible objects. In fact, the first identification of a quasi-stellar radio source was made in 1960, but at the time it was taken to be a visible star that had the rather unusual property of emitting radio waves.

1. The Discovery

The key to the 1963 discovery was the careful spectral analysis of three sources, in which several known spectral lines could be identified if one assumed very large redshifts, suggesting that these sources are not stars in our galaxy but rather extragalactic objects. The chief observational data available at the time were the following:

1. In photographs they look like faint stars; they emit in both the optical and the radio regions of the spectrum.
2. Their spectra show very large redshifts.
3. Their brightness varies rapidly, for example, by as much as 30 percent in a year.

These observations were, of course, made with the aid of the most sophisticated instruments of observation available at the time.

The first problem concerns the redshifts. If they are cosmological—that is, results of the overall expansion of the universe—their sources must be very far away, for example, 5 billion light years. There are, of course, other types of redshifts, and for a time some astronomers denied that these were cosmological, but that notion seems by now to have been pretty generally abandoned. This was not a major scientific controversy; it was a relatively short-lived disagreement about the interpretation of the data. It follows, then, that these sources must be extremely bright—perhaps 100 times as bright as our galaxy. Otherwise, given their enormous distances, we would not be able to see them.

Item 3, the variability of the sources, has been crucial in the minds of many astrophysicists. They have used a causal argument in an attempt to show that the relatively rapid variability in brightness implies that the sources are extremely compact. This conclusion was drawn in 1963, and it has been frequently repeated ever since, right up to the present. Moreover, since 1963, many other quasi-stellar radio sources—now usually called quasars or QSOs—have been discovered with *much* greater redshifts and *much* more rapid variation (on the order of days). According to the standard line of reasoning, they must be *much* brighter and *much* more compact. It is this causal argument on which I wish to focus attention.

2. The Argument

The argument in question is based on what might be called "the $c\Delta t$ size criterion," where c is the speed of light and Δt is the time in which the variation occurs. It goes as follows:

> An overall change in brightness can be achieved only by propagating signals throughout the region of variation.
>
> No signal can travel faster than light.
>
> ---
>
> The region of variation cannot be larger than distance light travels in its time of variation.

It should be added that the variation need not be periodic, and that it may be either an increase or a decrease in brightness.

I shall show that this argument is fallacious—indeed, egregiously fallacious. The question is why it has hung on for thirty years without noticeable dissent (the only exception I know of is Dewdney [1979]), during which time it has been applied to a wide variety of other fluctuating objects, for example, BL Lacertae objects (BL Lacs), pulsars, X-ray bursters, Seyfert galaxies, and active galactic nuclei (AGN), including that of our very own Milky Way. A number of examples are given in the appendix of this article. As a matter of fact, the fallacious character of the $c\Delta t$ size criterion argument was pointed out in the following year by Banesh Hoffmann (1964) in a brief article in *Science*. Like any other fallacious argument, this one can, of course, be made valid by furnishing additional premises; but if this strategy is to be adopted, we deserve at least a hint of what these other premises might be. A different response to Hoffmann (Terrell, 1964), offered in the same year and in the same journal, advanced a geometrical argument to which I shall return later.

My interest in these issues was first aroused by an article on BL Lacertae objects in which the following claim was made:

> A successful model must account for the operation of what may be the most powerful engine in the universe, and it must fit that engine into a compartment of trifling size. *A crucial test of all models* is the fastest variation in luminosity that can be

accommodated, since that period corresponds to the time required for a signal to traverse the emitting region. *Some models have had to be discarded already* because they do not allow rapid enough variations. (Disney and Véron, 1977, p. 39; emphasis added)

A similar claim occurs in the *Astronomy and Astrophysics Encyclopedia* (Maran, 1992): "At present observations only give upper limits on the sizes of these objects [the central engines]. . . . Some AGN [active galactic nuclei] are strongly variable; in these, *causality limits the size* to the distance light can travel in a characteristic variability time" (ibid., p. 8; emphasis added). I do not know whether this encyclopedia qualifies as a serious technical reference work or merely a coffee table display piece. Be that as it may, a similar argument is offered in the massive treatise *Gravitation* (Misner, Thorne, and Wheeler, 1973, p. 634), which is *very* far removed from the category of coffee table fluff.

It should be noted that in the quotation from Disney and Véron, the $c\Delta t$ size criterion is applied to the "emitting region" of the object in question, not to its overall size. The same point arises in the *Encyclopedia* quotation. This statement explicitly applies the $c\Delta t$ size criterion to the nuclei of galaxies; it is not used to ascertain the overall sizes of the galaxies themselves.

As things have turned out, astrophysicists now have a rather generally accepted model of quasars—matter falling into a black hole from an accretion disk—and it does satisfy the $c\Delta t$ size criterion. Although there are still technical problems to be solved, I am *not* rejecting the model; the object of my criticism is the *argument*. It might be said that the argument is irrelevant; the aim of the exercise is to construct a satisfactory model rather than to support a statement—a theory or hypothesis—as the conclusion of an argument. But even if model-building is the aim of astrophysics, one can reasonably ask what constraints should be placed on the model. Surely no model that involved violation of the law of conservation of angular momentum could be accepted; the same would be true, I should think, of a model that violated special or general relativity or the laws of optics. But even though many authors seem to claim the same status for the $c\Delta t$ size criterion, suggesting that it is a consequence of special relativity, it cannot be put into the same category. It is neither *a basic principle of,* nor does it *follow from,* special relativity. So whether we are dealing with theories and their supporting arguments or models and their constraints, the same fundamental issues concerning the $c\Delta t$ size criterion remain.

3. The Fallacy

In order to see the invalidity of the argument based on the $c\Delta t$ size criterion, it is essential to understand the distinction between genuine causal processes and pseudo-processes. Consider a simple example (see fig. 25.1). Suppose a large circular building, such as the Astrodome, is fitted out at its center with a rotating beacon that sends out a beam of white light. When the light is on, and the interior is otherwise dark, the beacon casts a white spot on the wall. As the beacon rotates, the spot of light moves around the wall. A pulse of light traveling from the beacon to the wall is clearly a causal process, and it transpires at the speed of light. No causal process can travel faster than light (in vacuo). Its causal character is revealed by the fact that it can transmit a mark; for example, if a red filter is

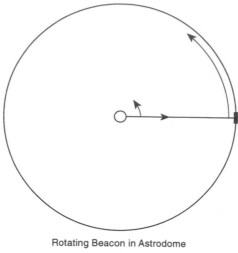

Rotating Beacon in Astrodome
Pulsar in Crab Nebula
$v \approx 4 \times 10^{13} \times c$

Causal vs. Pseudo-Processes

Figure 25.1

placed in its path anywhere between the beacon and the wall, the white light changes to red *and it remains red* from that point on until it reaches the wall.

The spot of light that moves around the wall is a pseudo-process. It is possible to mark the spot at any given place in its path—for example, by placing a piece of red cellophane at the wall where the spot passes—but when it travels past that point, it will revert to its white color. It will not continue to be red as a result of a single local intervention in its path. Pseudo-processes can be marked, but they do not transmit marks.

Suppose that our Astrodome has translucent walls, and that it is being observed at night by a distant observer. It will appear to get brighter and dimmer as the spot of light sweeps around the walls. Moreover, there is no finite limit on the speed at which the spot can travel. Imagine that as the beacon continues to rotate at the same rate, the size of the building increases. The time required for the spot to traverse the entire circumference will remain constant, but the distance traveled in that time will increase as the circumference does.

Let us consider, instead of the foregoing fictitious example, a real physical system. There is a well-known pulsar in the Crab Nebula that is believed to be a rotating neutron star that beams radiation toward us much as the fictitious beacon in the Astrodome beams radiation to the walls of the building. The pulsar rotates 30 times per second and it is located 6500 light years from us. Look at figure 25.1 again, but this time suppose that the beacon is the pulsar and that Earth is located at some point on the circumference of the circle. A light pulse would require 13,000 years to cross a diameter of the circle, but the spot of radiation requires one thirtieth of a second to sweep around the circumference. As this spot passes us, it is traveling at about $4 \times 10^{13} \times c$. Faster and more distant pulsars are known, but if 4×10^{13} is not a big enough factor to be convincing, I doubt that a few more orders of magnitude would do the trick.

One possible objection to the example of the Astrodome with translucent walls is that, while the *emitting region* has the dimensions of the whole building, the *source of emitted energy* is much more compact, namely, the beacon at the center. This source would satisfy the $c\Delta t$ size criterion. Applying the criterion to the energy source instead of the emitting region makes sense because, from the beginning, a major problem about quasars has been to explain how such prodigious quantities of energy could be radiated by highly compact sources. But even though it requires us to furnish a slightly more complicated example, this shift does not save the criterion. Imagine, instead of a beacon in a building with translucent walls, a celestial object surrounded by a cloud of atoms or molecules in a metastable excited state. Suppose that this central object emits a quick burst of radiation that propagates isotropically toward the surrounding cloud, and that this light causes a burst of radiation by stimulated emission. The central source emits a relatively small quantity of radiant energy; the major part of the energy radiated by the cloud resides in the cloud; the central light is only a trigger. Although it could not be seriously entertained as a model of a quasar, because its spectrum would be totally unsuitable, this physically possible example shows that the $c\Delta t$ size criterion does not necessarily apply even to the size of the energy source.

The foregoing examples are mine, but in an article, "The Quasar 3C 273," on one of the three discovered in 1963, Courvoisier and Robson (1991) appeal explicitly to the $c\Delta t$ size criterion, and they offer the following analogy to explain its application:

> As a simple example, consider a line of 10 light bulbs. If one wishes to decrease the total luminosity significantly, a large number of the bulbs must be switched off, say, at least six. To do this, one must send a signal that instructs the bulbs to turn off. . . . The dimming process will therefore take at least the time light needs to cross the distance from the center of the line of bulbs to the most distant bulb to be turned off. (p. 54)

In a letter (20 June 1991) to the editor of *Scientific American,* in which the article appeared, I wrote:

> Far from supporting their contention, this example clearly shows that the size of the array cannot be inferred from the time required for dimming. Suppose that the bulbs are arranged along an arc of a very large circle. At the center of this circle place a powerful laser on a pivot. Aim it at the line of bulbs, turn it on, and rotate it rapidly. This device can zap the entire group of 10 light bulbs in an arbitrarily small time interval. No part of the laser needs to travel at a speed greater than that of light. To decrease the amount of time required for any given speed of rotation of the laser, simply move the laser farther away. Better yet, set up ten lasers, one aimed at each bulb, and turn them on simultaneously, wiping out the whole array of bulbs instantaneously.

On 5 July 1991 I sent a postscript to the foregoing letter in which I added, "I thought you might be interested to note that the very argument I criticize occurs in [the July] issue as well. . . . In this case it is applied to the Great Annihilator, but it is precisely the same argument."

My letter was sent to one of the authors (Courvoisier); the entire text of his reply follows:

The flight time arguments brought by Professor Salmon are correct. They are indeed used in the description of light echo phenomena (for example in the context of SN 1987A). The relevance of these arguments in the context of quasars and AGN is, however, not convincing. The following points can be made:

1. The arguments we used can be applied to the distance between the laser or whatever control the apparatus and the bulbs.
2. One can imagine many kinds of particular geometry with alignments along the line of sight in which the light travel time arguments can be defeated. They all suffer from being peculiar and contrived. Consider e.g. one single line of bulbs aligned with the line of sight and have the switch signal start at the furthest bulb. Since the switch signal travels at about the speed of light along the wire, we will have the impression that the process of intensity decrease (or increase) takes less time than the length of the array divided by c.

The time of flight argument is not watertight and we know that very well; it is nonetheless a very reasonable estimate of sizes which does not presuppose specific geometries. (quoted by the kind permission of T. Courvoisier)

Courvoisier's first point—that the $c\Delta t$ size criterion can be applied to the operation of the laser I proposed—is true, but irrelevant. In the case of the quasars, we observe the fluctuation on the surface; we do not observe the mechanism that produces it. We must keep clearly in mind the fact that the entity emitting the radiation is a three-dimensional object, whereas all that we can observe is part of its two-dimensional surface. The problem is to infer the size of the object from its observed period of variation. Because we cannot observe the internal mechanisms, the $c\Delta t$ size criterion does not solve that problem.

In his second point Courvoisier complains that examples like mine are "peculiar and contrived." Regarding this criticism I have two responses. First, the example cited is theirs, not mine. Second, if someone produced an intricate and complex device that turned out to be a genuine perpetual motion machine of the first kind—one that could actually do work without any input of energy—I doubt that anyone would complain that it was contrived. It would indeed be a contrivance, but one that would be extraordinarily interesting and useful.

Courvoisier concludes by remarking that their argument is not "watertight" and they are perfectly aware of that fact. This constitutes an explicit recognition that the $c\Delta t$ size criterion does not have the status of a law of nature or a consequence thereof. The editors of *Scientific American* informed me that my letter and the coauthor's reply did not merit publication.

4. An Actual Counterexample

Enough of these fictitious setups. In 1986 I sent the following technical report to *Science:*

Charles V. Shank's article, "Investigation of Ultrafast Phenomena in the Femto-second Time Domain" (*Science,* 19 September 1986), contains a fascinating discussion of the generation and uses of extremely brief pulses of light. I was, however, astonished at what seems a glaring omission—i.e., any reference to the minute size of the apparatus that produces these pulses. Indeed, the article contains no hint of the miracle of miniaturization that has apparently been achieved.

My knowledge of this feature of Shank's work is not derived from direct acquaintance; it comes from an application of a principle of astrophysics. In discussions of such fluctuating sources as quasars, BL Lacs, X-ray bursters, and pulsars, appeal is often made to what might be called "the $c\Delta t$ size criterion." According to this criterion, an upper limit on the size of a source that fluctuates over an interval Δt is given by the product of Δt and the speed of light (3×10^{10} cm/sec). It is often presented as a rigorous consequence of special relativity, and hence as an inviolable law of nature.

For example, the very next issue of *Science* (26 September 1986) contains the article by K. Y. Lo, "The Galactic Center: Is It a Massive Black Hole?" (pp. 1394–1403). Writing about radiation from active galactic nuclei in general, he says, "Such radiation is sometimes found to vary on time scales as short as days, implying that the source is $< 10^{17}$ cm in extent" (p. 1395). Although Lo does not explicitly invoke the $c\Delta t$ size criterion—probably because it is too well known to require mention—that would appear to be the basis of his calculation. An upper limit on the size of a source that fluctuates in one day is about 2.6×10^{15} cm, and one that fluctuates in 100 days has the approximate limit given by Lo. In a 1982 report in *Science* on the same topic, M. Waldrop explicitly invoked the $c\Delta t$ size criterion, taking it as an unexceptionable law of nature (Waldrop, 1982). V. Trimble and L. Woltjer, in the recent survey article, "Quasars at 25" (*Science*, 10 October 1986, pp. 155–161) [they took 1960 as the date of birth], also seem to make repeated appeals to this size criterion (pp. 155–157).

Shank reports that optical pulses with durations less than 8 femtoseconds (1 fsec = 10^{-15} second) have been produced. Applying the $c\Delta t$ size criterion to the sources of such ultrabrief pulses we find that the upper limit on their size is 2.4×10^{-6} m. [This is roughly the length of a human chromosome; such apparatus would fit conveniently within a human cell.]

In all seriousness, I profoundly doubt that Shank's apparatus (including such equipment as tunable dye lasers) has dimensions of a couple of microns. So how are we to reconcile laser theory and astrophysics? The answer was given by Banesh Hoffmann in the pages of this journal more than 20 years ago. The $c\Delta t$ size criterion is not a law of nature; it is not a consequence of special relativity. What is its status? At best it is a heuristic device that has been used to assess the plausibility of physical theories pertaining to quasars, black holes, and other celestial objects.

This technical report was not published; it was returned to me without comment. The editors of *Science* apparently had no sense of humor. Also, I think, they had no sense of history. They seemed to see no problem in an Aristotelian brand of physics that has one set of laws for terrestrial phenomena and an entirely different set for celestial phenomena.

5. The Geometric Argument

Hoffmann (1964) pointed out that the surface of a sphere could brighten instantaneously as a result of a causal process that propagated uniformly from the center, reaching all parts of the surface simultaneously. A concrete illustration of this idea was published in the *New York Times* (Browne, 1993). It involves a new theory regarding the nature of

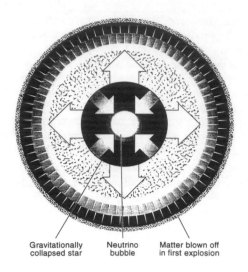

Gravitationally   Neutrino   Matter blown off
collapsed star   bubble   in first explosion

Supernova Explosion

Figure 25.2

supernova explosions. I am neither endorsing nor rejecting this theory; it is simply an example of Hoffmann's basic point. Standard supernova theory holds that when a star has used up almost all of its supply of hydrogen, a series of nuclear transmutations occurs creating iron and lighter elements. This is followed by a violent implosion in which heavier elements are created. According to the new theory, "a bubble of neutrinos forms in the implosion, lasts for 10 seconds, then ripples outward. As it reaches the surface a flash of light bursts out. Billions of miles out into space, shock waves from the star collide with gas ejected years before, generating radio signals that accompany supernova explosions." (See fig. 25.2.)

Given the obvious physical possibility of this sort of phenomenon, Terrell (1964) asks what a distant observer will see if it occurs. His answer is that the brightening will appear to occur not instantaneously but over a span of time. In a nice rhetorical flourish he appeals to relativity theory as the basis for asserting that the speed of light is finite (a fact we have known since Rømer's observations of the eclipsing of the moons of Jupiter in the seventeenth century). If we are observing a large spherical object, the light from the parts nearest to the observer will arrive earlier than light from the periphery because that light has farther to go to reach us (see fig. 25.3). Moreover, the difference in distance is roughly equal to the radius of the sphere, so the result is similar to the conclusion drawn from the $c\Delta t$ size criterion. I have noticed this argument in the semi-popular literature on quasars only once, namely, in Paolo Maffei's *Monsters in the Sky* (1980, p. 263).

Notice the relationship between the $c\Delta t$ argument and the geometrical argument. According to the former, actual instantaneous brightening is physically impossible. According to the latter, even if actual instantaneous brightening occurs, it will appear to be noninstantaneous. In fact, given certain particular geometrical configurations, these

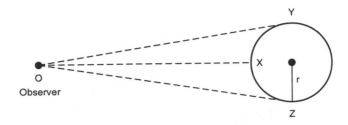

Light from a Luminous Sphere

Figure 25.3

two considerations can cancel each other out. The point (which I had not noticed before) was illustrated in Courvoisier's reply to my letter. Recall the case of a line of ten light bulbs arranged almost along our line of sight. A switch is flipped at the most distant bulb, and a signal travels along the line of bulbs at about the speed of light, turning on each bulb as it reaches it. Since the signal controlling the lights travels at about the same speed as the light from the bulbs, the light from all of the bulbs will reach us almost simultaneously, no matter how far apart the bulbs happen to be.

The striking feature of Terrell's geometrical argument is its dependency on the approximately spherical shape of the emitting object. Suppose, instead, that the emitting object is a flat disk oriented perpendicularly to our line of sight (see fig. 25.4). Let X be the center of the disk, Y a point on its edge, and O the position of the observer. The Pythagorean theorem, along with some trivial high school algebra, shows that the absolute difference between the length of OX and that of OY approaches zero as OX increases in length and XY remains fixed. Indeed, if we were looking at a ring with the same orientation instead of a disk, all of the light would take the same amount of time to reach us, no matter how large OX might be.

The question we must ask, therefore, concerns the shapes of objects that we find in the sky. Our own galaxy is a spiral; the ratio of its thickness to its diameter is approximately equal to that ratio in an ordinary phonograph record (see fig. 25.5). Of course, the edges are more ragged and there is a bulge at the center, but the approximation is pretty good. In order to display this shape, I looked through our home collection of old LP records

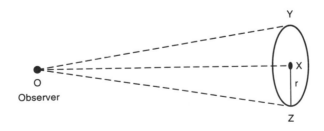

Light from a Luminous Disk

Figure 25.4

The Shape of the Milky Way

Figure 25.5

and serendipitously came upon "Cosmo's Factory" by the Creedence Clearwater Revival, a happy discovery given that we are interested in various types of engines in the cosmos, a clearing of the waters muddied by invalid arguments, and in a revival of credence in theories or models of such engines. Spiral galaxies are numerous in the universe, and there is no reason to think ours is particularly special. It is, of course, the homeland of humans, but this fact seems to me to have little cosmic importance.

Consider another familiar celestial object, the planet Saturn (see fig. 25.6). Although the planet itself is roughly spherical, the entire object, including the rings, is not. Now, imagine that the planet were to disappear, leaving only the rings visible to us. Imagine that Saturn began as a much more massive object than it is, and that it had undergone gravitational collapse, becoming a black hole. The result would be quite similar to the above-mentioned model astrophysicists currently favor for quasars—a black hole surrounded by an accretion disk, with matter from the disk falling into the black hole. Since a black hole is in principle invisible, what we see is a ring radiating prodigious quantities of energy. Of course, these rings are not all oriented in the same way with respect to our line of sight, and this is an important point. But not all quasars have the same rate of observed fluctuation, so the argument does not depend on any general assumption about the orientations of quasars.

6. Some Methodological Points

When quasars were first discovered, and for a long time thereafter, they presented extremely puzzling phenomena. No reasonably satisfactory model was available. Under such circumstances the most fruitful avenue for astrophysicists would be to conceive and develop as many models as possible, hoping to find one that at least comes close to fitting the facts. It would be counterproductive to impose artificial and unnecessary constraints,

Saturn

Figure 25.6

such as the $c\Delta t$ size criterion, and to rule out immediately any proposed model that would violate it—as reported by Disney and Véron in the quotation given earlier. Perhaps what actually happened was that a proposal was made—for example, that quasars are large, dense collections of stars (see Misner, Thorne, and Wheeler, 1973, p. 634)—that could be rejected on other grounds. In this case, for instance, there seemed to be no mechanism by which the brightening and dimming of members of the group could be orchestrated to produce the overall fluctuation that was observed. That is the real objection. It had nothing to do with the fact that the model violated the $c\Delta t$ size criterion.

As I noted earlier, a fluctuation violating the $c\Delta t$ size criterion could be produced by a huge shell of matter in an excited state that could be induced to radiate by stimulated emission from a central source. As we saw, however, this kind of entity would not yield a spectrum anything like those of observed quasars. Again, the $c\Delta t$ size criterion has nothing to do with the inadequacy of the model.

It may be that various models can be constructed that violate the $c\Delta t$ size criterion, and that every one of them can be rejected on completely different grounds. If that were to happen, it would be perfectly reasonable to treat the $c\Delta t$ size criterion as a plausibility claim. An astrophysicist might say, in effect, "I don't believe you can construct a satisfactory model that violates this criterion, but if you have any such model, let's take a look at it to see whether it's viable on other grounds. I seriously doubt that it will survive careful scrutiny." If this is what is involved, it seems to me, astrophysicists should say so explicitly, rather than invoking a criterion as a consequence of a law of nature when it is no such thing. And if the $c\Delta t$ size criterion is adopted as a plausibility principle, we may reasonably ask on what basis its plausibility rests. I have not found any answer to this question in the literature.

Perhaps the vast majority of authors feel queasy about appealing to plausibility considerations, thinking that such appeals fall short of scientific objectivity and rigor. Anyone who adopts a Bayesian approach to scientific confirmation can point directly to the fact that prior probabilities occur in Bayes's theorem; plausibility arguments can be identified with assessments of prior probabilities. According to the Bayesian approach, plausibility considerations are not only admissible; but also indispensable to scientific confirmation. Bayesianism has the important virtue of calling explicit attention to plausibility considerations, and to the grounds for their evaluations. When the *Astronomy and Astrophysics Encyclopedia,* cited earlier, says that "causality limits the size," this sounds like an inviolable principle. It would have been far better to say that models that violate the $c\Delta t$ size criterion do not stand much of a chance of meeting the other requirements that models of quasars must satisfy and why. Before astrophysicists had found fairly satisfactory models, the motto should have been, "Let a thousand flowers bloom, and let us look at them all."

7. The Rhetoric of the Argument

My attention was drawn to the $c\Delta t$ size criterion by articles in journals such as *Science* and *Scientific American.* I have subsequently found similar arguments in a number of other journals such as *Nature, American Scientist,* and *Physics Today.* These publications share an important characteristic, namely, that they are widely read by scientists and other scientifically literate people who want to find out what goes on in scientific areas outside their own specialties. *Nature* and *Science* are two of the most prestigious scientific journals in the world, but they are not confined to any single narrow specialty. *Science* and *American Scientist* are organs of scientific societies that purposely lack narrow disciplinary boundaries. But these are not publications, like *Time* and *Newsweek,* that are addressed to the general public. My complaint, then, concerns an apparent failure of scientists to level with their fellow scientists in other areas of research. Thoughtful astrophysicists apparently realize that the $c\Delta t$ size criterion has to be used with care, and only if certain conditions are fulfilled. Unfortunately it is difficult for 'outsiders' who have a genuine interest in the subject to discover any explicit formulation of the conditions that need to be taken into account.

In *The Remarkable Birth of the Planet Earth,* Henry M. Morris, the leading proponent of 'creation science' in the United States, says, "Cosmogony seems to be a sort of game that astronomers play, a tongue-in-cheek charade in which only the initiates know the rules and the spectators stand in awe" (1972, p. 57).The basis on which Morris makes this statement is, of course, entirely different from my concerns about the $c\Delta t$ size criterion and the geometrical argument. Nevertheless, it seems to me that bona fide scientists should do their best not to give potential ammunition to influential practitioners of pseudoscience.

8. Conclusion

A major theme of this essay, as suggested by its subtitle, is why no genuine scientific controversy emerged over the last 30 years regarding the sizes of variable celestial objects. I do not have an answer to this question. Even though some thoughtful astrophysicists are aware that the $c\Delta t$ size criterion has limited applicability, I fear that in general the connection between the time taken for variation and the size of the object may become a dogma—one whose basis for the rest of us remains obscure.

9. Late-Breaking Developments

During a few weeks (late September to mid-October 1993) I learned of two mechanisms that had recently been postulated as causes of fluctuations in the apparent brightness of celestial objects. The first of these is gravitational lensing. It has long been realized that a gravitational lens could intensify the radiation we receive from a distant source, but it has recently been suggested that brief pulses of intensified radiation from stars in the Large Magellanic Cloud—a nearby galaxy—were due to gravitational lensing by dark bodies in the halo of the Milky Way. The pulses were brief because the dark bodies passed relatively quickly between us and the more distant stars. (For technical reports, see Alcock et al. [1993]; Aubourg et al. [1993].) Alcock et al. found two such events by monitoring 3 million stars for more than three years; Aubourg et al. found one such event by monitoring 1,800,000 stars for one year. Since, however, as much as 90% of the matter in the universe may be dark, it seems too early to say much about the frequency and circumstances of fluctuations caused by this sort of gravitational lensing in the universe at large.

The second mechanism involves relativistic jets emitted by quasars. Conservation of angular momentum suggests that the trajectories of these jets will be spirals rather than straight lines. If the axis of the spiral makes quite an acute angle with our line of sight to the quasar, it follows from basic geometrical considerations that the main body of the jet will sometimes be traveling more directly toward us and sometimes more directly away from us. As a result, we will observe a brightening and dimming associated with the jet. I do not know whether this sort of fluctuation has any bearing on the size of the source of the jet. The technical report on this proposal is given in Schramm et al. (1993).

APPENDIX: MORE EXAMPLES FROM THE LITERATURE

This sample is given, not just for further documentation, but also to give the reader an appreciation of the variety of contexts in which the $c\Delta t$ size criterion is applied. These items are presented in chronological order.

Physics Today: Quasars

Many quasars vary their *optical* intensity on a time scale which is characteristically a tenth of a year.

Now it follows from this variability that quasars must be very compact. If they weren't compact, they couldn't vary in a tenth of a year. No object can double in brightness in a time much smaller than the light-transit time across an appreciable part of the object. So we conclude that one-tenth of a light year is a characteristic maximum dimension for the optical heart of a bright variable quasar. (Morrison, 1973, p. 25)

Physics Today: Black hole

Within the past year many observers have become convinced that Cygnus X-1 contains a black hole. The most recent evidence, reported at the December meeting of the American Astronomical Society in Tucson, is from an x-ray detector aboard a rocket; a group at the Goddard Space Flight Center reported seeing millisecond variations in intensity, suggesting a compact object. . . .

The most recent evidence that Cygnus X-1 contains a black hole . . . is that its x-ray output is flickering with variations as short as a millisecond, a behavior characteristic of a very small object. (GBL, 1974, pp. 17, 19)

Scientific American: Gamma-ray bursters

[T]heoretical reasons show it is plausible that a gamma-ray burster might contain a neutron star; certain observational facts make it probable that it does. One such fact is the very short time within which bursts change their intensity. Some bursts have been as short as .01 second, whereas a burst that occurred on March 5, 1979, rose in intensity in .0002 second. Since a source cannot significantly change brightness in a time shorter than the time it takes light to travel across the source region, the size of the March 5 burster must be smaller than .0002 light-second, or about 40 miles. There are few astronomical objects that meet the size limitations or have enough available energy to power a burst. A neutron star satisfies both of these requirements. (Schaefer, 1985, p. 55)

American Scientist: Quasars

For the 22 years since their discovery, quasars have occupied the attention, time, and resources of many of the world's astronomers. We are now essentially certain that they are the most luminous single objects in the universe and also very small—often significantly changing their vast output of energy within days, and in some instances within minutes. (This limits the size of the radiating region to the distance that light can travel in that time.) (Hutchings, 1985, p. 52)

Nature: BL Lacertae objects

Large-amplitude, rapid optical variability is a well-known identifying characteristic for BL Lacertae objects ('blazars'). Although large-amplitude variations on timescales ranging from days to decades have been well documented, considerable controversy surrounds the nature of microvariability, that is, optical variations on timescales significantly shorter than a day. Here we report observations of BL Lacertae in which rapid changes were detected in the total optical flux from this object. These variations occurred on timescales as short as 1.5 hours. Although their structure is complex, the minimum timescale for the variations may be used to place constraints on the size of the emitting region. (Miller et al., 1989, p. 627)

Sky & Telescope: Antimatter factory

From the time-scale of these variations and the finite speed of light, researchers argue that the radiation arises in a source less than 1 light-year across. ("Galactic Center Antimatter Factory Found at Last?," 1990, p. 363)

Nature: X-ray flare in quasar

The flaring timescale (Δt) provides, from the causality argument, an upper limit for the size of the emitting region $R < c\Delta t$. (Remillard et al., 1991, p. 591)

Astrophysical Journal: Gamma-ray emission from BL Lacertae object

Mrk 421 exhibits significant time variability in many wave-length bands, particularly in X-rays [references] and in optical wavelengths [references]. Most recently, Fink et al. [reference] observed a 20% change of the soft X-ray flux (0.1–2.4 keV) in 2 hr. The rapid variability reported in these references strongly suggests that Mrk 421 contains a compact object. (Lin et al., 1992, p. L61)

Science: Quasars, blazars, and gamma rays

We now know that there can be rapid variability in the gamma-ray emission of 3C279, which seems to require an emission site less than about a light-week away from the central black hole for 3C279. (Dermer and Schlickeiser, 1992, p. 1645)

Nature: Gamma-ray power

Photon-photon absorption limits the amount of energy that can escape from a bright source if the density of photons at different energies is high enough. Theorists have already had to cope with this limit in explaining the γ-rays from the weaker 3C273 source. The difficulty all depends on the size of the source region, which can be inferred from the timescale for the variability (a source can change only with a maximum rate determined by the light transit time over its dimensions). (Bignami, 1992, p. 299)

Nature: X-ray flares

The X-ray emission from active galactic nuclei is thought to emanate from volumes barely greater than the Solar System (10^{10} km across) around a supermassive black hole at the hub of each galaxy. Such compact sources are inevitably involved if the rapid variations in intensity are to be explained. (Baring, 1992, p. 109)

Scientific American: Inconstant cosmos

"When you look at the sky at high energies, it's an amazingly inconstant place," reflects Neil Gehrels, the project scientist for *GRO* [Compton Gamma Ray Observatory]. On time scales ranging from weeks to thousandths of a second, objects brighten and dim, flicker and oscillate. Such rapid changes imply that the sources of the radiation are minuscule on a cosmic scale (otherwise it would take far too long for a physical change to affect a large part of the emitting region). Yet those same objects are emitting tremendous quantities of energetic radiation. (Powell, 1993, p. 111)

American Scientist: Gamma-ray universe

Because gamma-ray bursts fluctuate over very brief periods of time (less than one-1,000th of a second in some instances), the region emitting some of the gamma rays must be quite small (less than 100 kilometers in diameter). . . .

Some active galactic nuclei release energy in all parts of the spectrum, from radio waves to gamma rays. They are the brightest objects in the universe. . . . Remarkably, some of these objects appear to be releasing most of their energy at gamma-ray wavelengths. One of these, a quasar identified as 3C 279, lies about 6 billion light-years away and may release as much as 10 million times more gamma rays than our own galaxy. Curiously, four months after 3C 279 was discovered it ceased flaring almost entirely. Such tremendous variation in output appears to be a common feature of these objects. In some cases they vary their output in less than a day, suggesting that the region of emission is relatively small (less than a light-day across). (Kniffen, 1993, pp. 344, 346)

Note

This essay was presented at a conference on scientific controversy at Vico Equense, Italy, in 1993. I should like to express my sincere thanks to the following individuals for helpful suggestions and discussions: Frederic Chaffee, T. Courvoisier, Donald Kniffen, James Small, and Raymond Weymann. Each of them would, I believe, have serious objections to the present essay; their generosity does not entail agreement with my main theses.

26

Dreams of a Famous Physicist

An Apology for Philosophy of Science

> Whether or not the final laws of nature are discovered
> in our lifetime, it is a great thing for us to carry on the
> tradition of holding nature up to examination, asking
> again and again why it is the way it is.
> —Steven Weinberg, *Dreams of a Final Theory*

Near the beginning of "The Importance of Scientific Understanding" (essay 5), I cited Steven Weinberg's *Dreams of a Final Theory* ([1992] 1994) to illustrate the attitudes of at least some late twentieth-century scientists toward scientific explanation. His view is nicely encapsulated in the foregoing epigraph, which is the final sentence of his main text (p. 275). It stands in sharp contrast to the dominant attitude in the early part of the century—concisely expressed by Karl Pearson—"Nobody believes now that science *explains* anything; we all look upon it as a shorthand description, as an economy of thought" ([1911] 1957, p. xi; emphasis in original). I was delighted to find such an outstanding example.

1. The Motivation

When *Dreams* was first published (1992) the United States Congress was considering the continued funding of the SSC (superconducting super collider), which was partly built at that time. Weinberg had testified at congressional hearings in 1987 in support of construction of this research facility, a particle accelerator 53 miles in circumference, at a projected cost of \$4.4 billion.[1] In 1992 he strongly urged continuation of the project. By 1994, when *Dreams* came out in paperback, Congress had already denied continued funding. The project was definitively canceled. An Afterword deploring that outcome was added to the paperback edition. In a later brief report, "Night Thoughts of a Quantum Physicist" (1995), he tells us that his hopes for the future of particle physics hinge on a European facility, now under construction, that will have one-third the power the SSC would have had and will be completed five years later than the SSC would have been.[2]

Ironically, in 1996 we learned that the General Accounting Office (a nonpartisan investigative agency for the Congress), after conducting an exhaustive four-year investigation, concluded that the successes claimed for "smart weapons" during the Persian Gulf war "were overstated, misleading, inconsistent with the best available data, or unverifiable." The American public and the U.S. Congress were persuaded by these deceptions to spend many billions of dollars on further development of such weapons, which are much more expensive than their "dumb" counterparts, but no more effective. The cost of the SSC is peanuts in comparison with the hundreds of billions that will be spent within the next couple of decades on "smart" weapons (see Weiner, 1996). The amount requested for the SSC in 1993 was much less than $1 billion.[3]

One could go on at length discussing the politics and economics of federal funding in the United States, but my focus in this essay is on problems in the philosophy of science. In reporting on his 1987 congressional testimony Weinberg begins by saying:

> My talk this afternoon will be about the philosophy of science, rather than about science itself. This is somewhat uncharacteristic for me, and, I suppose, for working scientists in general. I've heard the remark (although I forget the source) that philosophy of science is just about as useful to scientists as ornithology is to birds.[4]
>
> However, at just this time a question has arisen in the United States that will affect the direction of physics research until well into the twenty-first century, and I think that it hinges very largely on a philosophical issue. (1987, p. 433)

The philosophical issue in question involves scientific explanation and theoretical reductionism (which Weinberg prefers to call "objective reductionism").

2. Arrows of Explanation

When Weinberg writes of a final theory, he states clearly that we do not have any such thing in hand, and that we do not know what it will be like when we get it. Perhaps it will be some version of string theory; perhaps not. Furthermore, we do not know when we will have one. His claim is that the "explanatory arrows" we do have point to the existence of a final theory. This final theory will *explain* everything; it will enable us to *understand* the universe. The entire argument rests on explanatory relationships.

> There are arrows of scientific explanation that thread through the space of all scientific generalizations. Having discovered many of these arrows, we can now look at the pattern that has emerged, and we notice a remarkable thing: perhaps the greatest scientific discovery of all. These arrows seem to converge to a common source! Start anywhere in science and, like an unpleasant child, keep asking "Why?" You will eventually get down to the level of the very small. (1987, p. 435)

Weinberg's view is clearly and explicitly reductionistic. The fact that the explanatory arrows point to a unique final theory is not logically necessary; if true, it is a profound fact about our world. The explanatory arrows might converge toward several distinct theories, for example, if vitalism were true.[5] But Weinberg maintains that there is a final theory toward which particle physics is directed that will explain everything. Scientific explanation is so central to the scientific enterprise that, if one scientific theory could not explain another, "that would leave my whole generation of particle physicists with nothing to do" ([1992] 1994, p. 168). Weinberg's basic rationale for spending a projected $4.4 billion for construction of the SSC was the explanatory value of what would be learned by means of it. The fundamental goal is understanding the universe, and particle physics is uniquely qualified to pursue it. Since "explanatory arrows" are so central to the argument, it behooves us to try to figure out what scientific explanation is.

3. Against Philosophy?

My pleasure in discovering Weinberg's book was, I must confess, somewhat lessened when I came to the chapter titled "Against Philosophy," which seems to throw down the gauntlet. (We notice a barbed remark about philosophy of science in the opening paragraph of his 1987 lecture.) The chapter turns out, in fact, to be a polemic against positivism and relativism—one with which I largely agree. In an endnote he acknowledges this point but remarks, "I did not think that 'Against Positivism and Relativism' would be a very catchy title" ([1992] 1994, p. 304). Inasmuch as no sign indicating the existence of this endnote occurs in the chapter itself, a reader unfamiliar with current philosophy could easily gain the impression that contemporary philosophy of science consists of nothing but positivism and relativism. This would be a serious misconception.

In philosophy of science, positivism (known as *logical positivism* since the days of the Vienna Circle) has been dead for half a century. According to Weinberg, "positivism has preserved its heroic aura, so that it survives to do damage in the future. George Gale even blames positivism for much of the current estrangement between physicists and philosophers" (ibid., p. 176). If scientists are, indeed, still troubled by positivism, their philosophical concerns are severely anachronistic. An example of such anachronism on the part of eminent physicists occurs in a series of lectures by Stephen W. Hawking and Roger Penrose, published in *The Nature of Space and Time* (1996). Hawking says of Penrose, "He's a Platonist and I'm a positivist" (pp. 3–4). Penrose replies, "At the beginning of this debate, Stephen said that he thinks that he is a positivist, whereas I am a Platonist. I am happy with him being a positivist, but I think that the crucial point here is, rather, that I am a realist" (ibid., p. 134). It seems that many physicists have not heard the news of the demise of logical positivism.

Relativism is a different kind of issue. Serious philosophers, historians, and sociologists of science currently advocate relativistic doctrines, but they do not speak for the entire profession. I agree with Weinberg's sentiment, "It seems to me that we are discovering something real in physics, something that is what it is without any regard to the social or historical conditions that allowed us to discover it" ([1992] 1994, p. 188). The difference is that many philosophers, myself included, have offered philosophical arguments for our various forms of realism instead of merely expressing sentiments. (See "Why Ask, 'Why?'?" [essay 8]; Salmon, 1984b, chap. 8, provides a fuller account.)

Why is Weinberg sufficiently interested in philosophy of science to devote a disparaging chapter to it? Some indication may be found in this statement:

Ludwig Wittgenstein, denying even the possibility of explaining any fact on the basis of any other fact, warned that "at the basis of the whole modern view of the world lies the illusion that the so-called laws of nature are the explanations of natural phenomena." Such warnings leave me cold. To tell a physicist that the laws of nature are not explanations of natural phenomena is like telling a tiger stalking prey that all flesh is grass. The fact that we scientists do not know how to state in a way that philosophers would approve what it is that we are doing in searching for scientific explanations does not mean that we are not doing something worthwhile. We could use help from professional philosophers in understanding what we are doing, but with or without their help we shall keep at it. ([1992] 1994, pp. 28–29)

As a first response to this challenge, we should note that the quotation from Wittgenstein is taken from his *Tractatus Logico-Philosophicus,* first published in German in 1921, and in English translation in 1922. As already remarked, such views were widely held by philosophers and scientists early in the present century. Wittgenstein's comment is typical of the positivism that was philosophically outmoded by midcentury.

Charity should perhaps have led me to use an ellipsis in place of the next two sentences, the first an emotive expression, the second a colorful analogy with no argumentative import. We need not, however, deny ourselves the pleasure of Weinberg's rhetorical flourish.

The final two sentences go to the heart of the matter. "The fact that scientists do not know how to state in a way that philosophers would approve . . . does not mean that we are not doing something worthwhile." One can only agree, pointing out that "stating in a way that philosophers would approve" is the business of philosophers, not necessarily scientists. Of course, if scientists choose to do philosophy, we should not stand in their way, but we are entitled to evaluate their efforts critically. "We could use help from professional philosophers . . . but with or without their help we shall keep at it." Yes, do keep at it, Professor Weinberg; what you do is just great, and you do it wonderfully well. Keep on trying to explain the phenomena of nature; the explanations of scientists (not only physicists) are extremely interesting and valuable. If, however, you could use help from professional philosophers in accurately characterizing the nature of scientific explanation, a first step would be to consult their work on the subject.

One of the major philosophers to contribute to the theory of scientific explanation is Ernest Nagel, whose magnum opus (1961) Weinberg cites twice. The first citation points to the variety of senses of the word "why" ([1992] 1994, p. 26). Throughout his book Weinberg repeatedly emphasizes the importance of asking "why?" The second mentions an example of theoretical reduction (ibid., p. 40); Weinberg's third chapter, "Two Cheers for Reductionism," defends that doctrine. He offers no clue to Nagel's theory of scientific explanation, nor even to the fact that Nagel had such a theory. In any case, Nagel's book antedates Weinberg's by a full three decades. Recall that in 1961 the 3° cosmic background radiation had not been discovered and that, although it had been predicted much earlier by George Gamow, no one was hunting for it. When Penzias and Wilson discovered it by accident, they did not know what they had found; astrophysicists explained the nature and significance of the discovery. A philosopher writing on contemporary astrophysics without awareness of that phenomenon would look (and would be) ridiculous.

Another major contributor to the philosophy of scientific explanation is Karl Popper, who, in (1972), rejected the notion of an ultimate explanation (Weinberg, [1992] 1994, p. 230). Again, Weinberg gives no hint as to the nature of Popper's theory of scientific explanation, and this source antedates Weinberg's book by two decades. It was only in 1973, according to Weinberg, that an apparently insuperable problem regarding quarks—essential components of the "standard model"—was overcome through the discovery of asymptotic freedom (ibid., pp. 182–183). Anachronism again, but even more so. Popper's major work on scientific explanation came out in German in 1934 (imprint 1935)—the heyday of logical positivism, though Popper consistently distanced himself from that school—and in English translation with much added material in 1959.

Weinberg also quotes a 1962 remark by Paul Feyerabend that "the notion of scientific explanation developed by some [not *all*, but *some*] philosophers of science is so narrow that it is impossible to speak of one theory being explained by another" (ibid., p. 168). In an endnote unsignified in the main text, he acknowledges that Feyerabend was referring to the Vienna Circle logical positivists (ibid., p. 304). Both Feyerabend and Weinberg continue to flog dead horses (but see §6 of this essay). Weinberg's entire book contains not one reference to the work of Hempel, whose article "Aspects of Scientific Explanation" (1965b) elaborated what amounted to the received opinion on the subject—one might almost say "standard model," but Hempel offered three models—for a couple of decades. Nor does it mention any of the subsequent work that effectively overthrew the received view.

Weinberg's polemic is not actually directed at the failure of philosophers of science to provide adequate characterizations of scientific explanation; that issue is of little interest to him. He is not coy. At the outset of the chapter "Against Philosophy" he asks, "Can philosophy give us any guidance toward a final theory?" (ibid., p. 166). The short answer, to which we both agree, is NO, or at least very probably not. Why would anyone expect otherwise? "Physicists get so much help from subjective and often vague aesthetic judgments," he answers, "that it might be expected that we would be helped also by philosophy" (ibid.). His view seems to be that philosophy ought to contribute its share of mushiness, but he is disappointed:

> The value today of philosophy to physics seems to me to be something like the value of early nation-states to their peoples. It is only a small exaggeration to say that, until the introduction of the post office, the chief service of nation-states was to protect their peoples from other nation-states. The insights of philosophers have occasionally benefited physicists, but generally in a negative fashion—by protecting them from the preconceptions of other philosophers. (ibid.)

I would add that it could help to protect physicists from the philosophical errors of other physicists (see §5). A couple of paragraphs later he continues:

> This is not to deny all value to philosophy, much of which has nothing to do with science. I do not even mean to deny all value to the philosophy of science, which at its best seems to me a pleasing gloss on the history and discoveries of science. But we should not expect it to provide today's scientists with any useful guidance about how to go about their work or about what they are likely to find. (ibid., p. 167)

Weinberg finds philosophers who share his judgment: "Wittgenstein remarked that 'nothing seems to me less likely than that a scientist or mathematician who reads me should be seriously influenced in the way he [or she] works'" (ibid.). This is not an astonishing statement on the part of a philosopher. What I find missing from Weinberg's discussion is a reference to any statement whatever *by a philosopher* asserting that his or her work *should* have an effect on scientific practice. Indeed, upon extended reflection, I can think of only one such—namely, Popper's unfortunate claim:

> The main thing about the propensity interpretation is that *it takes the mystery out of quantum theory, while leaving probability and indeterminism in it.* It does so by pointing out that all the apparent mysteries would also involve thrown dice, or tossed pennies—*exactly* as they do electrons. In other words, it shows that quantum theory is a probability theory just as any theory of any other game of chance, such as the bagatelle board (pin board). (Popper, 1957, p. 68; emphasis in original)

The mysteries of quantum mechanics have not noticeably disappeared in the last forty years; if anything, they are now more acute than ever, as David Mermin has dramatically shown in "Quantum Mysteries for Anyone" (1981) and "Is the Moon There When Nobody Looks? Reality and the Quantum Theory" (1985).

Weinberg is admirably straightforward in expressing his intellectual and emotional orientation:

> It is only fair to admit my limitations and biases in making this judgment. After a few years' infatuation with philosophy as an undergraduate I became disenchanted. The insights of the philosophers I studied seemed murky and inconsequential compared with the dazzling successes of physics and mathematics. From time to time since then I have tried to read current work on the philosophy of science. Some of it I found to be written in a jargon so impenetrable that I can only think that it aimed at impressing those who confound obscurity with profundity. Some of it was good reading and even witty, like the writings of Wittgenstein and Paul Feyerabend. But only rarely did it seem to me to have anything to do with the work of science as I knew it.
>
> It may seem to the reader (especially if the reader is a professional philosopher) that a scientist who is as out of tune with the philosophy of science as I am should tiptoe gracefully past the subject and leave it to experts. I know how philosophers feel about attempts by scientists at amateur philosophy. But I do not aim here to play the role of a philosopher, but rather that of a specimen, an unregenerate working scientist who finds no help in professional philosophy. I am not alone in this; I know of *no one* who has participated actively in the advance of physics in the postwar period whose research has been significantly helped by the work of philosophers. ([1992] 1994, pp. 168–169)

What Weinberg is doing, in fact, is setting his own agenda for philosophy of science and complaining bitterly that philosophers of science have not fulfilled it. I will try to say a little later on about what I consider a reasonable agenda.

Unfortunately, I do not find Weinberg's disclaimers plausible. He is not just a poor working bloke who is doing his thing along with a bunch of other blokes engaged in the same job. The overall theme of his whole book is that physicists are succeeding in telling us why the world is the way it is, and there are strong indications that they will sooner or later come up with one unified theory that will be the ultimate explanation of the whole

universe. He pleads most eloquently that it would be wonderful to have such a theory—
so wonderful, in fact, that we ought to spend billions of dollars on the superconducting
super collider to take another major step in that direction. I agree with him completely on
the desirability of the SSC, and I agree with him on the enormity of the congressional
termination of the project.

4. Weinberg on Scientific Explanation

Since his major argument hinges on explanatory relations, let us have a look at some of
Weinberg's important comments. Taking his departure from a series of why-questions,
he says:

> The word "why" is notoriously slippery. The philosopher Ernest Nagel lists ten
> examples of questions in which "why" is used in ten different senses, such as "Why
> does ice float on water?" "Why did Cassius plot the death of Caesar?" "Why do human
> beings have lungs?" Other examples in which "why" is used in yet other senses come
> immediately to mind. . . .
>
> [I]t is a tricky business to say exactly what one is doing when one answers such a
> question. Fortunately it is not really necessary. Scientific explanation is a mode of
> behavior that gives us pleasure, like love or art. The best way to understand the nature
> of scientific explanation is to experience the particular zing that you get when someone
> (preferably yourself) has succeeded in actually explaining something. I do not mean
> that scientific explanation can be pursued without any constraints, any more than can
> love or art. In all three cases there is a standard of truth that needs to be respected,
> though of course truth takes different meanings in science or love or art. I also do not
> mean to say that it is not of any interest to try to formulate some general description of
> how science is done, only that this is not really necesssary in the work of science, any
> more than it is in love or art. (ibid., p. 26)

Many English words, including "explanation," partake of what is known as a *process-
product ambiguity.* "Photograph" is another. One can use this term to refer either to the
act of taking a photograph or to the material object consisting of a piece of paper with an
image on it. No confusion is apt to result. Notice, however, that in the preceding passage
Weinberg describes explanation as a "mode of behavior that gives us pleasure, like love
or art." He plausibly remarks that one can engage in all three of these activities without
being able to analyze them in precise detail. In other contexts, however, he speaks of
"one theory being explained by another" (ibid., p. 168); this can hardly refer to a human
activity such as love or art. It is difficult to conceive of one theory experiencing a
"particular zing" by being able to explain another. And the main theme of the book is the
"search for the ultimate laws of nature" (front cover, paperback edition), "the ancient
search for those principles that cannot be explained in terms of deeper principles" (ibid.,
p. 18). The "final theory" will explain theories that are less fundamental. Although the
description of scientific explanation as a human activity may be a psychologically
correct account of what humans do in seeking and finding explanations, it is logically
irrelevant to the question of what it means for one theory to explain another. Hans
Reichenbach's distinction between the context of discovery and the context of justifica-

tion would be of significant help here (Reichenbach, 1938, §1). Indeed, Weinberg immediately changes the subject; in the very next paragraph he continues:

> As I have been describing it, scientific explanation clearly has to do with the deduction of one truth from another. But there is more to explanation, and also less. Merely deducing one statement from another does not necessarily constitute an explanation, as we see clearly in those cases where either statement can be deduced from the other. . . . Explanation, unlike deduction, carries a unique sense of *direction*. ([1992] 1994, pp. 26–27)

Weinberg's point about the unique sense of direction of explanation is uncontroversial; I know of no philosopher or scientist who has claimed in any case that A explains B *and* B explains A. Moreover, I have never known anyone to argue that scientific explanation is no more than deduction. The most avid deductivist will insist that not all deductions qualify as scientific explanations; certain constraints must be added. There are, of course, cases in which two propositions A and B are logically equivalent, so that each entails the other, but in such cases we would deny that A explains B or B explains A, for the two propositions are just saying the same thing. To assert the contrary would be, in effect, to allow natural phenomena to explain themselves.

The unique directionality of explanation is essential to Weinberg's main thesis about the convergence of arrows of explanation, but he worries needlessly about it. He says, for example, that Newton's laws imply Kepler's laws and vice versa: "In formal logic, since Kepler's laws and Newton's laws are both true, either one can be said to imply the other. (After all, in formal logic the statement 'A implies B' just means that it never happens that A is true and B isn't, but if A and B are both true then you can say that A implies B and B implies A)" (1987, p. 435).[6] This quotation exhibits a simple confusion between material implication and logical entailment, a fundamental logical distinction that *must* be made in *any* introductory course in formal logic. This confusion leads Weinberg into unnecessary difficulty in claiming that more fundamental laws explain less fundamental laws, not vice versa. In cases such as Kepler's and Newton's laws, the logical entailment relation settles the question of fundamentality immediately and unambiguously. One look at the Newtonian Synthesis, as shown in figure 5.2, shows patently the direction of the explanatory arrows. This is just the sort of convergence that suggests an ultimate explanatory theory.

As a consequence of his confusion, however, Weinberg goes on to say:

> Talk of more fundamental truths makes philosophers nervous. We can say that the more fundamental truths are those that are in some sense more comprehensive, but about this, too, it is difficult to be precise. But scientists would be in a bad way if they had to limit themselves to notions that had been satisfactorily formulated by philosophers. ([1992] 1994, p. 27)

There are philosophical problems about fundamentality and comprehensiveness, but they do not arise from the source to which he has referred.

Leaving that issue aside, notice that in the final sentence Weinberg is again setting an agenda for philosophers, deploring their *supposed* inability to satisfy it. I cannot think of any serious philosopher who has ever suggested imposing such limitations on scientists. As a professional philosopher I would say that it is the business of scientists to employ the concepts they find useful, leaving it to the philosopher, normally later, to try to clarify these concepts and make them precise.

Continuing his discussion of the relationship between scientific explanation and deduction, Weinberg says:

> A scientific explanation can also be something less than a deduction, for we may say that a fact is explained by some principle even though we cannot deduce it from that principle. Using the rules of quantum mechanics we *can* deduce various properties of the simpler atoms and molecules and even estimate the energy levels of complicated molecules like the calcium-carbonate molecules of chalk. . . . But no one actually solves the equations of quantum mechanics to deduce the detailed wave function or the precise energy of really complicated molecules, such as proteins. Nevertheless, we have no doubt that the rules of quantum mechanics "explain" the properties of such molecules. This is partly because we can use quantum mechanics to deduce the detailed properties of simpler systems like hydrogen molecules and also because we have mathematical rules available that would allow us to calculate all of the properties of any molecule to any desired precision if we had a large enough computer and enough computer time. (ibid., pp. 27–28)

This paragraph exhibits once again the process-product ambiguity. It seems to me that one theory explains another, in the sense Weinberg intends, if the other follows logically from the first. We may not be able to carry out the human or computer activity of deducing the one from the other, but that means only that we may be unsure whether that relation holds between the theories. Weinberg shows that we can plausibly maintain that such a logical relation exists even in the absence of an actual derivation.

I have complained that Weinberg sets an inappropriate agenda for philosophy of science. It is time for me to offer a more suitable one. It contains two parts. First, I would suggest that there are certain foundational questions about science—such as the nature of explanation, confirmation, laws—that are intrinsically interesting but need not hamper the progress of science even while they remain unanswered. We have an excellent historical precedent. From the time that Newton and Leibniz invented the calculus until its arithmetization by Cauchy and others in the nineteenth century, this mathematical tool served science admirably, even though its foundations were in deplorable shape. The foundational studies led to profound investigations into the nature of arithmetic, and to the development of modern set theory. One can fairly say, I believe, that mathematicians were regularly proving theorems in analysis for a couple of centuries at least without any clear notion of what precisely constitutes a demonstration. Work in confirmation theory has analogously involved serious efforts to clarify the notion of empirical support for scientific theories, though it has not found the kind of success achieved in foundations of mathematics. I would add, as an aside, that Weinberg's lengthy discussion of the role of beauty in the evaluation of scientific theories would have benefited greatly from an awareness of Bayesian confirmation theory.

Second, I would paraphrase President John F. Kennedy's famous inaugural statement: Ask not what philosophy can do for science; ask rather what science can do for philosophy. My answer is that science can do a lot. Philosophers need to be aware of the substantive accomplishments of science in order to avoid serious philosophical error. An outstanding example is the discovery of non-Euclidean geometries and of their applicability to the physical world. Its bearing on the Kantian doctrine of synthetic a priori knowledge of space is epoch-making (see Salmon, 1975b, chap. 1). Another important case is the relevance of quantum mechanics to the doctrine of determinism, and its irrelevance to the problem of free will (see "Determinism and Indeterminism in Modern Science" [essay 2] and "Indeterminacy, Indeterminism, and Quantum Mechanics" [essay 17]). A third example is the theoretical work of Einstein and the experimental work of Perrin in the first decade of the twentieth century regarding the reality of unobservable entities such as atoms and molecules (see Salmon, 1984b, chap. 8).

What many physicists fail to realize, I think, is that philosophy of science is an intellectual discipline with its own subject matter, techniques, problems, and achievements. Weinberg does not fall into this category; he recognizes its existence but declines to study it systematically because he believes, rightly I think, that it won't help physicists find the final theory. He is a person with wide intellectual interests; in his preface he mentions his interest in history. I find it difficult to believe that reading general history—or even history of science—helps him to do physics. But I presume he enjoys reading history and probably many other subjects as well. Apparently he does not enjoy reading philosophy of science (for the most part). That's fine; I certainly do not fault him for it. His protestations to the contrary notwithstanding, however, that does not prevent him from *doing* philosophy, as his discussions of scientific explanation, theoretical reduction, empirical confirmation, and scientific realism clearly reveal. In fact, as we saw, in the opening portion of his 1987 lecture he frankly admitted that he was doing philosophy of science.

5. Explanatory Asymmetry

Bromberger's famous flagpole example, mentioned in "Explanatory Asymmetry" (essay 10), has appeared countless times in the literature on scientific explanation. It arose as a counterexample to the deductive-nomological (D-N) model of explanation propounded in the classic 1948 Hempel-Oppenheim article. It goes as follows. Given a vertical flagpole standing on a level patch of ground on a sunny day, and given the elevation of the sun at a particular time, we can invoke the law of rectilinear propagation of light to deduce the length of the flagpole's shadow from its height or to deduce the flagpole's height from the length of its shadow. Although both deductions fulfill the conditions to qualify as D-N explanations, only the first appears intuitively to have genuine explanatory force. The second appears to have none. In that essay I tried to show why the intuitive reaction is sound. If, however, it were inconvenient to ascertain the height of the flagpole in a more direct way, measurement of the length of the shadow and the elevation of the sun would, of course, enable one legitimately to *infer* its height. I will not rehash this trivial example here.

Consider a somewhat more sophisticated example. From the redshifts of light from distant galaxies, given the Doppler effect, we can infer that these galaxies are receding from us at high velocities. From the supposition that distant galaxies are receding from us at high velocities, again given the Doppler effect, we can infer that the light that reaches us from them is redshifted. If one asks why these galaxies are receding from us at these high velocities, it is essential to be clear on what kind of why-question is intended. Hempel carefully distinguished between *explanation-seeking* why-questions and *confirmation-seeking* why-questions. If the question is *why we believe* that the galaxies are receding—a confirmation seeking why-question—the observed redshift is a correct answer. This is an essential part of the evidence on which our belief is based, but we would never say that the actual recession occurs *because of* the redshifting of the light that reaches us on Earth. If the question is *why it happens,* the currently accepted answer is that it results ultimately from the "big bang." The scientific explanation of the fact that the galaxies are receding is one that involves a complex set of causal interactions and causal processes originating in the early history of the universe.

A commonsensical example clearly illustrates this basic distinction between explanation-seeking and confirmation-seeking why-questions. Suppose we read in a reliable newspaper that a prominent public official died. This constitutes good evidence that the death occurred; if asked *why we believe* it occurred, citing the newspaper report would be an appropriate answer. If asked, however, *why it occurred,* various possibilities are available. The correct answer might be an airplane crash. Notice that the newspaper account might contain *a report of the explanation* of death along with the report of the fact, but the newspaper account itself is not an explanation of the death. This example typifies a large class of cases—many to be found in serious scientific contexts—in which we have a record of some fact. Such records may be created by humans or by some other natural process. Tree rings, for example, record relative annual rainfall; they do not explain why given years were relatively wet or dry. Dendrochronologists study such tree-ring records to *infer* relative annual rainfall. The relative annual rainfall *explains* the tree-ring pattern. It is hard to see how anyone thinking seriously about such issues could confuse confirmation and explanation, but it happens.

In discussing the possibility of finding a fundamental physical theory that in principle explains everything, Weinberg gives considerable attention to string theories—to their strengths and shortcomings. He also treats, with well-founded misgivings, "a principle with a dubious status in physics, known as the *anthropic principle,* which states that the laws of nature should allow the existence of intelligent beings that can ask about the laws of nature" ([1992] 1994, pp. 219–220). He elaborates:

> The idea of an anthropic principle began with the remark that the laws of nature seem surprisingly well suited to the existence of life. A famous example is provided by the synthesis of the elements. . . . [I]n order for this process to account for the observed cosmic abundance of carbon, there must be a state of the carbon nucleus that has an energy that gives it an anomolously large probability of being formed in the collision of a helium nucleus and a nucleus of beryllium 8. [Such a state was found experimentally.] Once carbon is formed in stars, there is no obstacle to building up all the heavier elements, including those like oxygen and nitrogen that are necessary for all known forms of life. But in order for this to work, the energy of this state of the carbon nucleus must be very close to the energy of a nucleus of beryllium 8 plus the energy of a helium nucleus. If the energy of this state of the carbon nucleus were too large or too small, then little carbon or heavier elements will be formed in stars, and with only hydrogen and helium there would be no way that life could arise. The energies of nuclear states depend in a complicated way on all the constants of physics, such as the masses and

electric charges of the different types of elementary particles. It seems at first sight remarkable that these constants should take just the values that are needed to make it possible for carbon to be formed in this way. (ibid., pp. 220–221)

From the fact that we, as intelligent investigators, exist, we can *infer* that the universe is to some degree hospitable to intelligent life. In a certain context, says Weinberg—the confirmation context, I would say—this is just common sense. "Any scientist who asks why the world is the way it is must be living in one of the [logically possible] . . . universes in which intelligent life *could* arise" (ibid., p. 221). From the fact that we are here, however, we cannot conclude that that is *why* the universe is as it is—*why* it has the laws that it has. It is a sound principle on which to base an inference; it is an unacceptable basis for explanation. Its use for explanatory purposes is a scientifically unacceptable anthropomorphism. As Weinberg puts it, "The evidence that the laws of nature have been fine-tuned to make life possible does not seem to me very convincing" (ibid.). The phrase "fine-tuned to make life possible" strongly suggests, if it does not imply, the existence of an agent acting with a purpose.

Up to this point Weinberg's treatment of the anthropic principle is sound and illuminating, but later, after a discussion of a possible cosmological constant, he continues, "If all else fails, we may be thrown back on an anthropic *explanation*" (ibid., p. 226; emphasis added). He mentions Einstein's famous introduction and subsequent elimination of a cosmological constant in the equations of general relativity, but argues that some such cosmological constant may be admissible or even required. He also takes up the energy of the vacuum in contemporary quantum field theory. The result of combining the two he calls the *total cosmological constant*. It appears that this term must fall within fairly narrow limits if life as we know it is to be possible.

> To be specific, if the total cosmological constant were large and negative, the universe would run through its life cycle of expansion and contraction too rapidly for life to have time to appear. On the other hand, if the total cosmological constant were large and positive, the universe would expand forever, but the repulsive force produced by the cosmological constant would prevent the gravitational clumping together to form galaxies and stars in the early universe and therefore give life no place to appear. (ibid.)

From our own existence we can readily conclude by *modus tollens* that the value of the cosmological constant cannot be very different from zero. Weinberg further suggests that a value of the cosmological constant that is small and positive could furnish an answer to the well-known cosmological missing-mass problem as well as to problems concerning the age of the universe. From these interesting and important inferences Weinberg concludes, "[I]f such a cosmological constant is confirmed by observation, it will be reasonable to infer that *our own existence plays an important part in explaining why the universe is the way it is*" (ibid., p. 229; emphasis added). Weinberg does not accept this conclusion with enthusiasm:

> For what it is worth, I hope this is not the case. As a theoretical physicist, I would like to see us able to make precise predictions, not vague statements that certain constants have to be in a range that is more or less favorable to life. I hope that string theory really will provide a basis for a final theory and that this theory will have enough

predictive power to be able to prescribe values for all the constants of nature, including the cosmological constant. We shall see. (ibid.)

It seems to me that Weinberg's discussion of the anthropic principle ends up confusing explanation and inference. If there is a final theory that requires the value of one constant to be "put in by hand," the cosmological anthropic principle will do nothing to *explain why* that constant has the particular value that it has. As Weinberg realizes, not everything can be explained; the final theory explains but is not explained. The value of the cosmological constant (if the final theory requires insertion of such a term) would be an ultimate unexplained feature of the universe.

The most comprehensive text on the cosmological anthropic principle is Barrow and Tipler (1986). In the introductory chapter they distinguish three distinct anthropic principles. The first, the *Weak Anthropic Principle,* reads as follows:

(WAP): The observed values of all physical and cosmological quantities are not equally probable but they take on values restricted by the requirement that there exist sites where carbon-based life can evolve and by the requirement that the Universe be old enough for it to have already done so. (ibid., p. 16)

A little later they invoke Bayes's theorem to support WAP, thus showing that WAP is a principle of inference, not an explanatory principle at all (ibid., p. 17). This is the context in which Weinberg says that the principle is nothing but common sense.

In its generic form, the second principle, called the *Strong Anthropic Principle,* states:

(SAP): The Universe must have those properties which allow life to develop within it at some stage of its history. (ibid., p. 21)

Three distinct specific versions of this principle, which I will not take up separately here, are given (ibid., p. 22). Unlike WAP, SAP is a modal principle. The idea that explanation involves necessity is familiar; I have discussed it at length in "Comets, Pollen, and Dreams" (essay 3), "Deductivism Visited and Revisited" (essay 9), and "Scientific Explanation: Three Basic Conceptions" (essay 20); I find it unacceptable. It seems that their argument for it could be paraphrased as follows (see ibid., p. 15):

(1)   We represent a carbon-based form of life, and we are here.
(2)   It is necessary that if carbon-based life exists, certain constants must have values falling within fairly narrow ranges.

Therefore,

(3)   The values of certain constants necessarily fall within the above-mentioned fairly narrow ranges.

This argument has the form,

Necessarily, if P then Q.
P.

Therefore,

Necessarily, Q.

It is an obvious and well-known fallacy in modal logic. A time-honored example is

Necessarily, if John is a bachelor John is unmarried.
John is a bachelor.

Therefore,

Necessarily, John is unmarried.

This conclusion obviously does not follow. John's unmarried status may well be the result of some accident, such as a broken leg, which caused his previously scheduled wedding to be postponed to a future date.

The third principle, called the *Final Anthropic Principle,* reads:

(FAP): Intelligent information-processing must come into existence in the Universe, and, once it comes into existence, it will never die out.

Why anyone would believe it is beyond my comprehension. Barrow and Tipler close their introductory chapter with a warning:

[B]oth the FAP and the SAP are quite speculative; unquestionably, neither should be regarded as well-established principles of physics. In contrast, the WAP is just a restatement, albeit a subtle restatement, of one of the most important and well-established principles of science: that it is essential to take into account the limitations of one's measuring apparatus when interpreting one's observations. (ibid., p. 23)

In the teleological pre-Copernican universe expounded in Dante's *Divine Comedy,* for example, a cosmological anthropic principle holds a cardinal position. God created the world with Earth at its center as a place for humans to fulfill or fail to fulfill their spiritual potentialities. It is ironic, indeed, that such a powerful thinker as Weinberg should now be flirting with it. The clear conclusion, to my mind, is that explanatory asymmetries are as important to contemporary science as they are to contemporary philosophy of science. Philosophers of science have given extensive attention to the asymmetries of explanation, and these considerations should enable scientists and philosophers to deal effectively with issues such as those raised by the anthropic cosmological principle.

6. Explanations of Generalizations

In their classic 1948 paper on scientific explanation, Hempel and Oppenheim gave a precise explication of what later became known as the deductive-nomological model of scientific explanation of *particular facts,* but they explicitly acknowledged in footnote 33 (hereafter H-O fn. 33) that they could not offer an account of deductive explanations of general laws. The full text of note is as follows:

The precise rational reconstruction of explanation as applied to general regularities presents peculiar problems for which we can offer no solution at present. The core of the difficulty can be indicated by reference to an example: Kepler's laws, K, may be conjoined with Boyle's law, B, to [form] a stronger law K.B; but derivation of K from the latter would not be considered an explanation of the regularities stated in Kepler's laws; rather, it would be viewed as representing, in effect, a pointless "explanation" of Kepler's laws by themselves. The derivation of Kepler's laws from Newton's laws of motion and gravitation, on the other hand, would be recognized as a genuine explanation in terms of more comprehensive regularities, or so-called higher level laws. The problem therefore arises of setting up clear-cut criteria for the distinction of levels of explanation or for a comparison of generalized sentences as to their comprehensiveness. The establishment of adequate criteria for this purpose is as yet an open question. (Hempel and Oppenheim, [1948] 1965, p. 273)

Hempel did not address this question in his comprehensive essay "Aspects of Scientific Explanation" (1965b), nor did he take it up again elsewhere to the best of my knowledge. R. B. Braithwaite's *Scientific Explanation* (1953), which does not mention the Hempel-Oppenheim article, does not address this fundamental issue either. Braithwaite's theory is vulnerable with respect to the problem raised in this footnote. Nor does Karl R. Popper's *Logic of Scientific Discovery* (1959) take account of it, though it too is vulnerable. The theory of Michael Friedman (1974) would, if successful, have solved the problem. Unfortunately, as I explain in Salmon (1989, 1990, pp. 94–101), it is untenable.

Given the widespread acceptance of the view that scientific generalizations can be explained by derivation from broader generalizations, it is astonishing that H-O fn. 33, as well as the problem stated in it, has received virtually no attention in the literature. I must plead guilty to this charge, for I never took it up prior to 1989.

John Watkins, in *Science and Scepticism* (1984), though not mentioning H-O fn. 33, does address the problem: "It is rather remarkable that, although scientific theories are taken as the basic units by many philosophies and nearly all histories of science, there is no extant criterion, so far as I am aware, for distinguishing between a theory and an assemblage of propositions which, while it may have much testable content, remains a rag-bag collection" (1984, p. 204). He takes a theory to be a finite set of axioms together with all of its deductive consequences, and he assumes that a suitable distinction between the observational vocabulary and the theoretical vocabulary has been made. He also stipulates that any pure mathematics employed by the theory is to be axiomatized separately. As an answer to this problem, Watkins offers the *organic unity requirement.* An axiom system containing more than one axiom *fails* this requirement if it can be partitioned into two mutually exclusive and exhaustive nonempty subsets such that the set of testable consequences of the whole theory is equal to the set-theoretical union of the testable consequences of the two subsets.[7] The idea is that the axioms should work together to yield testable consequences that they cannot produce separately. With reference to H-O fn. 33, if we take Kepler's three laws K_1, K_2, K_3 and Boyle's law B as separate axioms, then the partition into Kepler's three laws and Boyle's law would violate the organic unity requirement. Moreover, if any axiom were a truth of logic or pure mathematics, the organic unity requirement would be violated.

It would, of course, be easy to *satisfy* this requirement by conjoining all of the axioms to yield a single axiom, somewhat along the lines suggested by H-O fn. 33. This move illustrates the fact that a given theory can be axiomatized in different ways. Two sets of axioms are equivalent if the deductive consequences of one are precisely the same as the deductive consequences of the other. Watkins attempts to block trivializing reaxiomatizations of theories by offering criteria for distinguishing *natural* from *unnatural* axiomatizations of a theory. An example is the requirement that no axiom should contain as a proper part any deductive consequence of the axiom set. This plausible requirement obviously suffices to dispose of the Hempel-Oppenheim example, which, as given, involves a single axiom that is the conjunction of the four separate laws. However, because many different tricks can be used in reaxiomatizations, stronger requirements are needed. Watkins offers a list of five:

(1) Independence: each axiom in the axiom set must be logically independent of the conjunction of the others.

(2) Nonredundancy: no predicate or individual constant may occur inessentially in the axiom set.

(3) Segregation: if axioms containing only theoretical predicates can be separately stated, without violating other rules, they should be.

(4) Wajsberg's requirement: an axiom is impermissible if it contains a (proper) component that is a theorem of the axiom set, *or becomes one when its variables are bound by the quantifiers that bind them in the axiom.* (See ibid., pp. 209–210, for the history of this requirement, including its title.)

(5) Decomposition: if the axiom set can be replaced by an equivalent one that is more numerous (though still finite) without violating the preceding rules, it should be. (ibid., pp. 208–209)

Condition (5) does the greatest part of the work; the preceding conditions are designed largely to prohibit its misuse. Condition (1), for example, prohibits increasing the number of axioms by adding theorems to the set of axioms. Condition (2) prohibits 'deoccamization' by the introduction of superfluous terms. Watkins justifies condition (3) "on the assumption, here taken for granted, that an axiomatization that enables us to go at once to a theory's fundamental assumptions is more perspicuous than one that has fundamental and auxiliary assumptions intermixed" (ibid., p. 209). Condition (4) blocks the example in H-O fn. 33, because $K_1.K_2.K_3.B$ contains $K_1.K_2.K_3$ as a proper part. If we were to rewrite it in prenex normal form, taking all of the quantifiers to the beginning of the conjunction, it would contain proper parts that, though not theorems because they are not closed, become theorems when bound by the appropriate quantifiers. This is prohibited by the italicized clause of condition (4). Conditions (4) and (5) mandate that the four component laws be taken as separate axioms.

Watkins considers condition (5) "uncontroversial": "One often wants to pinpoint as narrowly as possible those axioms in an axiom set that are specifically responsible for a particular theorem, especially in cases where the theorem has been falsified; and this means that one wants large, composite axioms to be broken up into small propositional units, indeed into the smallest propositional units, provided these units remain 'natural' . . . and are not dismembered artificially" (ibid., p. 212).

Having looked at the rules laid down by Watkins to determine what constitutes an appropriate axiomatization of a theory, we need to take a further step. Since, according to many philosophers—including Braithwaite, Hempel, Popper, and Watkins—an explanation of a generalization is a valid deductive argument, we must also consider the nature of the arguments. Let us begin by stipulating, first of all, that a given theory is suitably axiomatized in a way that would be natural according to Watkins's requirements, and, second, that no trivial deductions *of generalizations* are explanatory. Clearly the derivation of K from K.B is trivial. In general, we must conclude that the derivation of any proper or improper subset of the axioms of a theory is trivial. Similarly, the derivation of any conclusion that is logically equivalent to any subset of the axioms is trivial. In such cases the asymmetry of explanation breaks down. "All ravens are black" does not explain why all nonblack things are nonravens. As we saw in §4, Weinberg correctly remarks, "Explanation, unlike deduction, carries a unique sense of *direction. . . .* [A]lthough Newton derived his famous laws of motion [and gravitation] in part from the earlier laws of Kepler that describe the motion of planets in the solar system, we say that Newton's laws explain Kepler's, not the other way around" ([1992] 1994, p. 27). The words "in part" are crucial.[8] In this case, the entailment relation is not symmetric; Kepler's laws do not entail Newton's.

A modest proposal emerges from the foregoing considerations. Suppose that a theory has been axiomatized in accordance with Watkins's five conditions. Impose the explanatory asymmetry requirement. Let us then say that a deductive consequence of a subset of the axioms, all of which are essential to the derivation, explains that consequence only if that consequence does not entail the subset of axioms from which it was deduced. At this point Watkins's Decomposition Condition (5) performs important service. Perhaps this asymmetry requirement is sufficient.

A somewhat less modest proposal is to classify all arguments that use only truth-functional argument forms as trivial. This makes sense, I think, because truth-functional logic does not involve either the empirical content or the internal structure of statements that can be substituted for its variables. Of course, *modus ponens, modus tollens,* and other such forms are permissible in nontrivial arguments; it is just that to be nontrivial, a derivation must make essential use of other argument forms as well. Perhaps this is sufficient; however, I do not have a clear argument to show that it is.

The proposals discussed in this section are extremely tentative. Watkins's organic unity requirement makes good sense as a condition for theoretical unity. His conditions for natural axiomatization seem reasonable, though he stops short of claiming that they are sufficient to block all perverse reaxiomatizations (1984, p. 213). The requirement of asymmetry of entailment for explanations of generalizations is unexceptionable. The notion that strictly truth-functional derivations are trivial also seems correct. The problem we are confronting is fundamental to the theory of scientific explanation. If we cannot say with some precision what it means for one theory to explain another, our conception of scientific explanation is in pretty bad shape. It would mean that Feyerabend's 1962 remark, "[T]he notion of scientific explanation developed by some philosophers of science is so narrow that it is impossible to speak of one theory being explained by another," would apply to all theories of explanation.

7. Conclusions

1. In my personal experience, I have found many physicists who fail to realize that general philosophy of science is a genuine academic discipline with its own subject matter and its own agenda.[9] It attempts to clarify concepts such as scientific explanation

and scientific confirmation, and to grapple with doctrines such as reductionism and theoretical realism. Some physicists who are vaguely aware of the concerns of philosophy of science believe that, whereas students of sciences other than physics might benefit from instruction in philosophy of science, physics students absorb whatever may be worthwhile in the course of their training as physicists. Philosophers of science need to make serious efforts to convey the nature of their discipline to scientists (not only particle physicists). This does not mean that we should try to foist philosophical training upon those who have no interest in it or taste for it. I do not fault Weinberg for his disenchantment with philosophy of science.

2. Weinberg is aware that philosophy of science is an existing academic discipline, but he does not care for its agenda. He repeatedly sets his own agenda and complains that philosophers of science do not fulfill it. He does not seem to realize that there are other aims for philosophy of science, and that philosophers of science are not necessarily concerned to try to help with the search for a final theory. If they are, I agree that they are going about it in a very unproductive way.

3. My view is that those who regard the search for a final theory as a primary goal should be well-trained practicing physicists, and that extensive study of philosophy of science would be largely a waste of their time, unless they find pleasure or intellectual satisfaction in it for its own sake. I do not urge physicists to do philosophy if they are not interested, and I do not believe they will be less competent as physicists if they ignore philosophy altogether.

4. The principal moral I would draw is that, *if physicists insist on doing philosophy of science,* they should inform themselves as to what it is. This is my major quarrel with Weinberg's *Dreams.* He claims not to be doing philosophy of science, but he does, in fact, engage in it, and he does so with the open declaration of being unfamiliar with the great bulk of literature on the subject, especially literature that is less than three decades old. In addition to anachronism, his arguments suffer from a number of philosophical faults that have been pointed out in this essay.

5. To recapitulate these faults: Weinberg sets an inappropriate agenda for philosophy of science. He confuses logical entailment with material implication. He equivocates on the basis of the process-product ambiguity. He confuses confirmation-seeking why-questions with explanation-seeking why-questions. He makes himself a party, albeit reluctantly, to a modal fallacy connected with the cosmological anthropic principle.

6. Although this essay focuses on philosophy of science as it relates to physics, we should recall that essays 21–24 apply to anthropology and archaeology. A great many scientists in these fields are strongly receptive to philosophy of science, and make serious efforts to inform themselves about aspects that are relevant to their work.[10]

7. My final conclusion is a challenge to professional philosophers of science. If we are to have a "final theory" of scientific explanation—as if anything could be final in philosophy—we must come to terms with the problem set forth in H-O fn. 33. I have tried to provide a bit of progress in that direction, but I am not really convinced that the job is done. Scientists can, in all likelihood, distinguish intuitively between legitimate and illegitimate explanations of generalizations, but as philosophers we owe it to ourselves to try to codify explicitly the principles underlying the intuitions. In this and many other ways, our work on causality and scientific explanation is far from complete.

Notes

1. Like Weinberg, I use "billion" in the American sense to mean a thousand million.

2. "Night thoughts, of course, are what you get when you wake up at three in the morning and can't figure out how you are going to go on with your life" (Weinberg, 1995, p. 51).

3. The actual amount requested by the administration at the time was $640 million (Weinberg, [1992] 1994, p. 278).

4. If ornithologists could persuade Congress to invest huge sums for preservation of the natural habitats of birds, ornithology would have considerable value for birds.

5. Many biologists who have no truck with vitalism, including Ernst Mayr (1988, p. 475), strongly reject this sort of reductionism.

6. All parties to this discussion realize, of course, that Kepler's laws are not precisely true, and neither are Newton's, and that Newton's laws do not strictly entail Kepler's laws. However, the approximations we have adopted do not undermine the point of the example.

7. Truth-functional combinations of atomic observation statements, such as the conjunction of a testable consequence of Kepler's laws with a testable consequence of Boyle's law, do not qualify as testable consequences.

8. This marks a crucial difference between Weinberg (1987) and Weinberg ([1992] 1994).

9. In using the term "general philosophy of science" I mean to exclude such special fields as philosophy of space and time, philosophy of quantum mechanics, philosophy of biology, and philosophy of the social sciences. I believe that philosophers of science well qualified in such areas can make important contributions to the theoretical development of the scientific discipline in question.

10. When we arrived at the University of Arizona in 1973, archaeologists took the initiative of inviting Merrilee Salmon and me to participate in their seminars. Merrilee Salmon is presently recognized as a creative ongoing contributor to the discipline.

References

Achinstein, Peter. 1983. *The Nature of Explanation*. New York: Oxford University Press.

Alcock, C., et al. 1993. "Possible Gravitational Microlensing of a Star in the Large Magellanic Cloud," *Nature 365*, pp. 621–623.

Alston, William. 1971. "The Place of Explanations of Particular Facts in Science," *Philosophy of Science 38*, pp. 13–34.

Aristotle. 1928. *Posterior Analytics*, in W. D. Ross, ed., *The Works of Aristotle*, vol. 1 (Oxford: Clarendon Press).

Arnauld, Antoine. [1662] 1964. *The Art of Thinking (Port Royal Logic)*. Indianapolis: Bobbs-Merrill.

Asimov, Isaac. 1962. *The Genetic Code*. New York: New American Library.

Aubourg, E., et al. 1993. "Evidence for Gravitational Microlensing by Dark Objects in the Galactic Halo," *Nature 365*, pp. 623–625.

Baring, M. G. 1992. "Ignition of X-ray Flares," *Nature 360*, p. 109.

Barrow, John B., and Frank J. Tipler. 1986. *The Anthropic Cosmological Principle*. Oxford: Clarendon Press.

Basso, E. 1978. "Ethnographic Explanation." Unpublished ms. University of Arizona.

Bignami, G. F. 1992. "Gamma-ray Power from 3C 279," *Nature 355*, p. 299.

Binford, Lewis R. 1972. *An Archaeological Perspective*. New York: Harcourt, Brace, and Jovanovich.

————, ed. 1977. *For Theory Building in Archaeology*. New York: Academic Press.

Binford, L. R., and J. B. Bertram. 1977. "Bone Frequencies—and Attritional Process," in Binford, 1977, pp. 77–152.

Born, M. [1956] 1969. *Physics in My Generation*, 2d ed., revised. New York: Springer-Verlag.

Braithwaite, R. B. 1953. *Scientific Explanation*. Cambridge: Cambridge University Press.

Brill, A. A., ed. 1938. *The Basic Writings of Sigmund Freud*. New York: Random House.

Brody, B. 1975. "The Reduction of Teleological Sciences," *American Philosophical Quarterly 12*, pp. 69–76.

Bromberger, Sylvain. 1966. "Why-Questions," in Robert G. Colodny, ed., *Mind and Cosmos* (Pittsburgh: University of Pittsburgh Press), pp. 86–111.

Browne, M. W. 1993. "Strange, Violent Physics Born in the Death of Stars," *New York Times*, 27 April, p. B5.

Bunge, Mario. [1959] 1963. *Causality*. Cleveland: World Publishing Company.

Cairns-Smith, A. G. 1985. *Seven Clues to the Origin of Life*. Cambridge: Cambridge University Press.

Cajori, F., ed. 1947. *Sir Isaac Newton's Mathematical Principles of Natural Philosophy and his System of the World*. Berkeley: University of California Press.

Carnap, Rudolf. 1950. *Logical Foundations of Probability*. Chicago: University of Chicago Press.

———. 1966. *Philosophical Foundations of Physics*. New York: Basic Books.

———. 1974. *An Introduction to the Philosophy of Science*. Reprint, revised. New York: Harper Torchbooks.

Cartmill, Matt. 1980. "John Jones's Pregnancy: Some Comments on the Statistical-Relevance Model of Scientific Explanation," *American Anthropologist 82*, pp. 382–385.

Chaffee, Frederic H., Jr. 1980. "The Discovery of a Gravitational Lens," *Scientific American 243* (Nov.), pp. 70–88.

Church, A. 1940. "On the Concept of a Random Sequence," *Bulletin of the American Mathematical Society 46*, pp. 130–135.

Clark, Edwin Gurney, and William O. Mortimer Harris. 1961. "Venereal Diseases," *Encyclopedia Britannica 23*.

Clendinnen, F. J. 1992. "Nomic Dependence and Causation," *Philosophy of Science 59*, pp. 341–360.

Coffa, J. Alberto. 1973. "Foundations of Inductive Explanation." Ph.D. dissertation, University of Pittsburgh.

———. 1974. "Hempel's Ambiguity," *Synthese 28*, pp. 141–164.

Cohen, S. N. 1975. "The Manipulation of Genes," *Scientific American 233* (July), pp. 24–33.

Coles, John. 1973. *Archaeology by Experiment*. New York: Charles Scribner's Sons.

Copi, Irving. 1972. *Introduction to Logic*, 4th ed. New York: Macmillan.

Courvoisier, T., and E. I. Robson. 1991. "The Quasar 3C 273," *Scientific American 264* (June), pp. 50–57.

Crow, James F. 1979. "Genes That Violate Mendel's Rules," *Scientific American 240* (Feb.), pp. 134–146.

Darwin, Charles. 1859. *The Origin of Species*. London.

———. [1871] 1883. *The Descent of Man*. New York: D. Appleton & Co.

Deetz, James. 1967. *Invitation to Archaeology*. New York: American Museum Science Books.

Dermer, C. D., and R. Schlickeiser. 1992. "Quasars, Blazars, and Gamma Rays," *Science 257*, p. 1645.

Descartes, René. [1641] 1951. *Meditations*. Trans. Laurence J. LaFleur. New York: Library of Liberal Arts.

Dewdney, A. 1979. "A Size Limit for Uniformly Pulsating Sources of Electromagnetic Radiation," *Astrophysical Letters 20*, pp. 49–52.

Disney, M., and P. Véron. 1977. "BL Lacertae Objects," *Scientific American 237* (Aug.), pp. 32–39.

Dowe, Phil. 1992a. "An Empiricist Defence of the Causal Account of Explanation," *International Studies in the Philosophy of Science 6*, pp. 123–128.

———. 1992b "Process Causality and Asymmetry," *Erkenntnis 37*, pp. 179–196.

_____. 1992c. "Wesley Salmon's Process Theory of Causality and the Conserved Quantity Theory," *Philosophy of Science 59*, pp. 195–216.

_____. 1995. "Causality and Conserved Quantities: A Reply to Salmon," *Philosophy of Science 62*, pp. 321–333.

Dumond, Don E. 1980. "The Archaeology of Alaska and the Peopling of America," *Science 209*, pp. 984–991.

Earman, John, ed. 1983. *Testing Scientific Theories*, vol. 10, *Minnesota Studies in the Philosophy of Science*. Minneapolis: University of Minnesota Press.

_____. 1986. *A Primer on Determinism*. Dordrecht: D. Reidel.

_____. 1992. "Determinism in the Physical Sciences," in M. Salmon et al., *Introduction to the Philosophy of Science* (Englewood Cliffs, N.J.: Prentice-Hall), pp. 232–268.

Eddington, A. [1920] 1959. *Space, Time, and Gravitation*. New York: Harper & Brothers.

Eells, Ellery. 1991. *Probabilistic Causality*. Cambridge: Cambridge University Press.

Einstein, Albert. [1905] 1923. "On the Electrodynamics of Moving Bodies," in Albert Einstein et al., *The Principle of Relativity* (New York: Dover Publications), pp. 35–65.

Einstein, Albert, B. Podolsky, and N. Rosen. 1935. "Can Quantum-Mechanical Description of Physical Reality Be Considered Complete?," *Physical Review 47*, pp. 777–780.

Fair, David. 1979. "Causation and the Flow of Energy," *Erkenntnis 14*, pp. 219–250.

Feigl, Herbert, and May Brodbeck, eds. 1953. *Readings in the Philosophy of Science*. New York: Appleton-Century-Crofts.

Fetzer, James H. 1971. "Dispositional Probabilities," in Roger C. Buck and Robert S. Cohen, eds., *PSA 1970* (Dordrecht: D. Reidel), pp. 473–482.

_____. 1974a. "Statistical Explanations," in Kenneth Schaffner and Robert S. Cohen, eds., *PSA 1972* (Dordrecht: D. Reidel), pp. 337–347.

_____. 1974b. "A Single Case Propensity Theory of Explanation," *Synthese 28*, pp. 171–198.

_____. 1975. "On the Historical Explanation of Unique Events," *Theory and Decision 6*, pp. 87–97.

_____. 1976. "The Likeness of Lawlikeness," in Robert S. Cohen et al., eds., *PSA 1974* (Dordrecht: D. Reidel), pp. 377–391.

_____. 1977. "A World of Dispositions," *Synthese 34*, pp. 397–421.

_____. 1981. *Scientific Knowledge*. Dordrecht: D. Reidel.

_____. 1991. "Critical Notice: Philip Kitcher and Wesley C. Salmon, eds., *Scientific Explanation*, and Wesley C. Salmon, *Four Decades of Scientific Explanation*," *Philosophy of Science 58*, pp. 288–306.

_____. 1992. "What's Wrong with Salmon's History: The Third Decade," *Philosophy of Science 59*, pp. 246–262.

Feynman, Richard P., et al. 1963. *The Feynman Lectures on Physics*, 3 vols. Reading, Mass.: Addison-Wesley.

_____. 1985. *QED: The Strange Theory of Light and Matter*. Princeton: Princeton University Press.

Flannery, Kent. 1973. "Archeology with a Capital S," in C. L. Redman, ed., *Research and Theory in Current Archaeology* (New York: John Wiley and Sons), pp. 47–53.

Freud, Sigmund. [1900] 1938. *The Interpretation of Dreams*. In Brill, 1938, pp. 181–549.

_____. [1901] 1938. *The Psychopathology of Everyday Life*. In Brill, 1938, pp. 35–178.

Friedman, Michael. 1974. "Explanation and Scientific Understanding," *Journal of Philosophy 71*, pp. 5–19.

"Galactic Center Antimatter Factory Found at Last?" 1990. *Sky & Telescope 79*, p. 363.

Gamow, George. 1961. *The Atom and Its Nucleus*. Englewood Cliffs, N. J.: Prentice-Hall.

GBL. 1974. "Evidence Accumulates for a Black Hole in Cygnus X-1," *Physics Today 27* (Feb.), pp. 17, 19.

Giere, R. 1984. *Understanding Scientific Reasoning.* 2d ed. New York: Holt, Rinehart & Winston.

Good, I. J. 1961–62: "A Causal Calculus I-II," *British Journal for the Philosophy of Science 11*, pp. 305–318; *12*, pp. 43–51; errata and corrigenda, *13*, p. 88.

———. 1980. "Some Comments on Probabilistic Causality," *Pacific Philosophical Quarterly 61*, pp. 301–304.

———. 1983. *Good Thinking.* Minneapolis: University of Minnesota Press.

———. 1985. "Causal Propensity: A Review," in Peter D. Asquith and Philip Kitcher, eds., *PSA 1984*, vol. 2 (East Lansing, Mich.: Philosophy of Science Association), pp. 829–850.

Goodman, N. 1955. *Fact, Fiction, and Forecast.* Cambridge, Mass.: Harvard University Press.

Grayson, Donald K. 1980. "Vicissitudes and Overkill: The Development of Explanations of Pleistocene Extinctions," *Advances in Archaeological Method and Theory 3*, pp. 357–403.

Greeno, James G. [1970] 1971. "Explanation and Information," in Salmon, 1971, pp. 89–104.

Greenstein, J. L. 1963. "Quasi-Stellar Radio Sources," *Scientific American 207* (Dec.), pp. 54–62.

Griffin, J. B. 1956. "The Study of Early Cultures," in Harry L. Shapiro, ed., *Man, Culture, and Society* (London: Oxford University Press), pp. 22–48.

Grünbaum, Adolf. 1962. "Temporally Asymmetric Principles, Parity between Explanation and Prediction, and Mechanism versus Teleology," *Philosophy of Science 29*, pp. 146–170.

———. 1963. *Philosophical Problems of Space and Time.* New York: Alfred A. Knopf.

———. 1967. *Modern Science and Zeno's Paradoxes.* Middletown, Conn.: Wesleyan University Press.

———. 1973. *Philosophical Problems of Space and Time*, 2d enlarged ed. Dordrecht: D. Reidel.

Halley, E. [1687] 1947. "The Ode Dedicated to Newton by Edmund Halley." Trans. L. J. Richardson, in Cajori, 1947, p. xiv.

Hanna, Joseph. 1978. "On Transmitted Information as a Measure of Explanatory Power," *Philosophy of Science 45*, pp. 531–562.

Hanson, N. R. 1958. *Patterns of Discovery.* Cambridge: Cambridge University Press.

———. 1959. "On the Symmetry Between Explanation and Prediction," *Philosophical Review 68*, pp. 349–358.

Harré, Rom, and Edward Madden. 1975. *Causal Powers.* Oxford: Basil Blackwell.

Hawking, Stephen W., and Roger Penrose. 1996. *The Nature of Space and Time.* Princeton: Princeton University Press.

Hempel, Carl G. [1959] 1965a. "The Logic of Functional Analysis," in Hempel 1965a, pp. 297–330.

———. 1962a. "Deductive-Nomological vs. Statistical Explanation," in Herbert Feigl and Grover Maxwell, eds., *Scientific Explanation, Space, and Time*, vol. 3, *Minnesota Studies in the Philosophy of Science* (Minneapolis: University of Minnesota Press), pp. 98–169.

———. 1962b. "Explanation in Science and in History," in R. G. Colodny, ed., *Frontiers in Science and Philosophy* (Pittsburgh: University of Pittsburgh Press), pp. 7–33.

_____. 1965a. *Aspects of Scientific Explanation and Other Essays in the Philosophy of Science*. New York: Free Press.

_____. 1965b. "Aspects of Scientific Explanation," in Hempel, 1965a, pp. 331–496.

_____. 1966. *Philosophy of Natural Science*. Englewood Cliffs, N.J.: Prentice-Hall.

_____. 1977. "Nachwort 1976: Neuere Ideen zu den Problemen der statistischen Erklärung," in Carl G. Hempel, *Aspekte wissenschaftlicher Erklärung*. Berlin: Walter de Gruyter, pp. 98–123.

Hempel, Carl G., and Paul Oppenheim. [1948] 1965. "Studies in the Logic of Explanation," in Hempel, 1965a, pp. 245–296.

Hesslow, Germund. 1976. "Two Notes on the Probabilistic Approach to Causality," *Philosophy of Science 43*, pp. 290–292.

Hitchcock, Christopher Read. 1993. "A Generalized Probabilistic Theory of Causal Relevance," *Synthese 97*, pp. 335–364.

_____. 1995. "Discussion: Salmon on Explanatory Relevance," *Philosophy of Science 62*, pp. 304–320.

_____. 1996. "A Probabilistic Theory of Second Order Causation," *Erkenntnis 44*, pp. 369–377.

Hoffmann, B. 1964. "Fluctuating Brightness of Quasi-Stellar Radio Sources," *Science 144*, p. 319.

Holton, G., and S. Brush. 1973. *Introduction to Concepts and Theories in Physical Science*, 2d ed. Reading, Mass.: Addison-Wesley.

Hume, David. [1739–1740] 1888. *A Treatise of Human Nature*. Ed. L. A. Selby-Bigge. Oxford: Clarendon Press.

_____. 1740. "An Abstract of *A Treatise of Human Nature*," in Hume, [1748] 1955, pp. 183–198.

_____. [1748] 1955. *An Enquiry Concerning Human Understanding*. Indianapolis: Bobbs-Merrill.

Humphreys, Paul W. 1980. "Cutting the Causal Chain," *Pacific Philosophical Quarterly 61*, pp. 305–314.

_____. 1981. "Aleatory Explanation," *Synthese 48*, pp. 225–232.

_____. 1983. "Aleatory Explanation Expanded," in Peter Asquith and Thomas Nickles, eds., *PSA 1982*, vol. 2 (East Lansing, Mich.: Philosophy of Science Association), pp. 208–223.

_____. 1985. "Why Propensities Cannot Be Probabilities." *Philosophical Review 94*, pp. 557–570.

_____. 1989. *The Chances of Explanation*. Princeton: Princeton University Press.

Hutchings, J. B. 1985. "Observational Evidence for Black Holes," *American Scientist 73*, p. 52.

Irving, W., and C. Harington. 1973. "Upper Pleistocene Radiocarbon-Dated Artifacts from the Northern Yukon," *Science 179*, pp. 335–340.

Jeffrey, R. C. [1969] 1971. "Statistical Explanation vs. Statistical Inference," in Salmon, 1971, pp. 19–28. Originally published in Rescher, ed., 1969.

Kiev, Ari. 1964. "The Study of Folk Psychiatry," in A. Kiev, ed., *Magic, Faith, and Healing* (London: The Free Press), pp. 3–35.

King, John L. 1976. "Statistical Relevance and Explanatory Classification," *Philosophical Studies 30*, pp. 313–321.

Kitcher, Philip. 1976. "Explanation, Conjunction, and Unification," *Journal of Philosophy 73*, pp. 207–212.

————. 1981. "Explanatory Unification," *Philosophy of Science 48*, pp. 507–531.

————. 1985. "Two Approaches to Explanation," *Journal of Philosophy 82*, pp. 632–639.

————. 1989. "Explanatory Unification and the Causal Structure of the World," in Kitcher and Salmon, eds., 1989, pp. 410–505.

————. 1993. *The Advancement of Science*. New York: Oxford University Press.

Kitcher, Philip, and Wesley C. Salmon. 1987. "Van Fraassen on Explanation," *Journal of Philosophy 84*, pp. 315–330. Here reprinted.

————, eds. 1989. *Scientific Explanation*, vol. 13, *Minnesota Studies in the Philosophy of Science*. Minneapolis: University of Minnesota Press.

Kniffen, Donald A. 1993. "The Gamma-Ray Universe," *American Scientist 81*, pp. 342–349.

Kuhn, Thomas. 1962. *The Structure of Scientific Revolutions*. Chicago: University of Chicago Press.

Kyburg, H. E., Jr. 1970. "Conjunctivitis," in M. Swain, ed., *Induction, Acceptance, and Rational Belief* (Dordrecht: D. Reidel), pp. 55–82.

Laplace, Pierre-Simon, Marquis de. [1820] 1951. *A Philosophical Essay on Probabilities*. Trans. Frederick Wilson Truscott and Frederick Lincoln Emory. New York: Dover Publications.

LeBlanc, S. 1973. "Two Points of Logic," in C. L. Redman, ed., *Research and Theory in Current Archaeology* (New York: John Wiley and Sons), pp. 199–214.

Lehman, H. 1972. "Statistical Explanation," *Philosophy of Science 39*, pp. 500–506.

Lewin, W. 1977. "X-Ray Outbursts in Our Galaxy," *American Scientist 65*, pp. 605–613.

Lewis, D. 1973. "Causation," *Journal of Philosophy 70*, pp. 556–567.

Lin, Y. C., et al. 1992. "Detection of High-Energy Gamma-Ray Emission from the BL Lacertae Object Markarian 421 by the EGRET Telescope on the *Compton Observatory*," *Astrophysical Journal 401*, p. L61.

Lo, K. Y. 1986. "The Galactic Center: Is It a Massive Black Hole?" *Science 233*, pp. 1393–1403.

Locke, John. [1689] 1924. Ed. A. S. Pringle-Pattison. *An Essay Concerning Human Understanding*. Oxford: Clarendon Press.

Loeb, L. 1974. "Causal Theories and Causal Overdetermination," *Journal of Philosophy 71*, pp. 525–544.

Lucretius. 1951. *On the Nature of the Universe*. Trans. R. E. Latham. Baltimore: Penguin Books.

Mackie, John L. 1974. *The Cement of the Universe*. Oxford: Clarendon Press.

Madsen, W. 1964. "Value Conflicts and Folk Psychotherapy in South Texas," in A. Kiev, ed., *Magic, Faith, and Healing* (London: The Free Press), pp. 420–440.

Maffei, P. 1980. *Monsters in the Sky*. Cambridge, Mass.: MIT Press.

Maran, S. P., ed. 1992. *Astronomy and Astrophysics Encyclopedia*. New York: Van Nostrand.

Martin, Paul S., and Fred Plog. 1973. *The Archaeology of Arizona*. Garden City, N.Y.: Doubleday/Natural History Press.

Mayr, Ernst. 1988. "The Limits of Reductionism," *Nature 331*, p. 475.

Meehan, Eugene. 1968. *Explanation in Social Science—A System Paradigm*. Homewood, Ill.: The Dorsey Press.

Mellor, D. H. 1976. "Probable Explanation," *Australasian Journal of Philosophy 54*, pp. 231–241.

Mermin, N. David. 1981. "Quantum Mysteries for Anyone," *Journal of Philosophy 78*, pp. 397–408.

————. 1985. "Is the Moon There When Nobody Looks? Reality and the Quantum Theory," *Physics Today 38* (Apr.), pp. 38–47.

Mill, John Stuart. 1843. *A System of Logic*. London: John W. Parker.

Miller, H. R., et al. 1989. "Detection of Microvariability for BL Lacertae Objects," *Nature* *337*, p. 627.

Misner, C. W., K. S. Thorne, and J. A. Wheeler. 1973. *Gravitation*. San Francisco: W. H. Freeman.

Morris, H. 1972. *The Remarkable Birth of the Planet Earth*. San Diego: Creation-Life Publishers.

Morrison, P. 1973. "Resolving the Mystery of the Quasars?," *Physics Today 26* (Mar.), p. 25.

Mühlhölzer, F. 1994. "Scientific Explanation and Equivalent Descriptions," in W. Salmon and G. Wolters, eds., *Logic, Language, and the Structure of Scientific Theories* (Pittsburgh: University of Pittsburgh Press), pp. 119–138.

Nagel, Ernest. 1961. *The Structure of Science: Problems in the Logic of Scientific Explanation*. New York: Harcourt, Brace, and World.

Newton, Isaac. [1687] 1947. *Mathematical Principles of Natural Philosophy*. Trans. Florian Cajori. Berkeley: University of California Press. Originally published *Philosophiae Naturalis Principia Mathematica* (London).

Nye, Mary Jo. 1972. *Molecular Reality*. London: Macdonald.

Olmsted, D. L. 1971. *Out of the Mouth of Babes*. The Hague: Mouton.

Otte, Richard. 1981. "A Critique of Suppes' Theory of Probabilistic Causality," *Synthese 48*, pp. 167–190.

Pauling, L. 1959. *No More War*. New York: Dodd, Mead & Co.

_____. 1970. *Vitamin C and the Common Cold*. San Francisco: W. H. Freeman.

Pearson, Karl; [1911] 1957. *The Grammar of Science*, 3d ed. New York: Meridian Books.

Perrin, Jean. 1913. *Les Atomes*. Paris: Alcan.

_____. [1913] 1916. *Atoms*. Trans. D. L. Hammick. London: Constable & Co.

Popper, Karl R. 1935. *Logik der Forschung*. Vienna: Springer.

_____. 1957. "The Propensity Interpretation of the Calculus of Probability, and the Quantum Theory," in S. Körner, ed., *Observation and Interpretation* (London: Butterworths Scientific Publications), pp. 65–70.

_____. 1959. *The Logic of Scientific Discovery*. New York: Basic Books. Translation, with added appendices, of Popper, 1935.

_____. 1972. *Objective Knowledge: An Evolutionary Approach*. Oxford: Clarendon Press.

Powell, C. S. 1993. "Inconstant Cosmos," *Scientific American 268* (May), p. 111.

Prigogine, Ilya. 1984. *Order Out of Chaos*. Toronto: Bantam Books.

Quine, W. V. [1951] 1953. "Two Dogmas of Empiricism," in W. V. Quine, *From a Logical Point of View* (Cambridge, Mass.: Harvard University Press), pp. 20–46.

Radcliffe-Brown, A. R. [1933] 1967. *The Andaman Islanders*. New York: The Free Press.

_____. [1952] 1965. *Structure and Function in Primitive Society*. New York: The Free Press.

Railton, Peter. 1978. "A Deductive-Nomological Model of Probabilistic Explanation," *Philosophy of Science 45*, pp. 206–226.

_____. 1980. "Explaining Explanation: A Realist Account of Scientific Explanation and Understanding." Ph.D. dissertation, Princeton University.

_____. 1981. "Probability, Explanation, and Information," *Synthese 48*, pp. 233–256.

Redman, C. L., ed. 1973. *Research and Theory in Current Archeology*. New York: John Wiley and Sons.

Reichenbach, Hans. [1928] 1957. *The Philosophy of Space and Time*. New York: Dover Publications. Originally published as *Philosophie der Raum-Zeit-Lehre* (Berlin: Walter de Gruyter).

_____. 1938. *Experience and Prediction*. Chicago: University of Chicago Press.

———. 1946. *Philosophic Foundations of Quantum Mechanics*. Berkeley: University of California Press.

———. 1956. *The Direction of Time*. Berkeley: University of California Press.

Remillard, R. A., et al. 1991. "A Rapid Energetic X-Ray Flare in the Quasar PKSO558-504," *Nature 350*, p. 591.

Rescher, Nicholas. 1958. "On Prediction and Explanation," *British Journal for the Philosophy of Science 8*, pp. 281–290.

———, ed. 1969. *Essays in Honor of Carl G. Hempel*. Dordrecht: D. Reidel.

———. 1970. *Scientific Explanation*. New York: The Free Press.

Rice, G. E. 1975. "A Systematic Explanation of a Change in Mogollon Settlement Patterns." Ph.D. dissertation, University of Washington.

Rosen, Deborah A. 1978. "In Defense of a Probabilistic Theory of Causality," *Philosophy of Science 45*, pp. 604–613.

Ross-Ashby, W. 1956. *An Introduction to Cybernetics*. London: Chapman and Hall.

Russell, Bertrand. 1922a. *Our Knowledge of the External World*. London: George Allen & Unwin.

———. 1922b. "The Problem of Infinity Considered Historically," in Russell, 1922a, pp. 159–188.

———. 1927. *The Analysis of Matter*. London: George Allen & Unwin.

———. 1929. "On the Notion of Cause," in *Mysticism and Logic* (New York: W. W. Norton), pp. 180–208.

———. 1948. *Human Knowledge, Its Scope and Limits*. New York: Simon and Schuster.

Salmon, Merrilee H. 1982. "Models of Explanation: Two Views," in Colin Renfrew et al., eds., *Theory and Explanation in Archaeology* (New York: Academic Press), pp. 35–44.

Salmon, Merrilee H., and Wesley C. Salmon. 1979. "Alternative Models of Scientific Explanation," *American Anthropologist 81*, pp. 61–74. Here reprinted.

Salmon, Wesley C. 1965. "The Status of Prior Probabilities in Statistical Explanation," *Philosophy of Science 32*, pp. 137–146.

———. 1967. *The Foundations of Scientific Inference*. Pittsburgh: University of Pittsburgh Press.

———. 1968a. "The Justification of Inductive Rules of Inference," in I. Lakatos, ed., *The Problem of Inductive Logic* (Amsterdam: North-Holland), pp. 24–43.

———. 1968b. "Reply," in I. Lakatos, ed., *The Problem of Inductive Logic* (Amsterdam: North-Holland), pp. 74–97.

———, ed. 1970a. *Zeno's Paradoxes*. Indianapolis: Bobbs-Merrill.

———. 1970b. "Statistical Explanation," in R. Colodny, ed., *The Nature and Function of Scientific Theories* (Pittsburgh: University of Pittsburgh Press), pp. 173–231. Reprinted in Salmon, 1971, pp. 29–88.

———. 1971. *Statistical Explanation and Statistical Relevance*. With contributions by J. G. Greeno and R. C. Jeffrey. Pittsburgh: University of Pittsburgh Press.

———. 1974a. "Comments on 'Hempel's Ambiguity' by J. A. Coffa," *Synthese 28*, pp. 165–169.

———. 1974b. "Russell on Scientific Inference; *or*, Will the Real Deductivist Please Stand Up?," in G. Nakhnikian, ed., *Bertrand Russell's Philosophy* (London: Gerald Duckworth), pp. 183–208.

———. 1975a. "Theoretical Explanation," in S. Körner, ed., *Explanation* (Oxford: Basil Blackwell), pp. 118–145. Here reprinted as "Causal and Theoretical Explanation."

———. 1975b. *Space, Time, and Motion: A Philosophical Introduction*. Encino, Calif.: Dickenson Publishing Co.

_____. 1977a. "A Third Dogma of Empiricism," in R. Butts and J. Hintikka, eds., *Basic Problems in Methodology and Linguistics* (Dordrecht: D. Reidel), pp. 149–166. Here reprinted.

_____. 1977b. "An 'At-At' Theory of Causal Influence," *Philosophy of Science 44*, pp. 215–224. Here reprinted.

_____. 1977c. "Why Ask 'Why?'— An Inquiry Concerning Scientific Explanation," *Proceedings and Addresses of the American Philosophical Association 51*, pp. 683–705. Reprinted in Salmon, 1979a, pp. 403–425. Here reprinted.

_____. 1977d. "Hempel's Conception of Inductive Inference in Inductive-Statistical Explanation," *Philosophy of Science 44*, pp. 180–185.

_____, ed. 1979a. *Hans Reichenbach: Logical Empiricist*. Dordrecht: D. Reidel. Large portions were previously published in *Synthese 34* and *35*.

_____. 1979b. "Propensities: A Discussion Review," *Erkenntnis 14*, pp. 183–216.

_____. 1980. "Probabilistic Causality," *Pacific Philosophical Quarterly 61*, pp. 50–74. Here reprinted.

_____. 1981a. "Causality: Production and Propagation," in P. Asquith and R. Giere, eds., *PSA 1980*, vol. 2 (East Lansing, Mich.: Philosophy of Science Association), pp. 49–69. Here reprinted.

_____. 1981b. "Rational Prediction," *British Journal for the Philosophy of Science 32*, pp. 115–125.

_____. 1982a. "Causality in Archaeological Explanation," in Colin Renfrew et al., eds., *Theory and Explanation in Archaeology* (New York: Academic Press), pp. 45–55. Here reprinted.

_____. 1982b. "Comets, Pollen, and Dreams—Some Reflections on Scientific Explanation," in R. McLaughlin, ed., *What? Where? When? Why?* (Dordrecht: D. Reidel), pp. 155–178. Here reprinted.

_____. 1984a. *Logic*, 3d ed. Englewood Cliffs, N.J.: Prentice-Hall.

_____. 1984b. *Scientific Explanation and the Causal Structure of the World*. Princeton: Princeton University Press.

_____. 1985a. "Conflicting Concepts of Scientific Explanation," *Journal of Philosophy 82*, pp. 651–654.

_____. 1985b. "Scientific Explanation: Three Basic Conceptions," in P. Asquith and P. Kitcher, eds., *PSA 1984*, vol. 2 (East Lansing, Mich.: Philosophy of Science Association), pp. 293–305. Here reprinted.

_____. 1989. "Four Decades of Scientific Explanation," in Kitcher and Salmon, eds., 1989, pp. 3–219.

_____. 1990a. "Causal Propensities: Statistical Causality vs. Aleatory Causality," *Topoi 9*, pp. 95–100.

_____. 1990b. *Four Decades of Scientific Explanation*. Minneapolis: University of Minnesota Press. Reprinted from Kitcher and Salmon, eds. 1989, pp. 3–219.

_____. 1990c. "Scientific Explanation: Causation and Unification," *Crítica 22*, pp. 3–21. Here reprinted.

_____. 1994a. "Carnap, Hempel, and Reichenbach on Scientific Realism," in W. Salmon and G. Wolters, eds., *Logic, Language, and the Structure of Scientific Theories* (Pittsburgh: University of Pittsburgh Press), pp. 237–254.

_____. 1994b. "Causality without Counterfactuals," *Philosophy of Science 61*, pp. 297–312. Here reprinted.

_____. 1997. "Causality and Explanation: A Reply to Two Critics," *Philosophy of Science 64*, pp. 461–477.

Sanders, W. T., and B. Price. 1968. *Mesoamerica: The Evolution of a Civilization*. New York: Random House.

Sayre, Kenneth. 1977. "Statistical Models of Causal Relations," *Philosophy of Science 44*, pp. 203–214.

Schaefer, B. E. 1985. "Gamma-Ray Bursters," *Scientific American 252* (Feb.), p. 55.

Scheer, Robert. 1996. "Dumb Weapons," *Pittsburgh Post-Gazette*, 18 July 1996, p. A13.

Scheffler, I. 1957. "Explanation, Prediction, and Abstraction," *British Journal for the Philosophy of Science 7*, pp. 293–309.

Schiffer, M. B. 1975. "Archeology as a Behavioral Science," *American Anthropologist 77*, pp. 836–848.

Schramm, et al. 1993. "Recent Activity in the Optical and Radiofrequency Lightcurves of Blazar 3C 345: Indications for a 'Lighthouse Effect' Due to Jet Rotation," *Astronomy and Astrophysics 278*, pp. 391–405.

Scriven, Michael. 1958. "Definitions, Explanations, and Theories," in Herbert Feigl, Grover Maxwell, and Michael Scriven, eds., *Concepts, Theories, and the Mind-Body Problem*, vol. 2, *Minnesota Studies in the Philosophy of Science* (Minneapolis: University of Minnesota Press), pp. 99–195.

———. 1959. "Explanation and Prediction in Evolutionary Theory," *Science 130*, pp. 477–482.

———. 1962. "Explanations, Predictions, and Laws," in Herbert Feigl and Grover Maxwell, eds., *Scientific Explanation, Space, and Time*, vol. 3, *Minnesota Studies in the Philosophy of Science* (Minneapolis: University of Minnesota Press), pp. 170–230.

———. 1975. "Causation as Explanation," *Nous 9*, pp. 3–16.

Shank, C. V. 1986. "Investigation of Ultrafast Phenomena in the Femtosecond Domain," *Science 233*, pp. 1276–1280.

Skyrms, Brian, and William Harper, eds. 1988. *Causation, Chance, and Credence*. Dordrecht: Kluwer Academic Publishers.

Smith, A. H. 1958. *The Mushroom Hunter's Guide*. Ann Arbor: University of Michigan Press.

Stegmüller, Wolfgang. 1973. *Probleme und Resultate der Wissenschaftstheorie und Analytischen Philosophie*. Band 4, Studienausgabe Teil E. Berlin: Springer-Verlag.

Stein, Howard. 1983. "On the Present State of the Philosophy of Quantum Mechanics," in Peter Asquith and Thomas Nickles, eds., *PSA 1982*, vol. 2 (East Lansing, Mich.: Philosophy of Science Association), pp. 563–581.

Stevenson, Charles. 1944. *Ethics and Language*. New Haven: Yale University Press.

Suppes, Patrick. 1970. *A Probabilistic Theory of Causality*. Amsterdam: North-Holland.

———. 1984. *Probabilistic Metaphysics*. Oxford: Basil Blackwell.

Tannen, Deborah. 1991. *You Just Don't Understand*. New York: Ballantine.

Terrell, J. 1964. "Quasi-Stellar Diameters and Intensity Fluctuations," *Science 145*, p. 919.

———. 1977. "Size Limits on Fluctuating Astronomical Sources," *Astrophysical Journal 213*, pp. L93–97.

Trimble, V., and L. Woltjer. 1986. "Quasars at 25," *Science 234*, pp. 155–161.

Tuggle, H. D., A. H. Townsend, and T. J. Riley. 1972. "Laws, Systems, and Research Designs," *American Antiquity 37*, pp. 3–12.

van Fraassen, B. 1977. "The Pragmatics of Explanation," *American Philosophical Quarterly 14*, pp. 143–150.

———. 1980. *The Scientific Image*. Oxford: Clarendon Press.

———. 1983. "Glymour on Evidence and Explanation," in Earman, ed., 1983, pp. 165–176.

———. 1985. "Salmon on Explanation," *Journal of Philosophy 82*, pp. 639–651.

Venn, John. 1866. *The Logic of Chance*. London: Macmillan and Co.

von Bretzel, Philip. 1977. "Concerning a Probabilistic Theory of Causation Adequate for the Causal Theory of Time," *Synthese 35*, no. 2, pp. 173–190. Also published in Salmon, ed., 1979, pp. 385–402).

von Mises, R. 1964. *Mathematical Theory of Probability and Statistics*. New York: Academic Press.

von Neumann, John. 1955. *Mathematical Foundations of Quantum Theory*. Princeton: Princeton University Press.

von Wright, Georg Henrik. 1971. *Explanation and Understanding*. Ithaca, N.Y.: Cornell University Press.

Waldrop, M. 1982. "A Hole in the Milky Way," *Science 216*, pp. 838–839.

Watkins, J. W. N. 1968. "Non-inductive Corroboration," in I. Lakatos, ed., *The Problem of Inductive Logic* (Amsterdam: North-Holland), pp. 61–66.

———. 1984. *Science and Scepticism*. Princeton: Princeton University Press.

Watson, James. 1968. *The Double Helix*. New York: Atheneum Publishers.

Watson, Patty Jo, Steven LeBlanc, and Charles L. Redman. 1971. *Explanation in Archeology: An Explicitly Scientific Approach*. New York: Columbia University Press.

———. 1984. *Archeological Explanation: The Scientific Method in Archeology*. 2d ed. New York: Columbia University Press.

Wauchope, R. 1966. *Archaeological Survey of Northern Georgia with a Test of Some Cultural Hypotheses,* Salt Lake City: Society for American Archaeology.

Weinberg, Steven. 1977. *The First Three Minutes*. New York: Basic Books.

———. 1987. "Newtonianism, Reductionism and the Art of Congressional Testimony," *Nature 330* (3 Dec.), pp. 433–437.

———. 1988. "Weinberg Replies," *Nature 331* (11 Feb.), pp. 475–476.

———. [1992] 1994. *Dreams of a Final Theory*. New York: Vintage Books.

———. 1995. "Night Thoughts of a Quantum Physicist," *Bulletin of the American Academy of Arts and Sciences 49* (Dec.), pp. 51–64.

Weiner, Tim. 1996. "Stealth, Lies and Videotape," *New York Times*, 14 July 1996, p. E3.

Wichmann, Eyvind H. 1967. *Quantum Physics*, vol. 4, *Berkeley Physics Course*. New York: McGraw-Hill.

Wiener, N. 1961. *Cybernetics*. Cambridge, Mass.: MIT Press.

Wilson, Brian R., ed. 1970. *Rationality*. Oxford: Blackwell.

Winnie, John. 1977. "The Causal Theory of Space-Time," in John Earman, Clark Glymour, and John Stachel, eds., *Foundations of Space-Time Theories*, vol. 8, *Minnesota Studies in the Philosophy of Science* (Minneapolis: University of Minnesota Press), pp. 134–205.

Wittgenstein, Ludwig. 1922. *Tractatus Logico-Philosophicus*. London: Routledge & Kegan Paul.

Wright, Larry. 1976. *Teleological Explanations*. Berkeley: University of California Press.

Index